Diversity Dimensions in Mathematics and Language Learning

DaZ-Forschung

Deutsch als Zweitsprache, Mehrsprachigkeit und Migration

Herausgegeben von
Bernt Ahrenholz
Christine Dimroth
Beate Lütke
Martina Rost-Roth

Volume 24

Diversity Dimensions in Mathematics and Language Learning

Perspectives on Culture, Education and Multilingualism

Edited by
Annemarie Fritz, Erkan Gürsoy,
Moritz Herzog

With a foreword by Barbara Sarnecka

DE GRUYTER

OA Funding by Stiftung Mercator

ISBN 978-3-11-066104-0
e-ISBN (PDF) 978-3-11-066194-1
e-ISBN (EPUB) 978-3-11-066276-4
DOI https://doi.org/10.1515/9783110661941

Library of Congress Control Number: 2021934661

Bibliographic information published by the Deutsche Nationalbibliothek
The Deutsche Nationalbibliothek lists this publication in the Deutsche Nationalbibliografie; detailed bibliographic data are available on the Internet at http://dnb.dnb.de.

Typesetting: Integra Software Services Pvt. Ltd.
Printing and binding: CPI books GmbH, Leck

www.degruyter.com

Foreword

When I was a PhD student, the first study that I designed was about the meanings that young children attribute to number words. I put together some tasks, went to some preschools, and tested some children. But when I analyzed the data, I found nothing: no interpretable patterns, no confirmation or disconfirmation of my hypotheses. I took the results to my faculty advisors. "Look," I said. "Nothing. I guess I have to think of a new study and start over."

"Well, wait a minute," my advisors said, looking closely at the data. "It looks like half of these kids are bilingual. Did you exclude them?" I had not. My advisors told me to run the analyses again, using data only from monolingual children. I followed their instructions, and *voila*! Clear results. The results were the opposite of what I predicted, but I didn't care – results are results! By eliminating one dimension of variation (multilingualism) from the sample, I had turned an unpublishable study into a publishable one.

I believe that experiences like these are the reason why so many developmental researchers have shied away from studying diverse dimensions of mathematics and language learning. Research is difficult to begin with, and every dimension that we include just makes it more difficult. After my first study, I spent the next handful of years studying language and number development only in monolingual English-speaking, typically developing children. I knew that most of the world's children are not monolingual, and I knew that development takes many different paths besides the path we call normal. But I wanted to simplify my work however I could.

Then I got a job in Southern California. Suddenly, 85% of the preschoolers in schools I visited were bilingual or multilingual. So I had a choice: I could either exclude data from 85% of children, or I could try to study multilingual development.

I think a similar thing is happening to our whole field. Experimental psychologists have always created artificial situations to study; our analytical methods require experiments to be simpler than real life. But real life now in many of our communities is so diverse that when we abstract away from dimensions like multilingualism or atypical development, the picture that we end up studying seems utterly divorced from the reality that we all live in.

The authors of these chapters know well that studying realistic diversity is difficult, and that it would make things simpler if we pretended that all children grew up monolingual and developed along a typical path, which we could all study using one agreed-upon research method. But the world is not so simple. We must find ways to handle the diversity of mathematics and language learning

https://doi.org/10.1515/9783110661941-202

in our research because that is the only way to gain real wisdom about both the universals and the particulars of human development.

The editors of this book have done us all a service by bringing together scholars who use different methods and address different topics. Each of us has only a narrow band of expertise. We are trained in particular methods that limit the kinds of questions we can ask, and we have deep knowledge of only a small corner of the scientific literature. As the old joke says, a researcher learns more and more about less and less until eventually she knows everything about nothing. This unavoidable narrowness in our individual expertise means that a truly diverse picture of mathematics and language learning can only come from a whole community of researchers in conversation with each other. That is what the editors have created in this book.

I conclude this foreword with the hope that this book will be starting a conversation that continues for many years, inspiring new collaborations and new lines of research that lead us back to real wisdom. When I teach human development to undergraduates, I start by saying that all human beings on earth only differ from each other by 1–1.5% of their genes. Although diversity is fascinating, the most important truths of human experience are true for all of us – everyone wants to be safe, everyone wants to be loved, everyone wants to be heard. Even as our studies of mathematics and language learning become more sophisticated and more able to handle the diversity of real life, I suspect that they will keep bringing us back to certain home truths – that the human brain is amazingly plastic; that learning the representational systems of mathematics and language actually transforms the way we think; and that human development is simultaneously the most complicated research topic in the world, and also the most important.

Barbara Sarnecka

Introduction

In 2016, the three of us started working on a mathematics screening instrument for bilingual first-graders. It was a kick-off for an intensive time of learning, as we were relatively unfamiliar with each other's discipline. The one who was an expert in the development of arithmetic concepts was introduced to the linguistic aspects of multilingual learning; the one familiar with second-language acquisition learned about children's development of early numeracy; and the one who had advanced knowledge about early mathematics instruction encountered the diversity of linguistic influences on children's development. Thus, every one of us learned how important the others' expertise was for the own field, and vice versa, how important the own expertise was to the others' fields.

The book title *Diversity Dimensions in Mathematics and Language Learning* refers to children's diversity when acquiring fundamental linguistic and mathematical knowledge. This diversity comprises cultural properties such as the direction of reading and writing, or the structure of number words. Diversity dimensions can also refer to the instructional resources children contribute; for example, their experiences at home or the words they know to express numbers and number relations. And it may also include the diversity of home languages and how this diversity interferes with instruction, which in many educational environments may be monolingual.

However, this book can also be read in the light of the diversity of disciplines that engage with research on children's mathematics and language learning. The heterogeneity of involved disciplines, which comprise knowledge and theories from different traditions, will inevitably bring different perspectives to bear on mathematics and language learning that is not easily integrated. This book aims at collecting and interconnecting the various perspectives and insights gained by the different disciplines. Another consequence of this diversity is that this book has many contributors with varied approaches to the topic. We are proud and happy to have gathered these eminent researchers from different scientific backgrounds, who have contributed their expertise to this book.

To us, this book is the continuation and expansion of the learning process that started with our first meetings. In the following paragraphs, we would like to highlight the main insights gained while editing this book.

The first section "Perspectives on mathematics and language from different disciplines" represents the range of disciplines that are involved in mathematical and language learning: linguistics (Everett), psychology (Hartmann and Fritz; Dowker), neuroscience (Klein et al.), and mathematics education (Prediger). The focus is on fundamental theories and key findings arising from these disciplines and thus paves the path for the following sections. The variety of scientific

https://doi.org/10.1515/9783110661941-203

backgrounds shows also in the other sections, which are dedicated to one dimension of diversity each.

The second section "Language learning and mathematics development" focuses on the impact that language has on mathematical learning. Children's development of early numerical knowledge depends in part on their linguistic environment. Bahnmüller et al. provide an overview of the ties between mathematics and language. The way in which the labeling of numerosities affects the development of counting skills is highlighted in the chapter by Pixner and Dresen. Desoete et al. investigate the importance of children's learning opportunities at home.

Worldwide, there are more children learning mathematics who are multilingual than those who are monolingual: being multilingual therefore is the norm rather than the exception. The manifestation of multilingualism has varying effects on number representation as illustrated in the third section ("Multilingualism and mathematical learning"). Ashkenazi and Mark-Zigdon show how spatial number representation is affected by the languages spoken at home, while Klein reports findings on the acquisition of exact number representation in multilingual children. Multilingualism is often seen as a challenge for teachers. Martini et al. discuss the effect of the choice of cut-off values for mathematical difficulties among different home language groups.

Most research on mathematics and language learning focuses on children whose development proceeds within a normal range. However, there are children, who systematically differ from typical development due to specific impairments. The mathematical development of these children, focused in the fourth section "Vision and speech language impairments", has rarely been researched. Crollen et al. present findings on visually impaired children's mathematical development, while Schuchardt and Mähler investigated the mathematical skills of speech and language-impaired children. How the language of learning and instruction as second language affects mathematical learning is addressed by Moser, Opitz, and Schindler.

The final section "Language as a learning resource in school" attends to the instructional aspects of mathematics and language learning. While trying to bridge the gap between research and practice, the perspectives of the different disciplines are made visible: Moura et al. discuss the similarities of procession models for numbers and words informed by neuroscience. An important linguistic learning resource is children's mathematical vocabulary, whose development is the focus of Powell et al. The main topic in this section is word problems. Moschkovich and Scott demonstrate the pitfalls that word problems bring to learners. From a psychological perspective, Herzog et al. invert the usual process of mathematical modeling and explore how children write word problems when provided with given illustrations. Stephany takes a linguistic perspective and investigates the relation of text coherence and children's reading skills in the context of word problems. MacKay et al. address this challenge and illustrate how to prepare teachers for multilingual classrooms. The decision about when to intervene in a child's schooling is both complex and critical.

The various disciplines have contributed insights to the development of mathematics and language skills. It would seem to us, given the work presented in this book, that an imminent task for current and future researchers would be to integrate and interconnect these findings with the aim of forming a comprehensive theory of mathematical and language learning. Our common work on the mathematical screening for bilingual children worked as an initial spark for theorizing this interrelation and taking initial steps toward a theory. We hope to fuel this spark in our colleagues with the creation of this book.

Essen, September 2020
Annemarie Fritz, Erkan Gürsoy, and Moritz Herzog

Acknowledgments

The idea for this book was born during the very fruitful collaboration of the three editors on the validation of a math screening for first-grade students into Turkish and Arabic (Gürsoy et al., 2020). The math screening had originally been developed and standardized in German (Ehlert et al., 2020) with the aim of evaluating the children's learning prerequisites at the start of school and giving teachers orientation for their teaching activities.

When presenting the test to teachers in the context of further training courses for the prevention of numeracy difficulties, they often mentioned that not all of the children in their classes spoke German well enough to understand the test. However, these children spoke another language in which their learning prerequisites for mathematics could be evaluated. This problem, how to design a screening for early arithmetic concepts that is fair in regard to language, brought together a psychologist specialized in the development of arithmetic concepts and mathematical learning difficulties (Annemarie Fritz), a linguist specialized in multilingualism and German as second language (Erkan Gürsoy), and a mathematics educator who just started learning psychological aspects of mathematical learning (Moritz Herzog).

Our meeting not only led to a validation of the test in the languages spoken by many children as their first language. It also drew our attention to the importance of language for the acquisition of mathematics. Our own diversity as researchers highlighted the importance of integrating the different perspectives and approaches of the relevant disciplines. With this book we have tried to contribute to this integration by enriching the discussion on the relation of language and mathematics learning.

Our first thank goes to the editors of the book series Bernt Ahrenholz (†), Christine Dimroth, Beate Lütke, and Martina Rost-Roth for including our book concept and make our publication possible. We thank all of them sincerely.

We would like to thank Julie Miess of De Gruyter publishing, who helped a lot while preparing this publication. As editors, we are very proud that this book was published open access. We thank the MERCATOR foundation for providing substantial financial support for this book as an open-access resource.

The success of a book naturally depends first and foremost on the contribution of authors. The idea behind the book was to bring together experts in the field of childhood development, in both language and mathematics knowledge. Authors came together from different scientific fields – psychology, mathematics education, special education, multilingualism, and (second) language acquisition – to write papers drawing from various theoretical perspectives, engaging with literature in the field and reporting on research findings. The resulting

https://doi.org/10.1515/9783110661941-204

overview, comprising 20 chapters, allows for a profound insight into the topic. We would like to thank the numerous authors who quickly became interested in the book and enriched this book with their well-researched chapters. It was an honor to bring this product to completion.

Our very special thanks go to Barbara Sarnecka, who with enthusiasm and without hesitation complied with our request to write the Preface.

Caroline Long was responsible for language editing. She worked meticulously through the texts to achieve conceptual clarity and assist the authors to convey their messages with assurance.

Contents

V Language as learning resource in school

I Perspectives on mathematics and language of different disciplines

Caleb Everett

The diversity of linguistic references to quantities across the world's cultures

1 Introduction: Defining core concepts

It is challenging, perhaps impossible, to discuss "numbers" without bringing to bear particular assumptions of one's culture, language, theoretical bias, or some combination thereof. What are numbers, exactly? Are they innate concepts given to us by natural selection? Are they entities that exist in nature, awaiting discovery by the brains of humans and other animals? Are they cultural and linguistic constructs that have gradually accrued across the world's populations in different but constrained ways? Some scholars would offer affirmative answers to only one of the three preceding questions, while others might offer positive answers for all three. Volumes have been written on these possible perspectives and, perhaps, many of those volumes are of little relevance to those interested in more quotidian, and perhaps more significant, pedagogical concerns associated with numbers. Nevertheless, it is useful to have some basic agreement as to what we mean when we talk about learning numbers and the basic mathematical principles revolving around them – to have some shared understanding about what numbers even are. In this chapter I will focus on the last of the three questions above, outlining in basic form the cross-linguistic variation that exists vis-à-vis spoken number systems to illustrate how such systems have accrued in variable ways across human cultures – even if the relevant variations are constrained in some ways. The survey presented should, I hope, benefit scholars interested in mathematical pedagogy who are not entirely familiar with the extent of cross-cultural variation in the number systems of the world's languages.

Before embarking on the survey, though, allow me to establish the definitions of three terms that I will be using during its course. These definitions reflect my own theoretical predispositions, informed as they are by cross-cultural and cross-linguistic data. The three key terms and associated definitions I will employ are "quantical," "numerical," and "numbers." The definitions are grounded in other work, primarily Núñez (2017), though related terms and definitions have been presented by a variety of scholars. I begin with "numbers," which I define as verbal symbols representing precise quantities. Given that verbal symbols for precise quantities have primacy both ontogenetically and cross-culturally in our species, when contrasted to nonverbal symbols, I find it useful to interpret

https://doi.org/10.1515/9783110661941-001

them as the default form of numbers. This may seem odd in cultural contexts in which written symbols are sometimes interpreted as equally (or more?) basic, learnable units, but I believe that a focus on numbers-as-words is a useful reminder of the primacy of verbal symbols for representing precise quantities. Judging from the cross-cultural data, humans' most basic symbolic tools for manipulating quantities are verbal (Everett, 2017).

The distinction between "quantical" and "numerical" concepts is more recent and esoteric, but I believe it to be extremely useful and well motivated. For a fuller discussion of the merits of this distinction, I refer the reader to Núñez (2017). The chief motivation is that much research in psychology refers to basic and native "numerical" cognition, putatively shared by our entire species, that appears to be neither basic nor native once the extent of cross-cultural diversity in quantitative cognition is surveyed with sufficient care. Nevertheless, it is generally agreed that all humans do share some basic native capacities for quantity discrimination. For instance, humans can generally distinguish small quantities (1, 2, and 3) from each other precisely without training (as can the members of some other species). Humans can also approximately discriminate larger sets of items, for instance, eight sticks from sixteen sticks, presuming that the ratio between the sets is large enough. (This ability is also phylogenetically primitive – some have suggested it stretches back to the first vertebrates.) These basic quantitative reasoning skills are not apparently contingent on cultural scaffolding, but they are not really "numerical" in that they offer no means of delimiting, for instance, five from six items with consistency. Numbers like "five" and "six" do not simply follow from our native quantitative capacities; they must be crafted and honed by distinct cultural practices that rely on those capacities. These practices allow us to transfer our modest native exact quantity recognition into the realm of larger quantities. For such reasons, it is not particularly useful (from my perspective anyhow) to refer to native quantical abilities, shared with other species, as "numerical," or to liken them to a "number sense." Terms like "number sense" may give the false impression that we are somehow born with numbers in our heads or are wired to learn basic arithmetic (Dehaene, 2011). In the words of Núñez:

> Humans and other species have biologically endowed abilities for discriminating quantities. A widely accepted view sees such abilities as an evolved capacity specific for number and arithmetic. This view, however, is based on an implicit teleological rationale, builds on inaccurate conceptions of biological evolution, downplays human data from non-industrialized cultures, overinterprets results from trained animals, and is enabled by loose terminology that facilitates teleological argumentation. (2017:409)

Given my shared desire to avoid teleological argumentations where they are not warranted, and given this chapter's focus on cross-cultural variability, I adopt the terminological distinction proffered by Núñez, the distinction between "quantical" concepts and "numerical" concepts. The former term refers to humans' native, biologically endowed capacities for differentiating quantities in generally coarse ways. The latter term, "numerical," refers to exact, symbolic practices evident when humans use "numbers." Framed differently: The existence of quantical cognition is a necessary condition for the flowering of numerical cognition, but it is, critically, not a sufficient condition. Maintaining a distinction between "quantical" and "numerical" cognition is particularly useful as a background for discussing the extent of cross-cultural variability in the ways that people talk about quantities, and the potential relevance of that extensive variability to mathematical pedagogy. It is important to dissociate the universals of human quantical thought from the cross-cultural variability of numerical thought and numbers. This clear dissociation could positively impact efforts to more effectively convey numerical concepts to individuals across the world's cultures.

So, to be clear, this contribution aims to shed light on the diversity of numbers in the world's languages in the expression of numerical concepts, and also will survey some differences in how languages describe quantical concepts. Approaches to the pedagogy of arithmetic could only benefit, I hope, from an understanding of commonalities and differences in the ways the world's languages refer to such concepts. These could offer some insights into the best ways to approach, for instance, cross-culturally effective instruction strategies. (I leave it to the experts on pedagogy, however, to judge how the findings discussed here might benefit mathematical instruction across cultures.) At the least, such commonalities and differences can hopefully give the reader a better sense of just how typical or atypical our own linguistic strategies for encoding numerical and quantical concepts are, when considered in the light of the typological data. By examining an adequately representative sample of number systems in the world's languages we can, inter alia, better understand which numerical concepts are most easily acquired by the members of our species.

2 Cross-population differences in grammatical number

The grammars of the world's languages often refer to quantical concepts, what is commonly referred to as "grammatical number." Grammatical number refers to a variety of phenomena that denote distinctions between small precise quantities and large imprecise quantities (e.g., singular vs. plural), or between small precise quantities (e.g., singular vs. dual). Grammatical number markers take many forms, including noun suffixes and prefixes, verb suffixes and prefixes, and many more. In English, for example, suffixes are added to nouns to demonstrate whether there is one or more than one of an item or entity to which the speaker is referring. In the languages in which grammatical number exists, it serves overwhelmingly to distinguish between sets of exactly one (singular) and more than one (plural). In rarer cases grammar is also used to distinguish one, from two, from more than two items. Languages with that kind of grammatical number are said to have singular, dual, and plural marking. Rarer still are languages that have singular, dual, trial, and plural marking. So grammatical number is always used to designate sets of items (1, 2, 3, or many) that humans are capable of discriminating via their native quantical cognition, as defined above.

Grammatical number refers only to small quantities precisely, and to large quantities approximately. In this way its function is limited, but in another sense its function is very robust: Languages that have grammatical number often use it to obligatorily denote the quantity of reference, and this obligatory status means that it is extremely pervasive in speech. In this chapter alone there are hundreds of cases of grammatical number inflected on verbs and nouns. English learners, whether children or adults, must learn the ways of adding regular plural markings, not to mention irregular plural markers. They must also learn that some nouns, say, "deer," are not marked at all in the plural. More broadly, they learn that the quantity of referents is always relevant, even if only in approximate ways, during communication.

This is not the case in many of the world's cultures. In fact, in about 10% of the world's languages, there is no grammatical means of designating the number of referents to which a speaker is referring. For example, the Karitiâna language, on which I have done a fair amount of research, has no nominal plurality. Consider the following phrases from that language:

(1) *myjyp ambi*
three house
"Three houses."

(2) *y-ambi*
1st.Singular.Possessive-house
"My houses."

(3) *ombaky naokyt taso*
jaguar killed man
"The jaguar(s) killed the man/men."

(4) *yj-pyt ombaky*
our-hand jaguar
"Five jaguars."

As we see in (1) and (2), the word for house does not change even when there are many houses being referred to. The same is true of "jaguar," and "man," as seen in (3) and (4), because all the nouns in the language do not denote quantity distinctions.

There are many languages like Karitiâna scattered around the world. In a survey of data of 291 languages representing many distinct language families and geographic regions, the linguist Martin Haspelmath found that about 10% (n = 28) of the languages were like Karitiâna, with no nominal plurality evident in their grammars (Haspelmath, 2013). ("Nominal plurality" refers to cases in which the quantity of an item referred to by a noun is denoted in the grammar, typically with a suffix on the noun.) In another 19% (n = 55) of the languages, nominal plurality was found to be optional in all cases. So rather than saying something like, for instance, "three cars," one could say "three cars" *or* "three car," and either would be grammatically correct. There is a sense in which this is intuitive, as the *-s* suffix in a phrase like "three cars" is, after all, redundant, encoding information about plurality that is already contained in the preceding number word. In other cases the plural marking may prove quite informative. For example, the interpretation of clause (3) could vary significantly. Did one jaguar kill one man? Did one jaguar kill many men? Did many jaguars kill many men? Did many jaguars kill one man? In actuality, though, context and real-world prior information (e.g., that jaguars are fairly solitary creatures) help to constrain most cases of ambiguity. Speakers can communicate just fine without grammatical reference to things like plurality. One could make the case that grammatical number is most relevant for human nouns, since speakers tend to talk about human referents, and since humans can occur in varying group sizes (Everett, 2019). The global distribution of grammatical number types supports this intuition: Haspelmath (2013) found that about 7% (n = 20) of the world's languages have plural marking that is optional but can only be used to denote plural human referents. Furthermore, in about 14% (n = 40) of the sampled languages, plural marking is obligatory but is restricted to human nouns. And in 5% (n = 15) of the languages, plural marking occurs on all noun types but is optional for inanimate nouns.

Haspelmath's survey reveals, then, just how variable grammatical number marking is across the world's languages. Less than half of the languages in his sample, or 46% (n = 133), exhibit the kind of grammatical number marking evident in English and most European languages, in which multiple referents must be designated with plural-marked nouns in an obligatory manner. In over half of the world's languages, grammatical plural marking is either absent, or is optional, or is only obligatory for nouns that refer to human referents. This variability of grammatical plural marking is evident across diverse regions and language families.

One logical question that follows from the diversity of grammatical number is whether one's native language impacts how s/he becomes familiar with the distinction between the notions of "one" vs. "more than one." (This topic has been raised in contemporary discussions of "linguistic relativity"; see for example Everett, 2013.) Such an impact may seem implausible given that these are quantical concepts, native to all members of our species and countless others. Yet the question is not whether variation in grammatical number enables humans' simple capacity for tracking singularity or plurality, but whether it affects how a person habitualizes themselves to such distinctions during every-day events.[1] For instance, if a person speaks a language that only indicates plurality on human nouns, does this bias that person to pay attention to quantity more when speaking about or conceptualizing human referents? Perhaps not, but to my knowledge no experimental evidence has been brought to bear on the topic. There is now evidence, however, that distinctions in grammatical number can affect how adroitly children handle quantical concepts. Some of that evidence will be discussed below.

Grammatical duals are the formal means, often noun suffixes as in the case of plural markers, that languages use to denote precisely two referents. This dual marking is not extremely rare. For instance, in a recent survey of 218 languages, Franzon et al. (2018) find that grammatical duals occur in some form in 84 of the languages. In Everett (2019) I observe that these duals tend to be restricted in terms of geographic distribution and in terms of the language families in which they occur and are also restricted in terms of function. In most languages that use dual markers, they denote distinctions on human referents only. There are over 300 language families in the world (Bickel et al., 2017), and in the vast majority of these grammatical duals are not present. Still, grammatical

1 Note that this is just one of many issues that could be raised vis-à-vis the interaction of language, culture, and cognition. For more discussion on this topic, see Everett (2017) or Saxe (2012).

dual markers are more common cross-linguistically than some might assume, given that most of the world's most widely spoken languages lack grammatical duals. One notable exception to this trend is Arabic. Intriguingly, while Spanish and English and the vast majority of European languages lack a grammatical dual maker, Proto-Indo-European did apparently have one, as did ancient Greek and Sanskrit. And there are vestiges of the grammatical dual in English, notably in the words "either" and "both."

Despite their well-known tendency to have few numbers, as in a "one-two-many" system, some languages of Australia employ grammatical dual markers. Here are some examples from Dyirbal, taken from Dixon (1972: 51):

(5) *bayi Burbula miyandanyu*
"Burbula laughed."

(6) *bayi Burbula-gara miyandanyu*
"Burbula and another person laughed."

(7) *bayi Burbula-mangan miyandanyu*
"Burbula and several other people laughed."

In (7) we see that the suffix *-mangan* serves as a plural maker, denoting that multiple people are involved in the event. But this plural is only used to denote more than two people, since if there are precisely two people the *-gara* suffix is used as in example 6. (This kind of dual marker is called an "associative dual" since it refers to a specific person and exactly one other person.) While dual markers may tend to refer to human and pronominal referents, this is certainly not the case in all languages that use them. In the Sikuani language and various others, there is a suffix or other affix that refers to precisely two things. Consider these Sikuani words: *emairibü* "a yam" vs. *emairibü-nü* "yams" vs. *emairibü-behe* "two yams." The *-behe* suffix signifies that there are precisely two yams in question (Aikhenvald, 2014).

Grammatical trials are also evident in Franzon et al.'s (2018) survey. In that survey, 20 of the 218 languages have grammatical trials. However, the grammatical trial is evident in only one world region, Oceania. It is evident in clauses like the following example sentence from Moluccan:

(8) *duma hima aridu na'a*
house that we three own
"We three own that house" (Laidig & Laidig, 1990: 92)

The *aridu* pronoun is a first-person trial pronoun meaning "we three." Grammatical trials are generally restricted to pronouns, even more so than grammatical duals.

Given the distribution of grammatical number types, it seems fair to say that languages generally indicate a singular/plural distinction in their grammar, either with affixes attached to the noun, or with verbal affixes or other changes made to the verb that denote "agreement" with the number of items of a relevant noun. (Verbal affixes are prefixes or suffixes, in most cases, that are attached to a verb.) This singular/plural distinction is evident throughout most of the world's languages, but a substantive minority of languages do not make the distinction grammatically. Languages that refer to grammatical duals and trials are comparably rare, and the functional utility of these other categories tends to be limited.

Does the variation that exists in the world's grammatical number types impact how speakers of languages learn basic quantitative concepts like "precisely 2" and "precisely 3"? This may seem an odd suggestion given that quantical cognition allows us to differentiate 1 from 2, and 2 from 3. Yet simply because all humans are endowed with the capacity to differentiate these quantities, we cannot assume that they come to use them in the same ways and with the same dexterity, nor that the features of a language do not impact the ease with which the concepts are handled during childhood. To the contrary, there is now evidence that grammatical number has at least a modest effect on the ease with which quantical concepts are handled, at least in some contexts. English-speaking children tend to learn the word for 1 rapidly, when compared to Japanese and Mandarin speakers (Almoammer et al., 2013; Marušič et al., 2016). This may be due, at least in part, to the presence of grammatical number in English, which Mandarin and Japanese lack. Relatedly, speakers of one dialect of Slovenian that has a grammatical dual marker tend to learn the word for 2 earlier than speakers of the other languages for which comparable data are available. These include English, Russian, Japanese, and Mandarin (Marušič et al., 2016). While such results are consistent with a grammatical effect on the ease with which even quantical concepts are labeled and manipulated linguistically, the causal role of grammar is of course debatable given the host of cultural confounds entailed in such cross-cultural research.[2] One of the ways to circumvent this challenge is to examine

2 For example, cultures that rely heavily on trade may be more likely to refer frequently to distinctions between quantities, even small quantities (Everett, 2019). In such cases, the frequency of transactions requiring precise quantities could serve as a confounding explanation, perhaps explaining the observed differences in quantitative thought that could also correlate with linguistic differences.

groups that are relatively homogenous culturally, but differ in terms of one particular linguistic feature. Slovenian presents a critical test case, as dialects of Slovenian vary according to the presence of grammatical duals. Recent research with speakers of these dialects suggests that the kind of grammatical number that exists in a given Slovenian population impacts how and when Slovenian children learn to label and manipulate quantical concepts.

In dialects of Slovenian that employ a grammatical dual, it takes the form evident in (9).

(9)	*dva*	*rdeča*	*gumba*	*ležita*	*na*	*mizi*
	two	red.DUAL	button.DUAL	lie.DUAL	on	table

"Two red buttons are lying on the table" (Marušič et al. 2016: 2).

Note the pervasiveness of the grammatical dual in such a clause. The adjective ("red"), the noun ("button"), and the verb ("lie") are all inflected in a way that indicates the fact that there are precisely two buttons. Learning a language like this requires children to consistently refer to whether or not there are two, and precisely two, referents being discussed. This cognitive fixation might have some effect on the age at which children become comfortable with a more general ability to symbolically denote the notion of two. A research team led by Franc Marušič at the University of Nova Gorica, Slovenia, tested the hypothesis with young children between the ages of two and four. Their sample was large, involving nearly 300 children from three Slovenian regions. Eighty-three of these children were from Slovenska Bistrica, a region of Slovenia where the dual morphology evident in clause 9 is quite normal. Seventy-one represented Central Slovenia, another region in which the grammatical dual is used. One hundred fifty-eight children represented two other regions in which speakers do not generally use the grammatical dual: Metlika and Nova Gorica. Finally, a control population of 79 English speakers in San Diego was tested. The tasks involved in the work are common to research on the development of numerical cognition. A key task was the so-called Give-N task, in which children are tested on their familiarity with basic number words. For this variant of the task, the researchers gave kids 10 buttons and asked the kids (in Slovenian or English) the following question: "Can you put *N* in the box?" For example, "Can you put two in the box?" *N* refers to a number word. The results of the Give-N task were promising for the hypothesis, pointing to subtle but significant differences across the populations of Slovenian speakers. The researchers found that "overall, speakers of dual dialects were more likely to be 2-knowers than speakers of non-dual dialects" and reached the "2-knowing" stage at an earlier age (Marušič et al., 2016: 2). While the cross-population differences were not pronounced, they were consistent with the hypothesis that grammatical dual marking can impact how kids acquire basic numbers, even those associated with quantical concepts. These and other related findings led Marušič et al. (2016) to the following conclusion: "morphological marking of number in language facilitates learning of early number word meanings" (Marušič et al., 2016: 15).

We have seen in this section that languages vary in terms of how, and whether, they denote quantitative concepts grammatically. This survey has not been comprehensive, and for a fuller picture on grammatical number I refer the reader to Corbett (2000). Yet the survey was sufficient to demonstrate that variation in grammatical number is more substantive than some scholars may presume. Furthermore, I have highlighted recent work that suggests that variation in grammatical number, including the presence/absence of a grammatical dual, may impact when and how kids are able to symbolically represent quantical concepts like 2.

3 Cross-population differences in number words

There are many critical stages in the acquisition of basic numerical concepts. These include the well-known stage at which children master the cardinal principle, becoming fully aware that a set labeled by a word *N* corresponds to an exact quantity that is associated with the word *N* only. Relatedly, they learn the successor principle, becoming aware that each word in a sequence of number words refers to the quantity denoted by the previous number word plus exactly one more (Carey, 2009a, 2009b). Prior to the acquisition of these principles, kids are able to recite a list of number words but are unaware of the relationship between them. They merely recognize that number words, like the letters of the alphabet, come in a predictable order. Much debate remains as to how exactly kids acquire the cardinal and successor principles, but it is clear that cultural variation in finger counting and number words impinge on that acquisition. The presence of precise number words like "two" or "seven" (as opposed to "few" or "several") in a language appears critical to even more basic cognitive stages that do not rely exclusively on quantical capacities. For example, the mere recognition of one-to-one correspondence benefits from the presence of number words. There is some debate as to the extent of that benefit, but work among anumeric Nicaraguan homesigners, largely anumeric Munduruku indigenes, and totally anumeric Pirahã indigenes points in the same general direction: Number words are critical to scaffolding or at least enhancing the recognition of one-to-one correspondence for set sizes larger than 3–4 (Pica et al., 2004; Spaepen et al., 2011).

For differing views on the extent of the effects of an absence of number words in a culture, see Everett and Madora (2012) and Frank et al. (2008). (The Pirahã language is well known to lack precise number words (Everett, 2005).)

Much has been written about the cultures and languages with few or no number words, and admittedly the sparse studies carried out among the relevant groups leave room for multiple interpretations of a few key results. (See Frank et al. (2008) and Everett and Madora (2012) for one example of a disagreement in interpreting the experimental results among the Pirahã.) This is not surprising given that there is still debate on the acquisition of numerical concepts in cultures whose numerical cognition has been studied with thousands of studies, for example, Americans. (See, for instance, the differing views on some key topics by prominent researchers such as Carey (2009a), and Dehaene (2011).) But it is difficult to contest that number words are critical to the acquisition of very basic numerical concepts besides the cardinal and successor principles. This conclusion is, in a way, unsurprising. What is more contestable is whether current differences in types of number words impact numerical cognition. Setting aside the rare contemporary cases of anumeric or nearly anumeric cultures, then, what can we say about the vast majority of the world's 7000+ languages that have lexical numbers? Do cultures that rely on distinct kinds of number systems exhibit associated distinctions in how they think about and learn numerical concepts? The truly cross-cultural work on this topic is modest in scope, but it does hint that variation in number word systems yields some effects on basic numerical cognition.

Anecdotally, my own impression is that the extent of diversity in the world's number systems is underestimated by many scholars. In a detailed survey of 196 languages representing dozens of families and all major geographic regions, linguist Benard Comrie offers us a sense of that diversity. Twenty of these languages have "restricted" number systems, one of which is the aforementioned extreme case of Pirahã. Other restricted cases include Hup, which will be discussed below, and some other Amazonian and Australian languages. In New Guinea there are four languages from Comrie's (2013) survey that use an "extended body part" number system. In some of these cases, for example, Kobon, counting follows a trajectory up the arm (and back down the other side of the body in some languages). So the words for 1–5 are the same as the words for the fingers on the left arm, and then 6–12 are expressible via the words for the following body parts: wrist, middle of the forearm, the elbow (or, rather, the opposite side of the elbow), the upper arm, the shoulder, the collarbone, and then, lastly, the suprasternal notch (the indentation above the sternum). Such extended body part number systems, like restricted systems, have no number bases. In 172 of the 196 languages in Comrie's (2013) survey, there are bases.

Bases of verbal numbers are the key numbers around which larger numbers are structured, usually in a multiplicative fashion. For instance, English is base-10 or decimal because number words like "forty three" are constructed around "ten": *four x ten + three.*

According to Comrie's (2013) survey, 125 of the 196 languages examined have decimal bases, as in English, for numbers greater than 10. A smaller but sizable segment, 20 of the 196 languages, use vigesimal or base-20 numbers for higher quantities. Hybrid bases, which rely on a combination of decimal and vigesimal bases, are found in 22 of the languages. In total, then, 167 of 196 languages in the survey use some base that is derived from an obvious anatomical source. The existence of base-10 and base-20 systems owes itself, of course, to the fact that humans have 10 fingers and 20 fingers and toes. Taking Comrie's sample as a reasonable proxy for the world's languages, this means that about 85% of the world's languages likely rely on digitally based numbers, and most of the other extant number systems rely on anatomical features in some other way.

One base that is rarely attested but that has shaped much of western life, in an oblique manner at least, is the base-60 system that was once used in ancient Sumeria. This system has, over the last few millennia, worked its way into various aspects of our mathematical culture, for instance the use of 360-degree arcs evident in geometry and navigation. More fundamentally, due to its adoption by the Babylonians and Greeks, it ultimately came to shape how we define units of time. The minutes of the day are simply what one arrives at if hours are divided into 60 equal units and if we divide those units by 60 a second time we get, well, "seconds" (hours are an odd by-product of the ancient Egyptian sundials that divided the daylight into 12–10 units for when the sun was up, due to the decimal Egyptian language, plus one unit for dawn and one for dusk) (Everett, 2017). Base-60 systems are also attested in the ethnolinguistic literature, at least in the Ekari language of New Guinea:

(10) *èna ma gàati dàimita Mutò*
one and ten and Sixty
"Seventy one" (Drabbe, 1952: 30).

Interestingly, the most plausible account of the genesis of base-60 systems also points to the criticality of the fingers in the origins of numbers. An attested practice in some cultures is to count the 12 lines of the non-thumb joints of the inside of one hand with the five fingers of the other hand. (See image in Everett (2017: 80).) If each added finger is used to represent the 12 lines, then the total quantity represented by five fingers is 60 (Ifrah, 2000). So while the base-60 system we used for telling time is unrelated to the decimal system that developed in Indo-European languages, it shares with it manual origins.

There are roughly 400 Indo-European languages spoken in the world today, with English and other languages well represented across the globe as first and second languages. Proto-Indo-European, spoken somewhere in the vicinity of the Black Sea over 6,000 years ago, had a decimal system as evident by reconstructed words such as **dékmt*, "ten" and **duidkmti*, "twenty" (literally "two tens") or **trihdkomth*, "thirty" (literally "three tens"). Phonetic vestiges of such number words are still evident in descendant words, like the Portuguese word *dez* ("ten") or the word *decimal* itself, both of which bear some resemblance to **dékmt* (Everett, 2017). More critically, though, the structure of Portuguese numbers, English numbers, and numbers in other contemporary Indo-European languages still carry the structure of Proto-Indo-European numbers, whereby 10 is multiplied by smaller numbers to create larger number words. This decimal base is evident in the world's other largest language families today, including Niger-Congo, Austronesian, and Sino-Tibetan, which like Indo-European has over 400 languages and over a billion speakers. (The Niger-Congo and Austronesian families each have over 1,000 members, representing a sizable chunk of the world's 7,000+ languages.)

The manual/digital origins of number words are not simply evident in the preponderance of decimal and vigesimal number systems; they are also evident in the base-5 nature of number words less than 10 in many cultures. The critical nature of a word for 5 in constructing greater numbers is evident worldwide, and stems from the clear derivation of that number from counting with the fingers. For instance, the word for 5 in many languages is transparently derived from the word for "hand." In Proto-Austronesian, for example, the word for *hand* and the word for *five* were both *lima. The same correspondence is evident in very many unrelated languages, and the word for "five," once derived from the word for "hand," seems to kick-start the growth of larger number systems (Bowern & Zentz, 2012).

The digital foundations of numbers are even evident in some languages that have modest number systems, in words for precise numbers less than 5. In Hup and Dâw, two closely related languages of Amazonia, words for numbers are based around the kinship terms in the language. The word for 3, for instance, translates to "without a sibling" because 3 is odd. The word for 4 translates to "with a sibling," because it is even (Epps, 2006). These number words are not used by themselves, however, but alongside finger-counting strategies. So one needs to hold up four fingers and say "with a sibling" to fully denote the number four. Languages like Hup and Dâw drive home the general theme of this section: A survey of the world's spoken numbers suggests that languages vary tremendously in terms of the kinds of numbers they use, and in terms of the range of quantities denoted by those numbers. Yet there are also pervasive tendencies underlying this variability, and those tendencies point again and again to the ways in which finger counting is critical to the historical acquisition of numbers in diverse and unrelated cultural lineages.

Variation in kinds of cardinal numbers is just one of the sorts of variation in cultures' verbal representation of quantities. Ordinal numbers also vary in marked ways. In a recent survey of 321 languages, Stolz and Veselinova (2013) observe that over 10% do not have a distinct category of ordinal numbers. This is in contrast to languages like English, in which ordinal numbers are often denoted with a *-th* suffix, for example, fourth, fifth, sixth. In most languages there is some distinction between cardinal and ordinal numerals, however, and in most cases ordinal numbers are clearly derived from cardinal numbers as in the English examples just cited. Intriguingly, though, in almost two thirds of the languages surveyed by Stolz and Veselinova (2013), small ordinal numbers are treated differently. In many of these languages it is only the ordinal number for 1, as in English "first" (we do not say "oneth')" In some languages 2 also is denoted with a distinct ordinal number, as with English "second" (we do not say "twoth"). The cross-cultural variation in small ordinal numbers underscores that even basic reference to quantical concepts (quantities less than four) varies cross-culturally. This variation in the reference to quantical concepts, which was also evident in our discussion of cardinal numbers and grammatical numbers, is in some sense surprising. Languages vary extensively with respect to how they describe quantical concepts that all humans share and, as seen in cases like Slovenian, this variation has demonstrable effects on the age at which individuals become adept and using such "quantical" concepts. While linguists, anthropologists, psychologists, and others have long been aware of variation in terms of how languages denote numerical concepts, only relatively recently have we come to appreciate that that variation extends in key ways to quantical concepts. It is possible, however, that we still underestimate the ways in which languages vary vis-à-vis their expression of quantical concepts. In a very recent study involving data from nearly 6,000 dialects, I make the case that there is another key type of variation in number words for quantical concepts that has still not been explored systematically: The cross-cultural frequency in speech of words for 1 and 2 (Everett, 2019).

While the vast majority of the world's languages have words translatable as "one" and "two," this does not mean that those terms are used in the same ways or at the same rate. The exploration of their frequency seemed worthwhile for a few reasons. One reason is that the frequency of usage of number terms, even as small as "one" and "two," could well impact the rate and age at which children become practiced with basic quantitative concepts. This possibility is supported by the aforementioned work on grammatical duals, which suggests that the frequent grammatical reference to 2 facilitates to some degree children's refinement of certain facets of basic quantitative thought. While directly establishing the frequency in speech of words like "two" for most of the world's languages is not possible, there is one indirect way to test for frequency in speech. This way relies on a well-known fact about words: Highly frequent words tend to be reduced phonetically, that is, made shorter (Bybee, 2007). With this fact in mind, I examined the length of number words for "one" and "two" across the bulk of the world's languages. This was done via a database containing 40–100 commonly used words (phonetically transcribed) for the bulk of the world's languages (Wichmann et al., 2018). My work looked at 5,942 language varieties (dialects and mutually unintelligible languages), considering the average word length of all the words for each language. For each language variety, I then contrasted the word lengths for "one" and "two," respectively, with the average word length of all the other words in that language. Upon doing so, a very clear pattern emerged: The languages spoken by cultures with larger populations tend to have shorter words for "one" and "two," even after controlling for factors like the average word lengths of particular languages and the relatedness of languages. This pattern suggests strongly that larger populations tend to use number words more frequently than smaller populations. There are many factors that likely motivate this tendency across the world's culture, including greater frequency of number words in cultures relying on trade and industrialization.

This all may seem very intuitive and even trivial: Of course cultures vary in the degree to which they use number words, and in the frequency with which they use number words in practices like trade. Yet the key point is that such variation extends to number words for quantical concepts that are shared by all human populations. Previous work had suggested that quantical concepts, namely 1, 2, and 3, are less prone to being concretized in varied ways across cultures because they are native concepts (Franzon et al., 2018). Instead, I argue, they are treated pretty much like other quantitative concepts in terms of how they are referred to in speech. That is, they are prone to cross-cultural variation and are used with very different frequencies across the world's cultures – at least judging from the indirect word-length data. More broadly, the issue of the frequency of small number words raises yet another kind of cross-linguistic variation in numbers. This variation, like the variation in grammatical number types, may impact children's acquisition of numerical concepts. Work is required to explore this possibility.

In this section we have seen that there is an underlying manual basis of number systems but also an amazing diversity of number words overlaid over that manual basis. This includes diversity of several sorts: Diversity in number bases (despite their generally digital origins), diversity in the mere existence of number words (since some languages lack them), diversity in ordinal numbers, and diversity in the frequency with which numbers, even very small numbers, are used. This global diversity of number words impacts how kids acquire numerical concepts and even their facility with basic quantical concepts. All these factors are worth keeping in mind when considering how best to teach arithmetic across the world's cultures. The linguistic features of a given culture affect how the members of that culture learn even basic quantitative concepts.

4 Discussion and conclusion

While there are universal human quantitative capacities, each culture and language brings with it its own biases in terms of how it refers to quantical and numerical concepts. A greater awareness of the extant cross-cultural diversity of spoken numbers could, I hope, benefit those concerned with how best to teach basic arithmetic concepts. It is still very debatable just how much cross-cultural variation of numbers impacts how kids acquire numerical concepts. Yet, where relevant experimental evidence exists, it consistently suggests that such variation matters, often in marked ways. If people speak an anumeric language, this has marked effects on their ability to learn number concepts. If they speak a language with a grammatical dual, this seems to offer advantages to early numerical cognition. More commonly, cross-linguistic variation in the transparency of number bases may impact how kids acquire numbers. Some evidence suggests that Chinese children, for instance, outperform children from the UK, Russia, and other nations on mathematical tasks, and that this high performance is due in part to the greater transparency of the decimality of Chinese numbers (Rodic et al., 2015; though see Moschkovich, 2017). So, while languages tend to have decimal bases, the transparency with which decimality is expressed appears to affect the cross-cultural acquisition of numerical concepts.

All of this leaves us with two simple conclusions: (1) The cross-cultural variation of linguistic numbers impacts quantitative cognition, and (2) the cross-cultural variation of linguistic numbers is remarkable even if it is constrained by the typically digital origins of numbers. Both of these points seem worth bearing in mind as we adopt and refine pedagogical models for arithmetic instruction, if we are interested in the cross-cultural efficacy of those models.

References

Aikhenvald, Alexandra (2014): Number and noun categorisation: A view from north-west amazonia. In Storch, Anne, Dimmendaal, Gerrit J. eds.: *Number – Constructions and Semantics. Case Studies from Africa, Amazonia, India and Oceania*. Amsterdam: John Benjamins. 33–56.

Almoammer, Alhanouf et al. (2013): Grammatical morphology as a source of early number word meanings. *Proceedings of the National Academy of Sciences of the United States of America* 110, 18448–18453.

Bickel, Balthasar, Nichols, Johanna, Zakharko, Taras, Witzlack-Makarevich, Alena, Hildebrandt, Kristine, Rießler, Michael, Bierkandt, Lennart, Zúñiga, Fernando & Lowe, John B. (2017). *The AUTOTYP typological databases*. Version 0.1.0. doi: DOI 10.5281/zenodo.3667562.

Bowern, Claire & Zentz, Jason (2012): Diversity in the numeral systems of Australian languages. *Anthropological Linguistics* 54, 133–160.

Bybee, Joan (2007): *Frequency of Use and the Organization of Language*. Oxford: Oxford University Press.

Carey, Susan (2009a): *The Origin of Concepts*. Oxford: Oxford University Press.

Carey, Susan (2009b): Where our number concepts come from. *The Journal of Philosophy* 106, 220–254.

Comrie, Bernard (2013): Numeral bases. In Dryer, Matthew S., Haspelmath, Martin eds.: *The World Atlas of Language Structures Online*. Leipzig: Max Planck Institute for Evolutionary Anthropology.

Corbett, Greville (2000): *Number*. Cambridge: Cambridge University Press.

Dehaene, Stanislas (2011): *The Number Sense*. Oxford: Oxford University Press.

Dixon, Robert & Ward, Malcolm (1972): *The Dyirbal Language of North Queensland*. Cambridge: Cambridge University Press.

Drabbe, Peter (1952): *Spraakkunst Van Het Ekagi*. 'S-Gravenhage: Martinus Nijhoff.

Epps, Patience (2006): Growing a numeral system. *Diachronica* 23, 259–288.

Everett, Caleb (2013): *Linguistic Relativity: Evidence Across Languages and Cognitive Domains*. Berlin: De Gruyter Mouton.

Everett, Caleb (2017): *Numbers and the making of us: Counting and the Course of Human Cultures*. Cambridge, MA: Harvard University Press.

Everett, Caleb (2019): Is native quantitative thought concretized in linguistically privileged ways? A look at the global picture. *Cognitive Neuropsychology* 10.1080/02643294.2019.1668368.

Everett, Caleb & Madora, Keren (2012): Quantity recognition among speakers of an anumeric language. *Cognitive Science* 36, 130–141.

Everett, Daniel (2005): Cultural constraints on grammar and cognition in Pirahã. *Current Anthropology* 46, 621–646.

Frank, Michael, Everett, Daniel L., Fedorenko, Evelina & Gibson, Edward (2008): Number as a cognitive technology: Evidence from Pirahã language and cognition. *Cognition* 108, 819–824.

Franzon, Francesca, Zanini, Chiara & Rugani, Rosa (2018): Do non-verbal systems shape grammar? Numerical cognition and number morphology compared. *Mind & Language* 10.1111/mila.12183.

Haspelmath, Martin (2013): Coding of nominal plurality. In Dryer, Matthew S., Haspelmath, Martin eds.: *The World Atlas of Language Structures Online*. Leipzig: Max Planck Institute for Evolutionary Anthropology.
Ifrah, Georges (2000): *The Universal History of Numbers: From Prehistory to the Invention of the Computer*. Hoboken, New Jersey: Wiley.
Laidig, Wyn & Laidig, Carol (1990): Larike pronouns: Duals and trials in a central moluccan language. *Oceanic Linguistics* 29, 87–109.
Marušič, Franc, Žaucer, Rok, Plesničar, Vesna, Razboršek, Tina, Sullivan, Jessica, & Barnder, David. (2016) Does Grammatical Structure Accelerate Number Word Learning? Evidence from Learners of Dual and Non-Dual Dialects of Slovenian. *PLOS ONE 11*(8) https://doi.org/10.1371/journal.pone.0159208
Moschkovich, Judit N. (2017): Revisiting early research on early language and number names. *Eurasia Journal of Mathematics, Science & Technology Education* 13 (7b), 4143–4156.
Núñez, Rafael (2017): Is there really an evolved capacity for number. *Trends in Cognitive Sciences* 2, 409–424.
Pica, Pierre, Lemer, Cathy, Izard, Veronique & Dehaene, Stanislas (2004): Exact and approximate arithmetic in an Amazonian indigene group. *Science* 306, 499–503.
Rodic, Maya, Zhou, Xinlin, Tikhomirova, Tatiana, Wei, Wei, Malykh, Sergei, Ismatulina, Victoria, Sabirova, Elena, Davidova, Yulia, Tosto, Maria Grazia, Lemelin, Jean-Pascal & Kovas, Yulia (2015): Cross-cultural investigation into cognitive underpinnings of individual differences in early arithmetic. *Developmental Science* 18 (1), 165–174.
Saxe, Geoffrey (2012): *Cultural Development of Mathematical Ideas: Papua New Guinea studies*. Cambridge, UK: Cambridge University Press https://doi.org/10.1017/CBO9781139045360.
Spaepen, Elizabet, Coppola, Marie, Spelke, Elizabeth S., Carey, Susan E. & Goldin-Meadow, Susan (2011): Number without a language model. *Proceedings of the National Academy of Sciences* 108, 3163–3168.
Stolz, Thomas & Veselinova, Ljuba (2013): Ordinal numerals. In Dryer, Matthew, Haspelmath, Martin eds.: *The World Atlas of Language Structures Online*. Leipzig: Max Planck Institute for Evolutionary Anthropology.
Wichmann, Soren, Holman, Eric & Brown, Cecil (eds.) (2018). The ASJP database (version 18).

Julia Hartmann and Annemarie Fritz

Language and mathematics: How children learn arithmetic through specifying their lexical concepts of natural numbers

1 Introduction

When children are about 18 months old their speech output rapidly increases. It's like an explosion where about 10 new words are learned every day. It seems as if children suddenly understand how they can use language to interact with their surroundings. At 21 months of age the 100-word milestone in productive vocabularies is reached (Pine, 2005). Words are still mostly content words, used to refer to concrete objects and to describe the relationship between objects with expressions such as "car there," "mommy's mug," or "doggy sleep" being common. Around their second birthday they start to use words to describe the relationship of singular and plural. After being able to say and point out "car there" and doing so for all the cars seen at that moment, all of the sudden they say "car there, many car," pointing out all the cars observed (Barner et al., 2007). With the usage of natural quantifiers infants engage verbally with the world of numerical relationships. Soon after this development, children are able to describe objects as being "two." What seems like simply naming a group of things needs in fact the development of deep lexical concepts, which rely, on the one hand, on innate structures, and which, on the other hand, is learned from conversational interactions (Carey, 2009).

Being able to name the number of things seen in their surroundings means that infants refer to lexical concepts which are concrete and abstract at the same time. The twoness of something is concrete because of being unique and distinct from being "three" or "one"; on the other hand, it is abstract because it names and highlights just one feature of the objects seen. At the same time the word "two" has a whole bundle of different significations. We are, for example, referring to two cars meaning the magnitude, to the second car meaning the numerical order, and to two gallons of water describing a continuous substance. So, while "two" always has the same numerical value, it differs in shape, color, form, and size (Wiese, 2007). To integrate all these and even more information into one lexical concept requires about six years to develop as we will elaborate in the chapter.

https://doi.org/10.1515/9783110661941-002

Our aim in this chapter is to describe how lexical concepts for natural numbers develop. In order to address the complexity of the topic, we present interlinking sections, each dealing with a distinct though connected topic.

In Section 2, we provide a brief overview of the innate knowledge of number and magnitude, as it manifests in the child's numerical development. We focus on the innate core systems being the *approximate number system (ANS)* and the *object tracking system (OTS)*. In parallel, drawing on Carey (2009), we introduce "language" as the third core system and point out which linguistic structures seem to be fundamental for the numerical development.

In Section 3, we refer to the construct "bootstrapping," and explain this most important learning tool that is needed to integrate all numerical knowledge we presume is stored separately in each of the three core systems. To do so we introduce the *knower-level theory* (Le Corre & Carey, 2007), which explains how children gain the specific lexical knowledge, constituting the vocabulary of natural numbers up to "four."

In Section 7, we describe cognitive constraints as a key learning tool. Cognitive constraints function as a parallel working process, which helps children to organize their surroundings and order all global knowledge. This process, the functioning of cognitive constraints, may be likened to children building a mental closet with drawers for all different categories.

In Section 4, we present a model of early arithmetic development, where we focus on hierarchy in the development of numerical knowledge and align this hierarchy with age.

Considering the number of theories introduced and intertwined with each other each section ends with an interim conclusion summing up what has been stated so far. We close the chapter with an all-embracing conclusion bringing all knowledge components together.

2 Innate knowledge of number and magnitude

Over the last 20 years of research a lot has been discovered about innate structures of number and their magnitude. Whereas arithmetic, dealing with the properties and operations of number, is a culturally dependent and learned tool, knowledge of magnitude is an evolutionally old structure securing, for example, survival. Dehaene (1999) calls this knowledge of magnitude, the number sense. The two systems we rely on, as well as most of the animals, are called core systems. The first core system holds the knowledge of magnitude and is called the *approximate number system*. Here the approximately cardinal value of perceived sets is represented by a physical magnitude that is roughly proportional to the number of objects, or individuals, in the set being enumerated. What is currently known from the literature is that analog magnitude representations of number are available as early as six months of age. Number represented as analog magnitude is not unusual; dimensions like brightness, loudness, and temporal duration are also represented this way. In all cases the absolute distance of two entities of greater magnitudes is increasingly harder to discriminate and underlies a function of their ratio described by Weber`s psychological law of quantifying change (Sarnecka & Carey, 2006). With this system it is possible for infants and children to process the number of big sets and compare or distinguish these if the ratio between the sets is as big as 1:2, slowly decreasing to a 2:3 ratio. Xu and Spelke (2000) could show this ability to distinguish starting with seven-month-old infants. In their experiment the children were shown displays of eight dots. After a while the children lost their interest in displays showing eight or even more than eight dots. They recovered interest only when a novel display held at least 16 dots.

The second core system is called the *object tracking system*. This system processes mental representations of individual object-files as in "one," "one-one," and "one-one-one." The symbols in this system explicitly represent discrete objects. This innate input analyzer represents implicit numbers; each object-file corresponds one-to-one with its match in the world. Being nonverbal this system does not hold any numerical information and the models are just compared as being either equal or unequal.

Wynn (1992) found that children even form expectations about how these models should interact. Five-month-old children are able to perform transformations on small sets. In her experiment children observed a small set of one or two discrete objects. After familiarization a screen hid the objects from view. The children could then only see how one entity was added or removed. In both events the child saw the action without seeing the outcome. After removing the screen children reacted with more attention to the unexpected event, which led Wynn to the supposition that they had mentally also performed the adding or subtracting task and expected the situation to match their "new" working memory model. The argument is that children are able to mentally hold and match sets like this, and they therefore can use this information to distinguish entities according to quantity.

Feigenson et al. (2002) showed that children always chose the greater quantity, if numbers ranged between one and four. In their experiment they placed up to four cookies in two opaque cookie jars of the same size using different pairings. Being just 12 months of age children always chose the cookie jar having more cookies in it, if the ratio was 1 to 2 or 2 to 3. If the pairing was 1 to 4, 2 to 4, or 3 to 4 children chose by random, showing no differentiation. Here the limit to store up to three object-files simultaneously is reached, after which performance declines. While observing how the cookies are placed the system creates a Working Memory Model, which contains one object-file for each cookie. Afterward the system must keep track of whether the object seen at one point in time, is the same *one* as that object seen at the previous point in time. The decision the system makes dictates whether an additional individual file is established, and this guarantees that a mental model of a set of three crackers will contain three cracker symbols.

This outcome leads to the interim conclusion that pre-numerical set sizes are supported by *iconic mental representations.* Each object-file held in the short-term memory is an icon for its match in the real world (Wiese, 2007). Nonetheless children have no awareness of the fact that three object-files are equivalent to the numerical quantity of three. In fact, neither of the core systems, although storing numerical content, has the power to represent natural number (Carey, 2009). Both core systems share constraints that do not allow the child to distinguish between a single object and multiple objects precisely. There are no symbols, in these core systems for plural. There are no concepts in either of these systems of core concepts with the content, some, all, or the indefinite article "a" meaning "one" (Carey, 1999: 194). Based on this argument that the complex cognitive system, holding these two systems, is incomplete, a third system, that of language, is introduced as playing a pivotal role in the development of number knowledge.

The starting point for children is learning and recognizing number words in speech. Toddlers do this very early in their development, and while observing the surrounding speech patterns they quickly understand that number words are linguistically different from other adjectives. The usage of number words differs very much from descriptive adjectives; they are used not to denote something, but to describe a relation, for example, a set size or a sequential position (Wiese, 2007). The class of number words soon forms surface concepts, meaning that toddlers know that these words are used to somehow describe the quantity and relationship of quantities to each other, and in addition, that you need to point to entities while using these words. Number words are now the "placeholder." As expressed by Negen and Sarnecka (2012), "The placeholder symbols are the memorized count list and associated counting routine – without those, the number concepts themselves would not be created. In that sense, number-concept creation may depend heavily on language" (Negen & Sarnecka, 2012).

Moving forward in their development children sequentially integrate over the years the following critical pieces of information. Around their second birthday they integrate knowledge about the stable order of the number sequence. They understand singular and plural relationships and can describe them verbally with quantifiers such as *more* or *less*. In the year following, by the third birthday, they start integrating the knowledge that a number word matched to a point sequentially further along the number line is the same number used for a larger quantity. The vocabularies of spatial (e.g., between) and temporal prepositions like "before" or "after" are crucial in the early development of number comprehension. "Before" and "after" used in context of the number line refer to changing magnitudes and can be synonym to "less" and "more."

Children gradually improve their counting by applying the five counting principles introduced by Gelman and Gallistel (1978). The first principle is the one-to-one relation; the second, the stable order of the number line; and third, the cardinal principle, or the *last word rule*, referring to the number word of the last object counted, which is the answer to the question, "How many objects are there altogether?" As we will see, children knowing how to answer this question are not necessarily aware of the fact that this answer describes all the counted objects (Fritz et al., 2013). The fourth and fifth principle, respectively, acknowledge that counting can take place without a real counterpart *(abstraction principle)*, and that the starting point of counting is flexible *(order irrelevance principle)*.

In summary, it may be asserted that to learn the natural number system children start off with three separate systems: the ANS, the OTS, and language. None of the systems independently, as we have seen, holds enough numerical information to form stable concepts of natural number. The ANS has the power to form approximate representations of large magnitudes, the OTS stores iconic representations as object-files up to a number of three. Language has the power to discriminate between individual objects and sets. Chierchia (1998) formulates this to be a semi-lattice structure (Fig. 1) which is linguistically universal. This is the basis of all *language quantification systems*, which will be used by children to link core knowledge together into numerical concepts.

{a,b,c,d,...}

{a,b,c} {a,b,d} {b,c,d} {a,c,d} ...

{a,b} {a,c} {a,d} {b,c} {b,d} {c,d} ...

a b c d = At

Fig. 1: Semi-lattice structure (Chierchia, 1998).

Thus, for children there is the need to combine all three systems to construct new stable concepts representing the various features. Therefore, what starts as a surface lexical concept increases and at approximately four years of age holds information of a stable order, that each number word is used for exactly one corresponding magnitude; it is also the last word voiced while counting a set of objects, or the spaces on a number line, and it represents the quantity counted.

3 Bootstrapping

To combine all three systems, toddlers use what Carey calls bootstrapping. This process is used to not simply match information together, but to actually construct completely new mental concepts building on the foundation of innate structures.

One could think that the knowledge that the number word "three" refers to the sum of three individual objects is simply a problem of matching the number word and the iconic representation of three object-files correctly together. One might liken this process to grown-ups learning the number line in French and matching each number word to the corresponding magnitude. If this were the case, the development of numeracy knowledge would be a relatively quick process once the three core systems were in place as they hold all the information that is needed (Sarnecka & Carey, 2006). But in fact, the development is a slow and extended process, in which each child constructs new mental concepts. We will see this development in the *knower-level theory*.

Building new concepts on the foundation of innate structures is a unique human resource. We use this resource every time we learn a cultural tool like, for example, arithmetic. To do so Carey (2009) coined the term *conceptual-role bootstrapping* – bootstrapping as a metaphor, which means to get oneself out of a situation using existing tools. In the case of a toddler confronted with the task to differentiate, describe, and work with number and magnitude, it needs to find an answer to this dilemma from within its own set of innate resources. Early on children are confronted with number in speech. They soon know that number words refer to something special, since this group of adjectives does not work like "normal" adjectives, such as adjectives of comparison; for example, number words cannot be increased. To give an example you can talk about a small dog and then about a smaller one, but you cannot talk about two dogs and then about "twoer" dogs. Only number words and color words share this feature.

Although rooted in the same class of words and being linguistically more alike, talking of "three bricks" and "red bricks" still has different reference objects. Red builds a feature belonging to the brick; three does not belong to the brick in itself, but rather describes the relationship the bricks have to each other. A motherly demand, "Hand me those three bricks," challenges the child on a very high level. The child first of all needs to understand brick and the task of "giving something." But the real work starts with analyzing the adjective "three" and its grammatical and semantic features.

To truly understand the cognitive demand on the child, let us consider the situation the child shares with its mother and what information the child gets from its innate knowledge about the phrase "Hand me those three bricks." The child and its mother play with bricks; there is a group of three bricks lying on the floor next to the child. The mother then asks the child to give them to her.

1. The word "those" indicates that mother and child have a common focus while the mother points out which group of bricks she means, making sure her child pays attention to the right group.
2. When focusing on the group of bricks, the child's innate core structures respond and the OTS builds – in this particular case – an abstract mental representation of "one–one–one." The model matches the real situation (iconic representation).
3. The iconic representation of three is held securely in the working memory during the whole analyzing process.
4. In the word "bricks" the plural "s" is grammatically decoded.
5. The number word "three" has – dependent on the depth of the lexical concept of "three" – to be matched to the iconic representation.
6. The child might count out the number; or hand three bricks over directly.

The cognitive information up to step four comes from innate structures and emerges without much effort from the child. But actually, being able to hand over any exact number of things is a learning process that children need to work on for over a year. Earlier we stated that number words form placeholders, meaning children know the words and use them, but they have not assigned deep lexical meaning to them. These surface concepts will now be enriched step by step. Le Corre and Carey (2007) describe the process of gaining knowledge of the cardinal number by evoking the *knower-level theory*.

Toddlers understanding only a "giving task" will grab any amount and hand it over. They might insist that the number matches the magnitude asked for, thereby showing they understood that a quantity was asked for but may not be able to count out the correct number. These children are referred to as "grabbers." Afterward they start working out the cardinal number of "one." Being a *One-knower* means toddlers are capable of handing over exactly one object but would grab an arbitrary amount of bricks when asked for more than one. Once the child knows what *one* means, the child will then start working out cardinal values of *two* to *four*. This development means that in the next step the number word "two" and sets containing two objects intersect. Children are then capable of naming sets of two, plus after being asked for any set of two, they can respond with the correct amount. Having conceptualized the set number of "two," they start the process on "three" in the same manner. It takes about one year to develop cardinal knowledge up to a magnitude of "four."

Referring to our example this means that to meet the mother's request our child has to be at least a three-knower since in this case the iconic representation shows three and the number word is "three."

Sarnecka (2014) affirmed that the attainment of deep lexical concepts occurs via mapping number words to the corresponding object-files. She compared the amount of time children required to step up to conceptualize the next cardinal number in different languages. Doing this she could even show that children acquiring the next knower level took a longer time when reaching the boundary to the undefined plural. If languages have a singular/plural marking system like English or in fact German, children stay *one knower* for a longer time than children growing up with languages marking singular/dual/plural as used in Slovenian or Saudi Arabic. Here the children step up to know the meaning of *two* faster but take a longer time to step up to knowing the undefined plural starting with three. If languages do not mark singular/plural like, for example, Japanese, children need more time to even become a *one knower*.

So far, the two systems of need were *quantificational language* and the *OTS*. The natural limit of the OTS is the cardinal value of four. This limit means there are no more exact mental models as a reference. From this point on the process of counting forms the basis for attainment of exact cardinal number. Children having worked out the meaning of *one* to *three* have understood how quantities can be measured by counting them out. Yet not all higher magnitudes can be detected easily. It seems as if now the ANS needs to sharpen. Le Corre and Carey (2007) tested children on magnitudes raging between 6 and 10 and could show that even though children could give any exact number asked for through counting, the capability to semantically map higher magnitudes firmly to their corresponding number word still needs, on average, about six more months to develop. They differentiated children linguistically in groups of *non-mappers* and *mappers*. When presented with magnitudes raging between 6 and 10, non-mappers answer at random while mappers estimate closely to the correct cardinal number. Le Corre and Carey (2007) interpret this to mean that even though children have understood how to count out any number asked for, they still need about six months to actually map the approximate representation held by the ANS to the corresponding number word.

Drawing the evidence together, it can be concluded that to learn natural number children rely on three innate structures. These structures being the *analog magnitude system* for cardinal number higher than 4, the *object tracking system* with its parallel individuation models for cardinal number up to three at the most four and the *system of quantificational language*. To grasp the full meaning, they have to override the limitations of the two core systems representing numerical content and map all numerical features to the number words.

What seems most important is that not one system alone has the power to represent all these pieces of information, but that all three must be intertwined together to actually build new stable and enriched lexical concepts. Therefore, although it seems as if number words form the foundation for all this development, Sarnecka (2016) states that they, the number words, are only the scaffolding. Once the new concept is attained, the scaffolding is no longer necessary, and the number word becomes just one feature among all others in the deep conceptualization of natural number.

In addition to matching the initial cardinal knowledge to number words, children need to deepen their understanding of the relationship numbers hold to each other and the relationship with regard to the number line construct. Additionally, children are constrained by the kinds of hypotheses (referred to as cognitive constraints) that they use to work out not only cardinal aspects but also ordinal dimensions of number. How these constraints work and the extent of their influence on the deep lexical concepts of natural number will be discussed in the following section.

4 Cognitive constraints

The previous section dealt with the children's ability to work out initial aspects of quantity. This means, in essence, that they learned how number words can be used to count out small quantities by applying a one-to-one correspondence of number words to matching objects. They learned that the last counted number word answers the question "how many?" Having grasped these aspects of cardinality they still do not know how each number stands for a whole set representing the magnitude of a number. This aspect is gained through counting out quantities and responding to the demand to give an exact amount seen in the "give a number" task.

They have also learned so far that the number word sequence has a fixed order that has to be maintained. This preservation of order is the linguistic feature crucial to attributing ordinal number assignment. Ordinal assignment describes through language the position or rank of a number (Wiese, 2007) and requires the conceptual development of a mental number line.

Even before children are verbally well grounded in their knowledge of the order of numbers up to ten, and before they have built up cardinal knowledge about magnitude, they are capable of arranging quantities according to their amount. The innate system used here is the ANS, which leads children to estimate quantities. With this knowledge children can through a one-to-one-correspondence work comprehensively with quantities. This knowledge means that they can put them into a sequential arrangement following their magnitude or even equally split magnitudes. Linguistically they do not use number words to do this but rather quantificational language such as *more*, *less*, or the *same* (even) (Fritz & Ricken, 2008).

Keeping these aspects of innate knowledge derived from the ANS and the OTS (see section one) in mind, the children´s task is to integrate those concepts. They do this using cognitive constraints in organizing learning that are an innate foundation in itself. Sophian (2019) states that cognitive constraints underlie the children's task to learn very complex bodies of knowledge by assuming that they start with some expectations that simplify the learning task. With respect to vocabulary learning, one might understand the challenging problem children experience here. Each unknown word has a large number of possible interpretations. It can describe focal objects, parts of an object, characteristics, or actions of an object. To narrow down their chances Markman (1990) describes the whole-object constraint to be the first one that children rely on. This means that hearing any new word children tend to search for a referential object. Using the whole-object constraint they pick out objects that have not been labeled yet and guess the new word to name the whole object. With this nominal strategy children form distinct objects. The taxonomic constraint then leads children to the assumption that single words can be used to label a whole class of objects that are taxonomically related. A third constraint, the mutual exclusivity constraint, leads the children to assume that each object has only one name, so hearing a new word they search for an unfamiliar reference, which will then be labeled. If there is no unfamiliar object this constraint is the one that helps children to override the others. If all objects are named, children seek new referents for the new word, which may be one part of the object in question. It is in this way that constraints are used heuristically to get a learning process started (Sophian, 2019).

Congruent with the counting principles described in section one, Gelman (1991) suggests these principles function as constraints. She proposes that they also need to be overridden. Shipley and Shepperson's (1990) ideas of how constraints help children's development of natural number resemble the whole object constraint of Markman (1990). The tendency to focus on whole, distinct objects leads children to interpret numbers used in counting as tags in the counting sequence (Sophian, 2019). Wiese (2007) refers to this as the nominal dimension numbers hold. This means that each tag (number word) is used like a fixed label or name for the objects counted (iconic counting).

A central task in numerical development is that of reasoning about relations. Resnick (1992) describes four levels of numerical reasoning. Roughly speaking these four levels can be divided into two levels, “protoquantitative reasoning” and “quantitative reasoning” in the preschool years, and “mathematics of numbers” and “mathematics of operations” in the school years (Sophian, 2019). Since this chapter focuses on the preschool years levels III and IV will be omitted.

The protoquantitative schemata basically describe prenumercial relations of equality and inequality as well as less-than and greater-than. As seen earlier in this development, AMS and quantificational language work together.

When children have gained knowledge of the number word sequence and have developed counting abilities, they enter the world of quantitative reasoning and are able to think about ordinal relations. With matching quantity to number words “less-than” and “greater-than,” relations hold numerical content that can in the following years be used to order magnitudes due to their exact quantity. Starting off with the development of a mental number line, ordering numbers describes the relations numbers have with each other.

5 The mental number line

The mental number line comes into existence as soon as global knowledge of magnitude and counting combine. It allows children to understand the hierarchical arrangement of number words to represent a sequence of increasing numbers and, based on this understanding, the subsequent sequences of increasing quantities (Resnick, 1983), meaning that each number has a fixed position in the hierarchy of numbers. Integrating more and more number words with their magnitudes means therefore that more positions on the number word line become “numerical.” What has to be kept in mind here is that children order magnitudes due to their fixed position coming from the number words so there is as yet no numerical knowledge of distance relations between numbers. Children at this developmental stage do not know about the successor function, meaning that all neighboring numbers/magnitudes differ by exactly *one* (Fritz et al., 2013). Tasks asking children to write out a number line show that bigger numbers are positioned at narrower intervals (Fig. 2).

The placeholder structure which the line of number words had is thereby transformed into the mental number line, where each number word will by the end of kindergarten hold knowledge about set size and position.

Children now have an abstract representation of the number line where they can locate any given number, know the neighboring numbers, and use sections of the number line to solve arithmetic problems encountered in the early years. Fuson (1992) showed that children solved contextually rich arithmetic tasks by

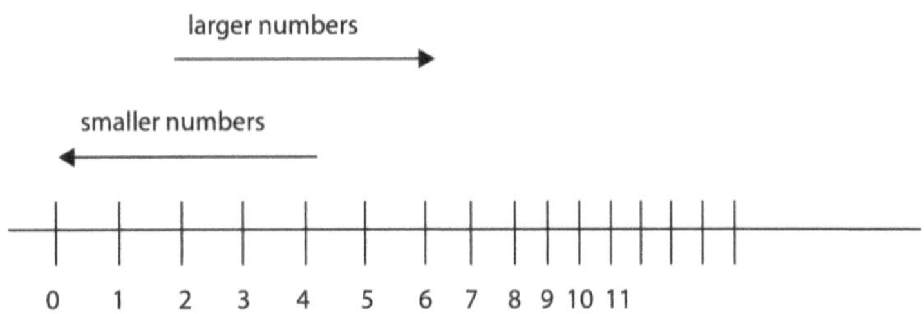

Fig. 2: Number line (Fritz et al., 2013).

"wandering up and down" the mental number line. Using the counting-all strategy, they first count out both subsets and then count the whole starting from one.

6 Second interim conclusion

The argument up to now shows that children construct cardinal and ordinal knowledge using innate structures. On the one hand they use the core systems described in section one; on the other hand, they use cognitive constraints that help them organize their surroundings and order the development of concepts.

In the preceding sections we have described two levels of learning that children consolidate in their first five years of life. The first level defines initial cardinal knowledge. Children have integrated magnitude and number word, and with this development are able to enumerate small sets of numbers. They answer questions concerning the magnitude by counting the whole set. The second step describes how ordinal aspects of numerical knowledge are formed and a mental number line is constructed. Third, children deepen their cardinal knowledge. By the age of around five they are full cardinal principle knowers, knowing about the magnitude of any number.

But having come this far in their numerical development children still lack deeper understanding of numerical relations. At this point they do not know about the part-part-whole relation or the complex ordinal relation that numbers on the mental number line always differ by a magnitude of one.

During the next two years, about ages six and seven, most children develop firm concepts of part-part-whole relations and concepts based on the relation of congruent intervals that exist between successive numbers on the number line. Sophian (2019) considers these firm concepts, part-part-whole relations and congruent intervals, to be the step from quantitative meaning to the mathematics of numbers that in her account seems to be closely related to success in schooling, since the disassociation of number from any physical referent is typically called on in school exercises.

7 Model of early arithmetic development

Taking these results into account, Fritz and Ricken (2008) (see also Ricken et al., 2013, 2011) framed a model of early arithmetic knowledge and its development. Theoretically confirmed by substantial research they could show empirically how numerical knowledge develops following six levels of arithmetic understanding. Each layer is thereby acquired separately and represents its own distinct numerical conceptual innovations. Though being hierarchical, the successive levels of understanding are not discrete but rather develop in "waves"; for example new knowledge is already present, while old knowledge is still in use (Fig. 3).

A model describing the competences that children acquire has the chance to become a didactically powerful tool, where children at the end of kindergarten and in grammar school can be tested for their actual conceptual knowledge. Since the model represents the conceptual ability that a child has, by mapping the child's competence against the model, help can be structured more easily, and children can then be guided on to acquire incrementally hierarchically higher mathematical concepts.

Each level thereby formulates the key competences children gain and will now be described in brief.

Level I: Count number and level II: The mental number line

These two levels form the foundation of any numerical learning and are basically described in the earlier sections of the chapter. they describe how children gain the fundamental knowledge and how these levels are intertwined in cognitive learning and linguistic processes, they take up most of the time of children entering the world of numerical relations.

From level III onward formal instruction is known to take over as the main influence in elaborating arithmetic knowledge and the chance to perform higher mathematics. Sophian (2019) calls this the mathematics of numbers, where "children move beyond quantitative reasoning" (p.162), where numbers no longer need references to physical amounts and where relations between numbers form the focus of learning.

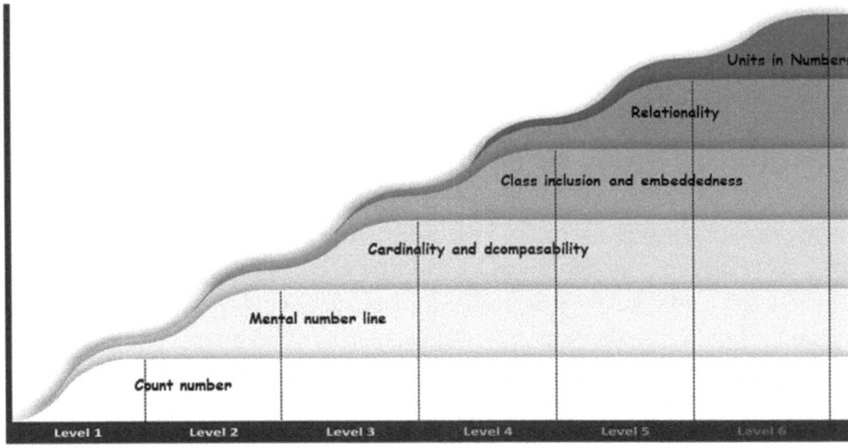

Fig. 3: Developmental model of early arithmetic learning.

Level III: Cardinality and decomposability

At the start of primary school (1st grade) children should be at level III or be "on the jump" from level II to level III. At level III the principle of *cardinal number* is developed. This basically means that a number word does not only stand for the ordinal position but that each number word unites all its counted elements. This does not happen automatically; instruction is required. To conceptualize the cardinal principle, children need to mentally integrate all counted elements into a whole. So, the question "How many are there?" is no longer being answered relying on the *last word rule* – the last number word counted indicates how many there are – but is grasped on a deeper level. The reasoning is as follows. If the quantity being counted holds eight elements, each single element is assigned a number word, and all together are assigned the mightiness, in this case "the eightness." With this mental integration, linguistic-numerical concepts now hold, alongside the feature *ordinal,* also the feature *cardinal.* Both features exist side by side, meaning that the task, which is more, seven or eight? can be answered and extended in two ways. Seven is less than eight, firstly because of its position on the number line, and then secondly, the amount seven holds fewer elements than the amount eight. Having mastered the step that number is a composite unit that consists of single elements, children also understand that numbers can be decomposed again. This very first understanding of part and whole forms is not only the basis for the next level but is also the key competence required to access effective calculating skills. While up to now all adding and subtracting tasks had to be worked out using the *counting all strategy*, children now have the ability to switch to the *counting on strategy* (Fuson, 1992). This strategy builds on the newly added knowledge that each partial quantity stands for a cardinal unit which is embedded in a whole (Fritz & Ricken, 2008). In order to solve an addition task, the first sum is counted out and the second sum is counted onward. The same strategy can now be used for a subtraction task. A subtrahend can be taken away from the whole as a unit. Also tasks where the second sum is to be completed, for example, “I need to go 9 steps to win this board game, my cube shows five. How many steps are missing?” can now be solved by counting on from the first partial quantity.

Level IV: Class inclusion and embeddedness

At level IV the most important concept is learned. They are able to work with the scheme of part-part-whole when given three partial quantities. This means they know that whatever the task is, the relationship of these three quantities is fixed in a triangular relationship and will not ever change. Now any task can be solved when two quantities are given, and in addition, children are able to create new triangles by shifting elements from one partial quantity to the other. Knowing this means children have grasped the concept of class inclusion which states that the connection of partial quantities and the whole can be expressed by the child. They now work at a more formalized cognitive level. With the understanding of the part-part-whole concept they have understood that numbers include other numbers and can be decomposed flexibly. The relation between parts and whole is determined. Thus, addition and subtraction problems, asking for the starting quantity, the exchange, or final quantity, are of equivalent difficulty.

Level V: Relationality

When children reach level V, they include all the knowledge gained on the previous levels resulting in a deep understanding of the concept of natural number. This means in essence that children have integrated ordinal and cardinal knowledge plus knowledge of their relationship in the part-part-whole composition. Numbers are now in position on the mental number line; they are representatives for both the mightiness and countable units in themselves. A newly added feature on this level is that of equidistance. This means that coming from the right of the mental number line on level II, children now know that the distance between any two neighboring numbers is exactly one. With this they can define by how many two quantities differ. This principle holds equally for distances of the same size that are equidistant independent of their position on the number line.

Level VI: Units in numbers

With reaching the final rung in the ladder in the developmental model of arithmetic learning children specify their part-part-whole concept that they gained on level IV. They already know that two partial quantities and their sum form a fixed triangle, where the partial quantities can be changed by shifting elements from one to the other partial quantity. What they now understand is that one whole can be divided into bundles of the same mightiness – 18 can, for example, be decomposed into three bundles each holding six elements. These bundles in themselves form now abstract units and are therefore countable as used in multiplication tasks. At the same time children know that each whole includes different sets of bundles of differing size (18: 3×6 or 2×9). The ability to bundle and unbundle numbers flexibly into sets of the same magnitude therefore shows that children have finally reached a stable, deep, and complex linguistic-numerical concept of natural number.

8 Composite conclusion

It can be concluded, based on the reasoning thus far, that learning arithmetic and gaining deep mathematical understanding is a process that not only takes about seven years to develop but is also dependent on different inputs.

There are, firstly, the innate structures that help children to construct cardinal and ordinal knowledge. The core systems from section one (levels l and ll) and the cognitive constraints from section two are fundamentally responsible for the childlike capability to construct concepts obtaining knowledge about "how many" and a ranking of the number line. The number word hierarchical order builds on the scaffolding that will finally be integrated into the concept itself.

This development takes place throughout the preschool years. With the transition from kindergarten to school, children leave the notion of "quantitative meaning" and enter the world of the "mathematics of number" (both Sophian, 2019).

Children's numerical concepts are now enriched due to formal instruction and become progressively more abstract. They no longer rely on a physical referent but start to work with numbers as abstract units that hold relations to each other. These relations reveal a successive order and describe firstly the establishment of cardinal knowledge and with this understanding the mightiness of a number. Numbers will no longer stand for the position in a string alone, but also for the whole set of all counted objects. Secondly, the expanded knowledge for the structure of the part-part-whole concept, describing the ability to compose and decompose quantities, is gained. Children now understand that quantities are flexible units. Thirdly, children acquire the concept of congruent intervals between all numbers on the number line and finally they gain the competence to bundle numbers into quantities of the same mightiness. These bundles then become new abstract composite units and are therefore countable.

All these different steps have been corroborated empirically and can be described in a model of early arithmetic development (Fritz & Ricken, 2008).

Although research on numerical development is still in early stages and is an ongoing process, it can be confidently stated that numerical reasoning in children underlies fundamental changes during childhood. All empirical data including those of competence and limitation suggest that numerical cognition is based on the *relational* character of numerical reasoning. During the developmental process it is the changes of what and how children relate "kinds of entities (unmeasured quantities, measured quantities, or abstract numbers), and in the kinds of relation among those entities they consider (equivalent relations, additive relations, or multiplicative relations)" (Sophian, 2019: 168).

Thinking about numbers as lexically complex concepts built via bootstrapping processes it can be emphasized that it needs the impact of both innate structures and culturally transmitted knowledge to build up firm numerical cognition.

Bibliography

Barner, David, Thalwitz, Dora, Wood, Justin, Yang, Shu-Ju & Carey, Susan (2007): On the relation between the acquisition of singular-plural morpho-syntax and the conceptual distinction between one and more than one. *Developmental Science* 10, 365–373.

Carey, Susan (1999): Sources of conceptual change. In Scholnick, Ellin K., Nelson, Katherine, Gelman, Susan A., Miller, Patricia, H. (eds.): *Conceptual Development: Piaget's Legacy.* Hillsdale, NJ: Erlbaum, 293–326.

Carey, Susan (2009): *The Origin of Concepts*. New York: Oxford University Press.
Chierchia, Gennaro (1998): Plurality of mass nouns and the notion of semantic parameter. In Rothstein, S. (ed.): *Events and Grammar*. London: Kluwer Academic Publishers, 53–113.
Dehaene, Stanislas (1999): *The Number Sense*. Cambridge: Oxford University Press.
Feigenson, Lisa, Carey, Susan & Hauser, Marc (2002): The representations underlying infants' choice of more: Object files versus analog magnitudes. *Psychological Science* 13 (2), 150–156.
Fritz, Annemarie, Ehlert, Antje & Balzer, Lars (2013): *MARKO-D: Mathematik und Rechenkonzepte im Vorschulalter – Diagnose*. Göttingen: Hogrefe.
Fritz, Annemarie & Ricken, Gabriele (2008): *Rechenschwäche*. München: Reinhardt.
Fuson, Karen C. (1992): Research on learning and teaching addition and subtraction of whole numbers. In Leinhardt, G., Putnam, R., Hattrup, R.A. (eds.): *Analysis of Arithmetic for Mathematics Teaching*. Hillsdale, NJ: Erlbaum, 53–187.
Gelman, Rochel (1991): Epigenetic foundations of knowledge structures: Initial and transcendent constructions. In Carey, Susan, Gelman, Rochel (eds.): *The epigenesist of Mind: Essays on Biology and Cognition*. Hillsdale, N.J.: Erlbaum, 293–322.
Gelman, Rochel & Gallistel, Charles R. (1978): *The Child's Understanding of Number*. Cambridge, MA: Harvard University Press.
Le Corre, Mathieu & Carey, Susan (2007): One, two, three, four, nothing more: An investigation of the conceptual sources of the verbal counting principles. *Cognition* 105 (2), 395–438.
Markman, Ellen M. (1990): Constraints children place on word meaning. *Cognitive Science* 14, 57–77.
Negen, James & Sarnecka, Barbara (2012): Number-concept acquisition and general vocabulary development. *Child Development* 83 (6), 2019–2027.
Pine, Julien M. (2005): Constructing a language: A usage-based theory of language acquisition. *Journal of Child Language* 32 (3), 697–702.
Resnick, Lauren B. (1983): Mathematics and science learning: A new conception. *Science* 220 (4596), 477–478.
Resnick, Lauren B. (1992): From protoquantities to operators: Building mathematical competence on a foundation of everyday knowledge. In Leinhardt, G., Putnam, R., Hattrup, R.A. (eds.): Analysis of Arithmetic for Mathematics Teaching. Hillsdale, NJ: Erlbaum, 373–430.
Ricken, Gabi, Fritz, Annemarie & Balzer, Lars (2011): Mathematik und Rechnen – Test zur Erfassung von Konzepten im Vorschulalter (MARKO-D). Ein Beispiel für einen niveauorientierten Ansatz. *Empirische Sonderpädagogik* 3, 256–271.
Ricken, Gabi, Fritz, Annemarie & Balzer, Lars (2013): *MARKO-D: Mathematik- und Rechenkonzepte im Vorschulalter – Diagnose (Hogrefe Vorschultests)*. Göttingen: Hogrefe.
Sarnecka, Barbara (2014): On the relation between grammatical number and cardinal numbers development. *Frontiers in Psychology* 5, 1132.
Sarnecka, Barbara (2016): How numbers are like the Earth (and unlike Faces, Loitering or Knitting). In Barner, D., Baron, A. (eds.): *Core Knowledge and Conceptual Change*. NY: Oxford University Press, 151–170.
Sarnecka, Barbara & Carey, Susan (2006): *The development of human conceptual representations*; UC Irvine, permalink https://escholarship.org/uc/item/9qv5d4tw.
Shipley, Elizabeth F. & Shepperson, Barbara (1990): Countable entities: Developmental changes. *Cognition* 34, 109–136.

Sophian, Catherine (2019): *Children's Numbers*. New York: Routledge.
Wiese, Heike (2007): The co-evolution of number concepts and counting words. *Lingua* 117, 758–772.
Wynn, Karen (1992): Addition and subtraction by human infants. *Nature* August 27, 749–750.
Xu, Fei & Spelke, Elizabeth S. (2000): Large number discrimination in 6-month-old infants. *Cognition* 74, B1–B11.

Elise Klein, Stefanie Jung, and Liane Kaufmann

A neuropsychological perspective on the development of and the interrelation between numerical and language processing

1 Introduction

Impairments in number processing (e.g., dyscalculia, acalculia) are often associated with impairments in language processing (e.g., dyslexia, aphasia) in both developmental disorders and adult neurological syndromes (e.g., Geary, 1993; Willmes, 2008). However, number processing deficits in aphasics are not necessarily reducible to the language impairment (Basso et al., 2005). Recent studies in mathematical experts even suggest the existence of two distinct, non-overlapping networks for mathematics and language (Amalric & Dehaene, 2018). This underscores the importance to understand the neuropsychological foundations of mathematics and language processing. The main aim of this chapter is to present the current neuropsychological literature of numerical cognition so that in a second step similarities and intersections with language processing can be briefly outlined.

2 Number processing and mental arithmetic

2.1 Models for numerical cognition

In the past four decades, various theoretical models have been proposed to provide a conceptual framework for the cognitive components involved in number processing and mental arithmetic (e.g., Campbell, 1994; Cipolotti & Butterworth, 1995; Dehaene, 1992; McCloskey, 1992; for an overview, see Deloche & Willmes, 2000). These models differ in the number and type of postulated representations and their interactions. The goal of all attempts was to develop a sufficiently detailed model to explain the numerical and arithmetical skills of adults with and without specific (learning) impairments. In the last two decades, the Triple-Code Model (TCM) of Dehaene and its elaborations (1992; Dehaene & Cohen, 1995, 1997; Dehaene et al., 2003) became the most influential model in numerical cognition because of its unique integration of behavioral and neurofunctional

https://doi.org/10.1515/9783110661941-003

aspects. However, it is important to note that the original TCM is based on adults (i.e., mature brain systems) only (Kaufmann et al., 2013).

2.2 The Triple-Code Model

One of the most important postulates of the TCM (Dehaene, 1992) is the distinction between a number magnitude representation on the one side and arithmetic fact retrieval from memory on the other side. Moreover, a visual number form representation in bilateral fusiform areas is assumed for recognizing Arabic digits. Fig. 1.

As regards the number magnitude representation, a bilateral fronto-parietal network around the intraparietal sulcus (IPS) subserves the representation and mental manipulation of numerical quantities (e.g., 28 + 52). In contrast, simpler tasks such as multiplication with one-digit numbers (e.g., 3 × 2) are solved by arithmetic fact retrieval in a left-hemispheric network including perisylvian language areas and the angular gyrus (Dehaene et al., 2003).

It is important to note that these representations can dissociate. For instance, patients suffering from a left-hemispheric stroke can present with a selective deficit of rote verbal knowledge (including multiplication facts, e.g., Zaunmuller et al., 2009) with preserved semantic knowledge of numerical quantities. On the other hand, patients with IPS lesions can show specific impairments of quantitative numerical knowledge (e.g., in subtraction), while knowledge of rote arithmetic facts is preserved (e.g., in simple multiplication; Dehaene & Cohen, 1997). These double dissociations suggest that numerical knowledge is processed in different codes within distinct cerebral areas.

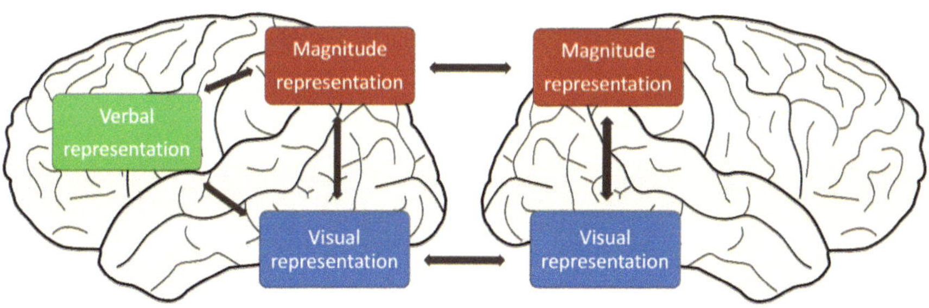

Fig. 1: Schematic Integration of Functional and Anatomical Assumptions of the Triple-Code Model (modified from Dehaene & Cohen, 1995). The brain regions postulated for the three codes (magnitude representation, verbal representation, visual representation) are projected onto lateral views of the left and right hemispheres. Arrows represent theory-based assumptions about transcoding pathways between codes rather than empirically substantiated white matter connections.

2.2.1 Neurofunctional correlates of number magnitude

The TCM incorporates evidence from studies on brain-lesioned patients, human functional neuroimaging, and primate neurophysiology in indicating that numerical cognition is subserved by a fronto-parietal network centered around the IPS. In particular, the IPS is dedicated to the mental manipulation of numerical quantities, if, for instance, it has to be decided which of two numbers is the numerically larger one. Here, numbers are coded as analogue magnitudes in an abstract notation-independent format (Piazza et al., 2007; but see Cohen Kadosh et al., 2007), each of them activating a small segment on a nonverbal, logarithmically compressed left-to-right oriented quantitative representation called "mental number line." The bilateral IPS are connected through transcallosal fibers, which enable the interplay between both hemispheres when semantic number magnitude is processed (Ratinckx et al., 2006). The IPS is active even in numerical tasks that do not necessarily require quantity processing (Eger et al., 2003, Klein et al., 2010) or that present numerical stimuli unconsciously (Naccache & Dehaene, 2001). In more complex numerical problems, the quantity-specific IPS is complemented by (pre)frontal areas involved in more general cognitive processes such as attention, working memory, or problem solving. However, the involvement of prefrontal cortices was only vaguely specified in the TCM. Based on an fMRI meta-analysis, Arsalidou and Taylor (2011) suggested a modification and extension of Dehaene's model specifying a refined picture of (pre)frontal functions and, thus, of supporting and domain-general functions implicated in solving arithmetic tasks.

Finally, the TCM considers the bilateral posterior superior parietal lobe (PSPL) to support magnitude processing via mental orientation of attention on the mental number line (Dehaene et al., 2003).

2.2.2 Neurofunctional correlates for arithmetic fact retrieval

Neurofunctional evidence on arithmetic fact retrieval primarily comes from studies that pursued the acquisition of arithmetic facts by means of drill training of difficult multiplication problems (e.g., $43 \times 9 = \_\_\_$; Delazer et al., 2003). A consistent finding was a switch from magnitude-related processing to verbally-mediated arithmetic fact retrieval with learning, reflected by a shift in activation from the bilateral fronto-parietal network of number processing to a left-hemispheric network including perisylvian language areas and angular gyrus (AG). The authors interpreted this to reflect a shift from quantity-based and working memory demanding computations to automatic retrieval of arithmetic facts from long-term memory. Consequently, it was argued that the left AG constitutes a key area for these retrieval processes (e.g., Dehaene et al., 2003).

In recent years mounting evidence suggested the additional involvement of long-term memory processes, subserved by the medial temporal lobe centered around the left hippocampus (Klein et al., 2016; Menon, 2016), while the role of the AG for arithmetic fact retrieval was challenged (e.g., Bloechle et al., 2016). The involvement of long-term memory areas in fact learning was corroborated by diffusion tensor imaging (DTI) data showing a significant increase of structural connectivity in fibers encompassing the left hippocampus but not the left AG following drill training of multiplication facts (Klein et al., 2019). Converging evidence for a central role of the hippocampus in mental arithmetic in general and arithmetic fact retrieval in particular comes from developmental research on functional connectivity in arithmetic fact learning (e.g., Rosenberg-Lee et al., 2018). For instance, hippocampal activation and functional connectivity increased with the acquisition of mathematical knowledge following a short-term training. However, the question whether the role of the hippocampus in fact learning is time-limited is not resolved yet. Recent neuropsychological single-case studies suggest that the hippocampus might be necessary only for the consolidation of arithmetic facts in memory rather than the retrieval of well-consolidated facts from memory (Delazer et al., 2019).

Taken together, it is assumed that the magnitude representation is supported by both cerebral hemispheres, while the verbal representation of arithmetic facts is situated in the left hemisphere only. Thus, only a bilateral parietal lesion would lead to a permanent impairment of the quantitative number representation.

2.3 Developmental models

Compared with the adult literature, there are far fewer studies investigating the neurofunctional correlates of number processing and calculation in children. A quantitative meta-analysis of these fMRI studies in children aged 14 years or younger is provided by Arsalidou et al. (2018). Interestingly, beyond brain areas reported to be associated with symbolic and non-symbolic mental arithmetic in adults (Arsalidou & Taylor, 2011), the authors identified additional brain areas (i.e., the insula and the claustrum) supporting number processing in children (Arsalidou & Taylor, 2018). These brain regions are not typically considered in numerical cognition models in adults.

Notably, neurofunctional trajectories of typically and atypically developing children may manifest differently at both the behavioral and the brain levels.

Below, we will focus on specific factors known to influence children's numerical development, namely, age, math proficiency, and notation.

2.3.1 Age effects

There is accumulating evidence that cerebral activation patterns in response to number processing are clearly *age dependent* (see the meta-analysis of developmental fMRI studies, Ashkenazi et al., 2013; Kaufmann et al., 2011). Compared with adults, children were found to activate more anterior (intra)parietal regions upon solving number magnitude comparison tasks (despite comparable accuracies and reaction times). A plausible explanation for the more anterior parietal activations (neighboring the postcentral gyrus, which is known to host sensory functions of the hand and fingers) is that children have a stronger need to rely on finger-based solution strategies (Kaufmann et al., 2011). Furthermore, when compared to adults, children are found to stronger recruit (pre)frontal brain regions that have been interpreted to reflect more effortful processing. Interestingly, upon investigating number processing in prematurely born six- and seven-year-old children that were just about to start formal education, Klein et al. (2018) found that gestational age predicted the frontal-to-parietal activation shift associated with number magnitude processing.

2.3.2 Effects of math proficiency

It is important to note that beyond the age of six or seven years (i.e., when formal schooling starts), age effects are inevitable confounded by math competency. Indeed, converging evidence suggests that with increasing age and schooling, cerebral activations associated with number processing become stronger in the posterior parietal cortex (including the IPS; symbolic number processing: Kaufmann et al., 2006; non-symbolic number processing: Ansari & Dhital, 2006; Cantlon et al., 2006; addition and subtraction:, Rivera et al., 2005). Hence, *math competency* might be a modulating factor for the ease with which children access number representations (e.g., Bugden & Ansari, 2011).

2.3.3 Notation effects

In children, different number notations are thought to be processed by distinct mental representations that become more overlapping with increasing age and math proficiency (Kucian & Kaufmann, 2009). Contrary to the assumption of an abstract number magnitude representation (TCM, Dehaene, 1992), typically developing children were found to process non-symbolic and symbolic magnitudes in distinct parietal and extra-parietal brain regions (Kaufmann et al., 2011): non-symbolic number processing yielded predominantly right (intra)parietal activations, located more anteriorly than the bilateral (intra)parietal activations associated with children's symbolic number processing.

With respect to calculation tasks, 9- to 12-year-old typically developing children revealed distinct activation pattern for symbolic (i.e., Arabic digits and number words) and non-symbolic (i.e., dot arrays) number formats upon solving simple subtraction tasks (Peters et al., 2016). While subtraction with symbolic formats yielded parietal activations (i.e., AG and supramarginal gyri), subtraction with non-symbolic format led to parietal and extra-parietal activations (i.e., middle occipital and superior parietal lobes, superior frontal gyrus and insula). Most likely, these differential activation patterns reflect differences in strategy use.

3 Neuropsychology of numerical cognition

The scientific interest in numerical deficiencies is quite young. It was only at the beginning of the twentieth century that Henschen (1919) coined the term "acalculia" by systematically describing acquired numerical deficits in brain-damaged patients. As regards innate numerical impairments in children, it took another 50 years before Kosc (1974) first introduced the term "dyscalculia." Notably, research on developmental dyscalculia and acquired acalculia has been separated from the beginning without systematic joint consideration and evaluation of the existing empirical evidence.

3.1 Acquired acalculia

Acquired acalculia designates the loss or impairment of the formerly intact ability to deal with numbers and/or to calculate following acquired brain pathology. In particular, up to 50% of patients with left-hemispheric and up to 30% of patients with right-hemispheric lesions suffer from acquired acalculia (Basso et al., 2005). Acalculia is often, but not necessarily, associated with other cognitive impairments such as language disturbances (Ardila & Rosselli, 2002, for a review).

The symptomatology of acalculia is not homogeneous (see Ardila & Rosselli, 2002; Domahs & Delazer, 2005, for reviews). This led to clinically oriented classifications into subtypes being proposed already in the early stages of research on acalculia (e.g., however, see the single-route model proposed by McCloskey, 1992). Indeed, several case studies in the 1990s clearly showed that specific subcomponents of numerical cognition can be impaired differentially.

Generally, the symptomatology of acquired acalculia can be divided into (i) impairments of a (quantitative) magnitude representation and (ii) calculation impairments (oral and/or written), which can be further subdivided into (a) impairments of arithmetical fact retrieval and (b) impairments of procedural as well as conceptual arithmetic knowledge. Additionally, (iii) transcoding impairments are frequently observed in patients with left-hemispheric lesions. Importantly, these capacities can be affected independently (Domahs & Delazer, 2005 for a review). This led Dehaene (1992) to formulate the multi-componential TCM, which assumes that numerical knowledge is processed in different formats within distinct cerebral areas.

3.1.1 Impairments of the (quantitative) magnitude representation

Problems with processing abstract numerical magnitudes have been observed less frequently than language-related errors. The postulated bilateral (i.e., redundant) representation of numerical quantity in the IPS makes it less vulnerable to focal brain damage. There are only a few patients reported in the literature suffering from unilateral IPS lesions of the language-dominant hemisphere who revealed problems in dealing with abstract quantities (Dehaene & Cohen, 1997; Delazer & Benke, 1997; Lemer et al., 2003). In these cases, counting or reciting multiplication tables from memory as well as reading aloud of numbers or transcoding of numerical symbols like dot patterns was preserved. In contrast, number magnitude comparison of Arabic digits was affected as well as number bisection (e.g., "What is the numerical middle between the two outer numbers 23 and 27?"). Simple subtraction tasks, which are not assumed to be solved via arithmetic fact retrieval from long-term memory, led to erroneous or no responses. Approximate calculation was impossible. Since number magnitude comparison of regularly placed sets of dots was impaired as well, the patients' problems with numerical magnitudes were not notation specific.

Delazer et al. (2006) reported a case of bilateral posterior cortical atrophy. The patient made errors in both a production and a verification version of the number bisection task (e.g., Is the middle number of this triplet also the numerical middle?" e.g., 23_26_29"; "23_25_29"). Similarly, approximate calculation tasks in which the less deviating alternative had to be selected quickly from two wrong solutions to an arithmetic problem were impaired.

Single-case studies also show that numerical knowledge can dissociate from non-numerical knowledge at the level of semantic processing. For instance, Cipolotti et al. (1991) reported a patient who was completely dysgraphic and dyslexic for all kinds of material, but showed preserved oral performance for words except for numerals above four. In particular, the patient could not discriminate Arabic digits from meaningless shapes or numerals from nonwords, nor could she produce the direct "neighbors" of a number word presented auditorily or do numerical magnitude comparisons for number words above four.

3.1.2 Impairments of arithmetical fact retrieval

Impairments of arithmetic fact retrieval are seen much more frequently. They comprise errors in highly overlearned simple addition and subtraction problems with numbers below 20, and in simple multiplication for arithmetical tables up to 9×9, which are typically retrieved from declarative long-term memory in healthy adults. Erroneous responses to simple multiplication tasks in acalculia tend to be from the same multiplication table or a close entry from another table (i.e., within-table errors, McCloskey et al., 1985; e.g., "$7 \times 8 = 48$" or "$7 \times 8 = 54$,") and not from a more distant entry or even a non-table response. Problems with arithmetic fact retrieval are not necessarily related to higher error rates only; they can also manifest in substantially longer response times, indicating the use of calculation routines or strategies in case of hampered or impossible fact retrieval (Warrington, 1982).

Zaunmuller et al. (2009) reported a patient with a severe impairment of multiplication fact retrieval following a left-hemispheric lesion of the middle cerebral artery. After a customized arithmetic fact drill training over 30 days, the patient's multiplication performance improved significantly, accompanied by a shift of activation to the contralesional right AG.

However, some studies challenge the view that simple problems are always retrieved from memory (e.g., LeFevre et al., 1996). Also Domahs and Delazer (2005) pointed out that a precise definition of what constitutes a number fact is not available. Multiplications with zero or with 10 are considered to be rule-based and have been shown to dissociate from "proper" multiplication problems (Pesenti et al., 2000).

3.1.3 Transcoding impairments

In acquired aphasia and acalculia alike, lexical and morphosyntactic errors can often be observed when number words are presented auditorily or visually.

These errors are sometimes accompanied by errors due to impaired verbal working memory. Number words are special because, unlike other long compound words (e.g., "football world championship"), semantics does not help to solve problems with the strict word order. In number words, all combinations of the elements of the number word dictionary are principally conceivable.

The two clinically most relevant transcoding pathways are between Arabic digits and spoken number words. When reading Arabic digits aloud, each of the three steps (identification of the digit chain, subsequent mental transformation into a sequence of words following fixed rules, utterance of the number word) can be disturbed.

In pure alexia, the initial (encoding) phase is impaired, while multi-digit numbers still can be compared with respect to their magnitude, because the non-dominant hemisphere can also encode Arabic digits and perform the magnitude comparison.

In lexical errors, a wrong element is selected from the same lexical number word class (ones 0–9; "teens" 11–19; decades 10–90), while the morphosyntactic structure (number word frame) of the target number word is preserved (e.g., 56 –> "seventy-six" or 411 –> "four hundred twelve"). In syntactical errors, an incorrect syntactic frame is generated which is filled with the "correct" number words corresponding to the digits in the Arabic numeral (e.g., 56 –> "five hundred and six").

When writing Arabic digits to dictation, not only lexical (e.g., "seventy-six" –> 56) and syntactical errors occur (e.g., "five-hundred-and-six" –> 56), but also errors, in which the numerical word is transcoded in sections ("term-by-term") and the additive composition principle of multi-digit numbers is not correctly applied ("three-hundred-and-sixty-eight" –> "30068").

Despite serious transcoding problems when verbalizing even single-digit numbers, a good understanding of the same numbers can be achieved by activating number-related semantic associations (e.g., 1945 –> "Hitler gone"). Patients with more severe naming disorders use the better-preserved up-counting technique for smaller numbers. Alternatively, fingers are used to show the number or the index finger is used to "imaginatively" write the Arabic numeral on a surface or in the air.

Acalculia and aphasia are often associated. Nevertheless, number processing problems and aphasia are not per se congruent (for an overview, see Willmes, 2008). Problems in reading and writing numbers occur mainly together, but also can be dissociated. Aphasic patients are typically better at choosing the larger number from a pair of Arabic multi-digit numbers than from a pair of spoken or written number words.

Likewise, various transcoding errors are found in neuropsychological conditions different from aphasia. For instance, visual processing disorders like hemianopia or visual neglect can result in leaving out the leftmost digits (e.g., Hécaen et al., 1961) even without a similar problem in semantically adequate texts.

3.1.4 Advantages and limitations of single-case studies

One of the main limitations of single-case studies is that the data collected cannot necessarily be generalized to the broader population, so they might seem less meaningful. Criticism of generalizability, however, is of less relevance when the intention is one of theory discovery and model falsification as outlined by Shallice (1988) for the case of double associations or dissociations of cognitive processes. In this situation, qualitative knowledge of a specific case may be generalized to significant segments of the population (Kennedy, 1979). However, regarding statistical generalization, caveats are necessary. For instance, if statistical procedures initially designed for data from statistically independent replications are applied to data from a single subject (e.g., to differentiate reliably between classical, strong, or weak dissociations), assumptions of independence can become questionable (Willmes, 1990). Here, statistical and psychometric aspects have to be applied carefully, for example, by using additional normative data together with the binomial model for criterion-referenced measurement (Sergent, 1988).

In this vein, single-case designs can be used to apply, to build, and to a lesser extent, to test a theory as well as in the study of unique cases because they allow a more detailed data collection and can be conducted on rare cases where large samples of similar participants are not available.

3.2 Developmental dyscalculia

Developmental dyscalculia (DD) is an innate learning disability hampering the typical development of numerical and arithmetical competencies in children (Kaufmann & Von Aster, 2012; Kucian & von Aster, 2015) that can persist into adulthood. It results in a failure to achieve adequate proficiency in arithmetic despite normal intelligence, scholastic opportunity, emotional stability, and sufficient motivation (e.g., Shalev & Gross-Tsur, 2001). Prevalence rates are rather high, ranging from 3.5% to 6.5% (Butterworth & Kovas, 2013). With respect to the etiology of DD the *core deficit hypothesis* (suggesting a deficit in the core representation of number magnitude information) has long been the dominant view (e.g., Butterworth, 2005; Butterworth et al., 2011). Even though the heterogeneity of DD has long been observed empirically (e.g., Temple, 1991 for a differentiation of fact and procedural dyscalculia), the existence of subtypes of DD was acknowledged only recently in theoretical models on numerical development (e.g., Kucian & Aster, 2015; Kaufmann & Aster, 2012 as well as Kaufmann et al., 2013 for critical reviews).

At the brain level, the core deficit hypothesis implies that compared with typically developing children those with DD have deviant (intra)parietal fMRI responses (Ashkenazi et al., 2013; Kaufmann et al., 2011). Indeed, both over- and underactivation of the IPS have been reported, which has been interpreted to reflect compensatory effortful functioning and deficient recruitment of number-relevant sites (i.e., deficient neuronal representation of numerosity), respectively. Furthermore, additional activation differences were found in (pre)frontal and occipital brain areas that were interpreted as reflecting domain-general compensatory mechanisms (Ashkenazi et al., 2013; Kaufmann et al., 2011; Peters & De Smedt, 2018; see also McCaskey et al. (2018) for a longitudinal developmental fMRI study of number processing).

It is important to note that beyond the above-mentioned core deficit of number magnitude representation, math learning difficulties may also be caused by other dysfunctional systems. Alternative accounts are, for example, (i) deficient mapping processes between number symbols and their internal magnitude representations (Rousselle & Noël, 2007; Rubinsten & Henik, 2005), (ii) domain-general deficiencies such as attention (Ashkenazi & Henik, 2010) and working memory (Rotzer et al., 2009; Toll et al., 2011), or (iii) a combined deficit of representing and manipulating numerical magnitude information (Ashkenazi et al., 2013). According to Ashkenazi et al. (2013), the latter hypothesis is especially apt to explain comorbid learning disorders (i.e., reading and math learning difficulties).

Fig. 2 provides a schematic overview of hypothetical subtypes of DD that is based on the TCM (Dehaene et al., 2003) and its postulate of three distinct types of (number) magnitude representations supported by distinct brain regions (i.e., numerical/analogue magnitude subtype: IPS, verbal subtype: left hemisphere language regions, spatial attentional subtype: posterior superior parietal lobe). According to this hypothetical model (Kaufmann & Aster, 2012), the cognitive core deficiencies of these three subtypes of DD could not only be used to develop diagnostic marker tasks but could also provide an empirically validated framework to design efficient and tailored intervention tools.

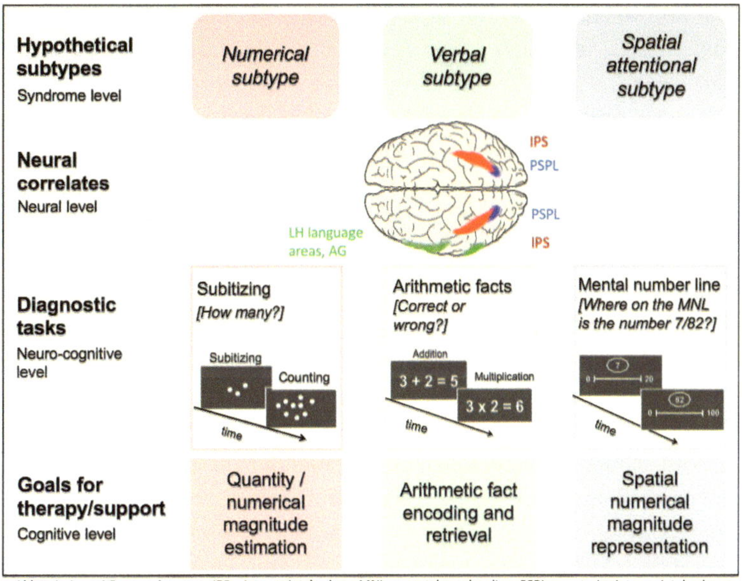

Abbreviations: AG = angular gyrus, IPS = intraparietal sulcus, MNL = mental number line, PSPL = posterior intraparietal sulcus

Fig. 2: Schematic representation of three distinct hypothetical subtypes of DD (differentiated at the behavioral and neural level, and assigned to specific diagnostic marker tasks and intervention goals).

4 Language processing

4.1 Models for language processing

Various cognitive components involved in language processing have been identified (e.g., language comprehension, reading and writing at the level of single words and in the context of sentences or functional communication) and described in various theoretical models (for language comprehension: e.g., the parallel interface model: Friederici, 2013, for reading: Coltheart, 1978; Caramazza et al., 1985; for writing: e.g., Ellis, 1982; 1988; Morton, 1980; for sentence production: Garrett, 1988; Levelt, 1993). These models have been elaborated on both typical and erroneous developmental trajectories of language processing (i.e., slip-of-the-tongue or slip-of-the-pen phenomena in the different language domains), and need to be differentiated from models describing impaired language processing in brain-injured patients (Ellis, 1982).

In principle, two types of models can be distinguished. First, in *serial models*, language processing is interpreted as a serial sequence of processing steps.

The functional architecture of serial models is often graphically displayed by box-and-arrow graphs, in which a box represents a processing module, and an arrow represents the route in the direction of processing. Crucially, modules and routes are assumed to be affected and treated selectively (Ellis & Young, 2013 for a review). Second, *interactive models*, also referred to as connectionist models, adopt hierarchically structured networks that can be activated in parallel (Dell, 1988; Dell et al., 1997; Dijkstra & de Smedt, 1996). These networks are characterized by specific nodes and edges through which activations spread in all directions within the network.

Serial and interactive models are often interpreted as contradictory. Nevertheless, there are also language models that appear as hybrid (mixed) models (e.g., Levelt, 1993).

Recently, an association between behavior and neural correlates has been taken into account in models of language development in children. These models predominantly focused on theoretic descriptions of children's observable language behavior. Neural correlates played at best a very subordinate role and the description of the neural basis of language development still remains difficult (Friederici, 2011). However, based on neuroimaging findings, domain-general aspects underlying language processing have been identified as being already established in infants. In contrast, the domain-specific aspects emerge only gradually until they are fully established in young adulthood (Skeide & Friederici, 2016).

4.2 Neuropsychology

In patients with acquired language disorders, speech or language is significantly impaired in one (or several) of the communication modalities (auditory comprehension, verbal expression, reading and writing, or functional communication).

4.2.1 Acquired language disorders

The generic term "acquired language disorders" covers various types of acquired functional impairments such as aphasia, alexia, and agraphia. The difficulties of people with aphasia can range from occasional trouble finding words to losing the ability to speak, read, or write while intelligence is unaffected. Alexia and agraphia can emerge as central and peripheral impairments of reading and writing. Central impairments often occur in association with other acquired language disorders such as aphasia (e.g., Delazer & Bartha, 2001) and may affect reading and writing in all domains (i.e., letters and digits in the context of acalculia) and output modalities [i.e., oral spelling, handwriting, and typing (Beeson & Rapcsak, 2015; Rapcsák, 2002)]. Peripheral impairments, in contrast, are related to the level of visual processing (alexia) or graphomotor planning and the execution of writing (Purcell et al., 2011).

The broad spectrum of symptoms as well as the combined and isolated occurrence of various speech and language deficits requires a precise understanding of typical language processes and their neural correlates in order to develop appropriate intervention approaches.

4.2.2 Developmental dyslexia and dysgraphia

Developmental dyslexia, like DD, belongs to the specific learning disabilities that develop in childhood and can persist into adulthood, and that cannot be explained by a child's chronological age, education, or IQ (APA, 2013). Both learning disabilities occur together more often than expected coincidentally (Wilson et al., 2015).

Developmental dyslexia describes the disability to sufficiently learn reading and/or writing despite an (above-) average intelligence and appropriate education. It is the most common subtype of learning disabilities with prevalence rates between 4 and 7% in early childhood, depending on the criteria used (Landerl et al., 2009). Crucially, a gender discrepancy is noticeable with boys being more often affected than girls. Empirical data clearly indicate that dyslexia increases an individual's risk for school dropout, low educational achievement, and unemployment (Esser et al., 2002).

Children (and adults) with dyslexia have a deficit in the mechanism of phonological awareness, the ability to recognize, identify, and manipulate syllables and phonemes within spoken language (Shaywitz & Shaywitz, 2012). They also struggle with handwriting aspects for various reasons, such as graphomotor planning and grapheme transcription (e.g., Kandel et al., 2017), correct spelling (e.g., Cidrim & Madeiro, 2017, for a review), and writing fluency (e.g., Sumner et al., 2013; but see Martlewm, 1992). They also have difficulties in recognizing and correcting errors (e.g., Horowitz-Kraus & Breznitz, 2011, for reading).

Etiological explanations for developmental dyslexia are twofold at least. On the one hand, difficulties due to basic deficits in (rapid) auditory (Tallal, 1980) and visual processing that arise from an impairment of the visual magnocellular system (Stein & Walsh, 1997) are described. On the other hand, deficits in attentional (e.g., Facoetti et al., 2001) and automatization processes (e.g., Nicolson & Fawcett, 2005) are discussed. Advocates of the latter theory argue that if cerebellar regions and the neural systems involved are affected (which are assumed to support automatization of basic articulatory and auditory skills), handwriting difficulties may arise (Nicolson & Fawcett, 2011). Liberman (1973) and later Snowling (2000) related dyslexia to a deficit in phonological awareness. Although there is supporting empirical evidence for all of these theories, not all cognitive domains are simultaneously affected and several subtypes of developmental dyslexia are identified (Heim et al., 2008).

Brain imaging techniques identified dysfunctional neuronal networks in dyslexia mainly in left-hemispheric temporo-occipital, parieto-temporal and frontal brain regions (Ashkenazi et al., 2013). Both hypo- and hyperactivations have been reported (see Vandermosten et al., 2012 for a review), in terms of either reduced performance or compensatory strategies. Parts of this network are also associated with dyscalculia, for instance, the fusiform gyrus as well as the angular gyrus, which may explain comorbidity of both learning disorders (i.e., reading and math learning difficulties).

5 Connectivity

Both number and language processing are cases of distributed and connected processing in the human brain. As such, both arithmetic (e.g., Dehaene & Cohen, 1997) and language processing (e.g., Hickok & Poeppel, 2004) rely on widespread, separate, and overlapping networks in the human brain. This underscores the importance of white matter fiber connections between the specialized grey matter brain areas.

Analogous to the visual system with a dorsal "where" and a ventral "what" stream (Mishkin et al., 1983), it is assumed that language is processed in a dual system connecting Broca's and Wenicke's areas for language processing (Hickok & Poeppel, 2004, 2007). The dorsal "where" stream along the arcuate fasciculus is dedicated to the mapping of auditory input to frontal articulatory networks (Hickok & Poeppel, 2007), while the ventral "what" stream along the external/extreme capsule system is involved in language comprehension (Saura et al., 2008). The functional link between the arcuate fascicle and language dates back to Wernicke's (1874) suggestion that a lesion of association fibers connecting the sensory and motor speech areas leads to a disconnection syndrome characterized by a failure to repeat verbal information ("conduction aphasia"). This is in line with more recent models, which propose that the arcuate fascicle connects dorsally brain regions involved in sensorimotor processes supporting speech production and speech perception (Hickok & Poeppel, 2007; Rauschecker & Scott, 2009), that is, superior temporal gyrus with premotor cortex and posterior inferior frontal gyrus. In contrast, the ventral pathway connects Wernicke's area with Broca's area through the extreme and external capsule subserving meaning (Weiller et al., 2011).

Only in 2013, a first study investigated whether the fronto-parietal network from numerical cognition can be integrated into this framework of dorsal and ventral processing pathways (Klein et al., 2013). Using probabilistic fiber tracking, the authors showed that magnitude- and fact retrieval–related processing are indeed subserved by two largely separate networks, both of them following dorsal and ventral pathways. Nevertheless, even though distinct anatomically, these networks operate as a functionally integrated circuit for mental calculation. In 2016, the TCM was complemented by adding these neuro-structural connections between cortical areas related to magnitude processing and arithmetic fact retrieval (Klein et al., 2016). These amendments suggest that the general principles associated with dorsal ("doing") and ventral ("understand what you are doing") processing streams, which seem to be instrumental for various domains (Rijntjes et al., 2012), can be adapted to number magnitude and fact retrieval processing in mental arithmetic.

So far, the work on the structural connectivity of these networks is primarily based on adults. Therefore, it is still not clear how any of these networks *develop* during childhood. Based on the work of Perani and colleagues (2011) for the case of language, the development of numerical competencies may most likely go hand-in-hand with brain maturation. In particular, studies from infancy to adulthood show later maturation of the dorsal compared to the ventral pathway (for an overview, see Friederici, 2012). While newborns have an adult-like ventral pathway at birth, the dorsal pathway that connects the temporal cortex to Broca's area develops later and is not fully matured at the age of seven (Perani et al., 2011). At this age, children still struggle with syntactically complex sentence processing. Therefore, it has been suggested that the degree of language competencies crucially depends on maturation of the dorsal pathway (Friederici, 2012). This would be in line with the idea children understand and process new numerical tasks first ventrally ("understanding what you are doing"), before more automated sequences of dorsal procedural processing can be established ("doing").

Taken together, the two neuro-cognitive networks of magnitude representation and verbal representation are (1) functionally distinct and dissociable, (2) anatomically largely distinct, but they (3) nevertheless work together in a functionally integrated way. However, it is controversially discussed whether they (4) are exclusively processed dorsally (Amalric & Dehaene, 2018) or follow general principles of dorsal and ventral processing (Klein et al., 2016).

6 Similarities in number and language processing

The question of an association between aphasia and acalculia reflects an important topic in cognitive neuroscience: To what extents are higher cognitive functions such as numerical cognition based on language? The spectrum of assumptions ranges from the mediation of arithmetic by lexical and syntactic linguistic processes to the assumption that in adulthood, arithmetic can be independent of language. In the latest amendment to the TCM, Amalric and Dehaene (2018) argued that numerical and arithmetical processing are independent of "sentence-level language processing" and "general semantic thinking." In contrast to this view, Chomsky (2017; but see Corballis, 2017) argued that all thinking is verbal/linguistic thinking.

In particular, Amalric and Dehaene (2018) suggested that the behavioral dissociation between mathematical and linguistic skills is accompanied by a major neural dissociation between brain regions associated with math and regions involved in language processing and semantics. This idea would be in line with neuropsychological findings on differences between numbers and language (e.g., Jung et al., 2015). In a single-case study, Jung et al. (2015) observed impairments in writing letters while number writing was preserved. Single letters (e.g., "E") do not necessarily carry semantic information. Also, a random assembly of letters (e.g., "EHKU") does not necessarily produce a word that carries semantic meaning. In contrast, any number as well as any random assembly of numbers (e.g., "3," "3826") conveys meaning, namely, the numerical magnitude of the respective numbers. Hence, the authors suggested that semantic magnitude information of numbers can facilitate their processing. Importantly, in the therapy of the same patient, semantic cues were then added to each letter to support the retrieval of these letters. This additional semantic information indeed successfully facilitated letter writing.

In contrast, Chomsky (2017) argued that numerical knowledge is an adjunct of human linguistic ability. This view assumes a close link between the domains of language and number processing. Indeed, there are several theoretical accounts that emphasize the importance of language for the phylogenetic and ontogenetic development of the human faculty of number processing (e.g., Carey, 1998; and already, Henschen, 1919). For instance, Wiese (2007) concluded that "it is language that opened the way for numerical cognition," suggesting that "it is no accident that the same species that possesses the language faculty as a unique trait should also be the one that developed a systematic concept of number" (p. 758). Furthermore, both letters and numbers are linked to verbal linguistic information and both natural language and number processing share the need for a recursion operation that creates embedded tree structures (Hauser et al., 2002). There is also evidence that language and number processing share a number of crucial functional characteristics at the brain level such as an overlapping ventral network, at least for specific cognitive processes such as semantic classification (Willmes et al., 2014).

Probably, neither of the two proposals is sufficiently detailed to explain the numerical performance of both healthy and cognitively impaired participants. Rather, we suggest that it may be the type of the specific problem at hand and its difficulty that determines which system is used to solve the task or how we translate our thoughts into language. Very simple arithmetic problems may be retrieved verbally as facts from long-term memory or solved applying overlearned procedures (e.g., "rule of three"). More difficult arithmetic problems may be tackled with different strategies, which do not always have to be linguistic. Nevertheless, future studies are needed to provide evidence in favor of or against one of these accounts.

Taken together, there is an ongoing debate on the building blocks of number and language processing: on the one hand, there is the view that number processing and arithmetic are independent of language processing (Amalric & Dehaene, 2018) and, on the other hand, there is the view that all thinking is linguistic thinking (Chomsky, 2017).

References

Amalric, Marie & Dehaene, Stanislas (2018): Cortical circuits for mathematical knowledge: Evidence for a major subdivision within the brain's semantic networks. *Philosophical Transactions of the Royal Society B: Biological Sciences* 373, 1740. doi:10.1098/rstb.2016.0515.

American Psychiatric Association. (2013). Diagnostic and statistical manual of mental disorders (DSM-5®). American Psychiatric Pub.

Ansari, Daniel & Dhital, Bibek (2006): Age-related changes in the activation of the intraparietal sulcus during nonsymbolic magnitude processing: An event-related functional magnetic resonance imaging study. *Journal of Cognitive Neuroscience*. MIT Press 18 (11), 1820–1828.

Ardila, Alfredo & Rosselli, Mónica (2002): Acalculia and dyscalculia. *Neuropsychology Review* 12 (4), 179–231. doi:10.1023/A:1021343508573.

Arsalidou, Marie, Pawliw-Levac, Matthew, Sadeghi, Mahsa & Pascual-Leone, Juan (2018): Brain areas associated with numbers and calculations in children: Meta-analyses of fMRI studies. *Developmental Cognitive Neuroscience* 30, 239–250.

Arsalidou, Marie & Taylor, Margot J. (2011): Is 2+2=4? Meta-analyses of brain areas needed for numbers and calculations. *NeuroImage* 54 (3), 2382–2393. doi:10.1016/j.neuroimage.2010.10.009. http://dx.doi.org/10.1016/j.neuroimage.2010.10.009.

Ashkenazi, Sarit, Black, Jessica M., Abrams, Daniel A., Hoeft, Fumiko & Menon, Vinod (2013): Neurobiological underpinnings of math and reading learning disabilities. *Journal of Learning Disabilities* 46 (6), 549–569.
Ashkenazi, Sarit & Henik, Avishai (2010): A disassociation between physical and mental number bisection in developmental dyscalculia. *Neuropsychologia* 48 (10), 2861–2868.
Basso, Anna, Caporali, Alessandra & Faglioni, Pietro (2005): Spontaneous recovery from acalculia. *Journal of the International Neuropsychological Society* 11 (1), 99–107. doi:10.1017/S1355617705050113.
Beeson, Pélagie M. & Rapcsak, Steven Z. (2015): Clinical diagnosis and treatment of spelling disorders. In *The Handbook of Adult Language Disorders*. Psychology Press, 133–154.
Bloechle, Johannes, Huber, Stefan, Bahnmueller, Julia, Rennig, Johannes, Willmes, Klaus, Cavdaroglu, Seda, Moeller, Korbinian & Klein, Elise (2016): Fact learning in complex arithmetic – the role of the angular gyrus revisited. *Human Brain Mapping* 37 (9), 3061–3079. doi:10.1002/hbm.23226.
Bugden, Stephanie & Ansari, Daniel (2011): Individual differences in children's mathematical competence are related to the intentional but not automatic processing of Arabic numerals. *Cognition* 118 (1), 32–44. doi:10.1016/j.cognition.2010.09.005. http://dx.doi.org/10.1016/j.cognition.2010.09.005.
Butterworth, Brian (2005): The development of arithmetical abilities. *Journal of Child Psychology and Psychiatry* 46 (1), 3–18.
Butterworth, Brian & Kovas, Yulia (2013): Understanding neurocognitive developmental disorders can improve education for all. *Science* 340 (6130), 300–305.
Butterworth, Brian, Varma, Sashank & Laurillard, Diana (2011): Dyscalculia: From brain to education. *Science* 332 (6033), 1049–1053. doi:10.1126/science.1201536.
Campbell, Jamie I.D. (1994): Architectures for numerical cognition. *Cognition* 53 (1), 1–44.
Cantlon, Jessica F., Brannon, Elizabeth M., Carter, Elizabeth J. & Pelphrey, Kevin A. (2006): Functional imaging of numerical processing in adults and 4-y-old children. *PLoS Biology* 4 (5), 844–854. doi:10.1371/journal.pbio.0040125.
Caramazza, Alfonso, Miceli, Gabriele, Silveri, Maria Caterina & Laudanna, Alessandro (1985): Reading mechanisms and the organisation of the lexicon: Evidence from acquired dyslexia. *Cognitive Neuropsychology* 2 (1), 81–114.
Carey, Susan (1998): Knowledge of number: Its evolution and ontogeny. *Science* 282 (5389), 641–642.
Chomsky, Noam (2017): Language architecture and its import for evolution. *Neuroscience and Biobehavioral Reviews* 81, 295–300. doi:10.1016/j.neubiorev.2017.01.053. https://doi.org/10.1016/j.neubiorev.2017.01.053.
Cidrim, Luciana & Madeiro, Francisco (2017): Studies on spelling in the context of dyslexia: A literature review. *Revista CEFAC*. SciELO Brasil 19 (6), 842–854.
Cipolotti, Lisa & Butterworth, Brian (1995): Toward a multiroute model of number processing: Impaired number transcoding with preserved calculation skills. *Journal of Experimental Psychology: General* 124 (4), 375–390. doi:10.1037/0096-3445.124.4.375.
Cipolotti, Lisa, Butterworth, Brian & Denes, Gianfranco (1991): A specific deficit for numbers in a case of dense acalculia. *Brain* 114 (6), 2619–2637.
Cohen, Kadosh, Roi, Cohen Kadosh, Kathrin, Kaas, Amanda, Henik, Avishai & Goebel, Rainer (2007): Notation-dependent and -independent representations of numbers in the parietal lobes. *Neuron* 53 (2), 307–314. doi:10.1016/j.neuron.2006.12.025.

Coltheart, Max (1978): Lexical access in simple reading tasks. In Underwood, Geoffrey (ed.): *Strategies of Information Processing*. San Diego, CA: Academic Press, 151–216.
Corballis, Michael C. (2017): The evolution of language: Sharing our mental lives. *Journal of Neurolinguistics* 43, 120–132. doi:10.1016/j.jneuroling.2016.06.003. http://dx.doi.org/10.1016/j.jneuroling.2016.06.003.
Dehaene, Stanislas (1992): Varieties of numerical abilities. *Cognition* 44 (1–2), 1–42.
Dehaene, Stanislas & Cohen, Laurent (1995): Towards an anatomical and functional model of number processing. *Mathematical Cognition* 1 (1), 83–120.
Dehaene, Stanislas & Cohen, Laurent (1997): Cerebral pathways for calculation: Double dissociation between rote verbal and quantitative knowledge of arithmetic. *Cortex* 33 (2), 219–250. doi:10.1016/S0010-9452(08)70002-9.
Dehaene, Stanislas, Piazza, Manuela, Pinel, Philippe & Cohen, Laurent (2003): Three parietal circuits for number processing. *Cognitive Neuropsychology* 20 (3–6), 487–506. doi:10.1080/02643290244000239.
Delazer, Margarete & Bartha, Lisa (2001): Transcoding and calculation in aphasia. *Aphasiology* 15 (7), 649–679.
Delazer, Margarete & Benke, Thomas (1997): Arithmetic facts without meaning. *Cortex* 33 (4), 697–710.
Delazer, Margarete, Domahs, Frank, Bartha, Lisa, Brenneis, Christian, Lochy, Aliette, Trieb, Thomas & Benke, Thomas (2003): Learning complex arithmetic – An fMRI study. *Cognitive Brain Research* 18 (1), 76–88. doi:10.1016/j.cogbrainres.2003.09.005.
Delazer, Margarete, Karner, Elfriede, Zamarian, Laura, Donnemiller, Eveline & Benke, Thomas (2006): Number processing in posterior cortical atrophy – A neuropsycholgical case study. *Neuropsychologia* 44 (1), 36–51. doi:10.1016/j.neuropsychologia.2005.04.013.
Delazer, Margarete, Zamarian, Laura, Benke, Thomas, Wagner, Michaela, Gizewski, Elke R. & Scherfler, Christoph (2019): Is an intact hippocampus necessary for answering 3 × 3? – Evidence from Alzheimer's disease. *Brain and Cognition* 134 (April), 1–8. doi:10.1016/j.bandc.2019.04.006.
Dell, Gary S. (1988): The retrieval of phonological forms in production: Tests of predictions from a connectionist model. *Journal of Memory and Language* 27 (2), 124–142.
Dell, Gary S., Schwartz, Myrna F., Martin, Nadine, Saffran, Eleanor M. & Gagnon, Deborah A (1997): Lexical access in aphasic and nonaphasic speakers. *Psychological Review* 104 (4), 801–838. doi:10.1037/0033-295X.104.4.801.
Deloche, Gerard & Willmes, Klaus (2000): Cognitive neuropsychological models of adult calculation and number processing: The role of the surface format of numbers. *European Child and Adolescent Psychiatry* 9 (2), 27–40. doi:10.1007/s007870070007.
Dijkstra, Ton & Smedt, Koenraad de (1996): Computer models in psycholinguistics: An introduction. In Dijkstra, T., and de Smedt (eds.): *Computational psycholinguistics*. London: Taylor and Francis, 3–23.
Domahs, Frank & Delazer, Margarete (2005): Concepts, procedures, and facts. *Psychology* 47 (1), 96–111. http://www.pabst-publishers.de/psychology-science/1-2005/ps_1_2005_96-111.pdf.
Eger, Evelyn, Sterzer, Philipp, Russ, Michael O., Giraud, Anne-Liese & Kleinschmidt, Andreas (2003): A supramodal number representation in human intraparietal cortex. *Neuron* 37 (4), 719–726.
Ellis, Andrew W. (1982): Spelling and writing (and reading and speaking). In Ellis, Andrew W. (ed.): *Normality and Pathology in Cognitive Functions*. London: Academic Press, 113–146.

Ellis, Andrew W. (1988): Normal writing processes and peripheral acquired dysgraphias. *Language and Cognitive Processes* 3 (2), 99–127.
Ellis, A. W., & Young, A. W. (2013): *Human cognitive neuropsychology: A textbook with readings*. Hove, East Sussex: Psychology Press
Esser, Günter, Wyschkon, Anne & Schmidt, Martin H. (2002): Was wird aus Achtjährigen mit einer Lese-und Rechtschreibstörung. *Zeitschrift für Klinische Psychologie und Psychotherapie; Forschung und Praxis* 31 (4), 235–242.
Facoetti, Andrea, Turatto, Massimo, Lorusso, Maria Luisa & Mascetti, Gian Gastone (2001): Orienting of visual attention in dyslexia: Evidence for asymmetric hemispheric control of attention. *Experimental Brain Research* 138 (1), 46–53.
Friederici, Angela D. (2011): The brain basis of language processing: From structure to function. *Physiological Reviews* 91 (4), 1357–1392. doi:10.1152/physrev.00006.2011.
Friederici, Angela D. (2012): Language development and the ontogeny of the dorsal pathway. *Frontiers in Evolutionary Neuroscience* 4 (FEB), 1–7. doi:10.3389/fnevo.2012.00003.
Friederici, Angela D. (2013): *Kognitive Strukturen des Sprachverstehens*. Berlin, Heidelberg: Springer.
Garrett, Merrill F. (1988). Processes in language production. In F. Newmeyer (Ed.), Linguistics: The Cambridge Survey (pp. 69–96). Cambridge: Cambridge University Press. doi:10.1017/CBO9780511621062.004
Geary, David C. (1993): Mathematical disabilities: Cognitive, neuropsychological, and genetic components. *Psychological Bulletin*. doi:10.1037/0033-2909.114.2.345.
Hauser, Marc D., Chomsky, Noam & Tecumseh Fitch, W. (2002): The faculty of language: What is it, who has it, and how did it evolve?. *Science* 298 (5598), 1569–1579.
Hécaen, Henri, Angelergues, René & Houillier, S. (1961): The clinical varieties of acalculias during retrorolandic lesions: Statistical approach to the problem. *Revue Neurologique* 105, 85–103.
Heim, Stefan, Tschierse, Julia, Amunts, Katrin, Wilms, Marcus, Vossel, Simone, Willmes, Klaus, Grabowska, Anna & Huber, Walter (2008): Cognitive subtypes of dyslexia. *Acta Neurobiologiae Experimentalis* 68 (1), 73–82.
Henschen, Salomon Eberhard (1919): Über Sprach-, Musik- und Rechenmechanismen und ihre Lokalisationen im Großhirn. *Zeitschrift für die gesamte Neurologie und Psychiatrie* 52 (1), 273–298. doi:10.1007/BF02872428.
Hickok, Gregory & Poeppel, David (2004): Dorsal and ventral streams: A framework for understanding aspects of the functional anatomy of language. *Cognition* 92 (1–2), 67–99.
Hickok, Gregory & Poeppel, David (2007): The cortical organization of speech processing. *Nature Reviews Neuroscience* 8 (5), 393–402. doi:10.1038/nrn2113.
Horowitz-Kraus, Tzipi & Breznitz, Zvia (2011): Error Detection Mechanism for Words and Sentences: A comparison between readers with dyslexia and skilled readers. *International Journal of Disability, Development and Education* 58 (1), 33–45.
Jung, Stefanie, Halm, Katja, Huber, Walter, Willmes, Klaus & Klein, Elise (2015): What letters can "learn" from Arabic digits – fMRI-controlled single case therapy study of peripheral agraphia. *Brain and Language* 149 (July 2015), 13–26. doi:10.1016/j.bandl.2015.06.003.
Kandel, Sonia, Lassus-Sangosse, Delphine, Grosjacques, Géraldine & Perret, Cyril (2017): The impact of developmental dyslexia and dysgraphia on movement production during word writing. *Cognitive Neuropsychology* 34 (3–4), 219–251.
Kaufmann, Liane & Aster, Michael von (2012): Diagnostik und Intervention bei Rechenstörung. *Deutsches Ärzteblatt International* 109 (45), 767–778. doi:10.3238/arztebl.2012.0767.

Kaufmann, Liane, Koppelstaetter, Florian, Siedentopf, Christian, Haala, Ilka, Haberlandt, Edda, Zimmerhackl, Lothar-Bernd, Felber, Stefan & Ischebeck, Anja (2006): Neural correlates of the number-size interference task in children. *Neuroreport*. Europe PMC Funders 17 (6), 587.

Kaufmann, Liane, Mazzocco, Michèle M., Dowker, Ann, Aster, Michael von, Göbel, Silke M., Grabner, Roland H., Henik, Avishai et al. (2013): Dyscalculia from a developmental and differential perspective. *Frontiers in Psychology* 4 (Aug), 1–5. doi:10.3389/fpsyg.2013.00516.

Kaufmann, Liane, Wood, Guilherme, Rubinsten, Orly & Henik, Avishai (2011): Meta-analyses of developmental fMRI studies investigating typical and atypical trajectories of number processing and calculation. *Developmental Neuropsychology* 36 (6), 763–787. doi:10.1080/87565641.2010.549884.

Kennedy, Mary M. (1979): Generalizing from single case studies. *Evaluation Quarterly* 3 (4), 661–678. doi:https://doi.org/10.1177/0193841X7900300409.

Klein, Elise, Korbinian, Moeller, Nuerk, Hans Christoph & Willmes, Klaus (2010): On the neuro-cognitive foundations of basic auditory number processing: An fMRI study. *Behavioral and Brain Functions* 6 (1), 42.

Klein, Elise, Moeller, Korbinian, Glauche, Volkmar, Weiller, Cornelius & Willmes, Klaus (2013): Processing pathways in mental arithmetic-evidence from probabilistic fiber tracking. *PLoS ONE* 8, 1. doi:10.1371/journal.pone.0055455.

Klein, Elise, Moeller, Korbinian, Huber, Stefan, Willmes, Klaus, Kiechl-Kohlendorfer, Ursula & Kaufmann, Liane (2018): Gestational age modulates neural correlates of intentional, but not automatic number magnitude processing in children born preterm. *International Journal of Developmental Neuroscience* 65 (October 2017), 38–44. doi:10.1016/j.ijdevneu.2017.10.004. http://dx.doi.org/10.1016/j.ijdevneu.2017.10.004.

Klein, Elise, Suchan, Julia, Moeller, Korbinian, Karnath, Hans Otto, Knops, André, Wood, Guilherme, Nuerk, Hans Christoph & Willmes, Klaus (2016): Considering structural connectivity in the triple code model of numerical cognition: Differential connectivity for magnitude processing and arithmetic facts. *Brain Structure and Function* 221 (2), 979–995. doi:10.1007/s00429-014-0951-1.

Klein, Elise, Willmes, Klaus, Bieck, Silke M., Bloechle, Johannes & Moeller, Korbinian (2019): White matter neuro-plasticity in mental arithmetic: Changes in hippocampal connectivity following arithmetic drill training. *Cortex* 114, 115–123. doi:10.1016/j.cortex.2018.05.017. https://doi.org/10.1016/j.cortex.2018.05.017.

Kosc, Ladislav (1974): Developmental dyscalculia. *Journal of Learning Disabilities* 7 (3), 164–177.

Kucian, Karin & Aster, Michael von (2015): Developmental dyscalculia. *European Journal of Pediatrics*. Springer 174 (1), 1–13.

Kucian, Karin & Kaufmann, Liane (2009): A developmental model of number representation. *Behavioral and Brain Sciences*. Cambridge University Press 32 (3–4), 340–341.

Landerl, Karin, Fussenegger, Barbara, Moll, Kristina & Willburger, Edith (2009): Dyslexia and dyscalculia: Two learning disorders with different cognitive profiles. *Journal of Experimental Child Psychology* 103 (3), 309–324. doi:10.1016/j.jecp.2009.03.006.

LeFevre, Jo-Anne, Bisanz, Jeffrey, Daley, Karen E, Buffone, Lisa, Greenham, Stephanie L & Sadesky, Gregory S (1996): Multiple routes to solution of single-digit multiplication problems. *Journal of Experimental Psychology: General*. American Psychological Association 125 (3), 284.

Lemer, Cathy, Dehaene, Stanislas, Spelke, Elizabeth & Cohen, Laurent (2003): Approximate quantities and exact number words: Dissociable systems. *Neuropsychologia* 41 (14), 1942–1958. doi:10.1016/S0028-3932(03)00123-4.

Levelt, Willem J M. (1993): *Speaking: From intention to articulation*. 1, Cambridge(MA): MIT press.
Liberman, Isabelle Y. (1973): Segmentation of the spoken word and reading acquisition. In *Bulletin of the Maryland: Orton Society, Inc.* JSTOR, 65–77.
Martlewm, Margaret (1992): Handwriting and spelling: Dyslexic children's abilities compared with children of the same chronological age and younger children of the same spelling level. *British Journal of Educational Psychology*. Wiley Online Library 62 (3), 375–390.
McCaskey, Ursina, von Aster, Michael, Maurer, Urs, Martin, Ernst, Ruth O'Gorman, Tuura & Karin, Kucian (2018): Longitudinal brain development of numerical skills in typically developing children and children with developmental dyscalculia. *Frontiers in Human Neuroscience*. Frontiers 11, 629.
McCloskey, Michael (1992): Cognitive mechanisms in numerical processing: Evidence from acquired dyscalculia. *Cognition* 44 (1–2), 107–157.
McCloskey, Michael, Caramazza, Alfonso & Basili, Annamaria (1985): Cognitive mechanisms in number processing and calculation: Evidence from dyscalculia. *Brain and cognition*. Elsevier 4 (2), 171–196.
Menon, Vinod (2016): *Memory and Cognitive Control Circuits in Mathematical Cognition and Learning. Progress in Brain Research*. 1st edn. 227, Elsevier B.V., doi:10.1016/bs.pbr.2016.04.026. http://dx.doi.org/10.1016/bs.pbr.2016.04.026.
Mishkin, Mortimer, Ungerleider, Leslie G. & Macko, Kathleen A. (1983): Object vision and spatial vision: Two cortical pathways. *Trends in Neurosciences* 6 (C), 414–417. doi:10.1016/0166-2236(83)90190-X.
Morton, John (1980): The logogen model and orthographic structure. In Frith, U. (Ed.): *Cognitive Approaches in Spelling*. London: Academic Press, 117–134.
Naccache, Lionel & Dehaene, Stanislas (2001): Unconscious semantic priming extends to novel unseen stimuli. *Cognition* 80 (3), 215–229. doi:10.1016/S0010-0277(00)00139-6.
Nicolson, Roderick I & Fawcett, Angela J (2005): Developmental dyslexia, learning and the cerebellum. In *Neurodevelopmental Disorders*. Springer, 19–36.
Nicolson, Roderick I & Fawcett, Angela J (2011): Dyslexia, dysgraphia, procedural learning and the cerebellum. In *Cortex: A Journal Devoted to the Study of the Nervous System and Behavior*. Elsevier Masson SAS.
Perani, Daniela, Saccuman, Maria C, Scifo, Paola, Anwander, Alfred, Spada, Danilo, Baldoli, Cristina, Poloniato, Antonella, Lohmann, Gabriele & Friederici, Angela D (2011): Neural language networks at birth. *Proceedings of the National Academy of Sciences*. National Acad Sciences 108 (38), 16056–16061.
Pesenti, Mauro, Depoorter, Nathalie & Seron, Xavier (2000): Noncommutability of the N+ 0 arithmetical rule: A case study of dissociated impairment. *Cortex*. Elsevier 36 (3), 445–454.
Peters, Lien & Bert, De Smedt (2018): Arithmetic in the developing brain: A review of brain imaging studies. *Developmental Cognitive Neuroscience*. Elsevier 30, 265–279.
Peters, Lien, Polspoel, Brecht, Op de Beeck, Hans & De Smedt, Bert (2016): Brain activity during arithmetic in symbolic and non-symbolic formats in 9–12 year old children. *Neuropsychologia*. Elsevier 86, 19–28.
Piazza, Manuela, Pinel, Philippe, Le Bihan, Denis & Dehaene, Stanislas (2007): A magnitude code common to numerosities and number symbols in human intraparietal cortex. *Neuron* 53 (2), 293–305. doi:10.1016/j.neuron.2006.11.022.
Purcell, Jeremy, Turkeltaub, Peter E, Eden, Guinevere F & Rapp, Brenda (2011): Examining the central and peripheral processes of written word production through meta-analysis. *Frontiers in Psychology*. Frontiers 2, 239.

Rapcsák, Tamás (2002): On minimization on stiefel manifolds. *European Journal of Operational Research*. Elsevier 143 (2), 365–376.
Ratinckx, Elie, Nuerk, Hans Christoph, van Dijck, Jean Philippe & Willmes, Klaus (2006): Effects of interhemispheric communication on two-digit arabic number processing. *Cortex* 42 (8), 1128–1137. doi:10.1016/S0010-9452(08)70225-9.
Rauschecker, Josef P. & Scott, Sophie K. (2009): Maps and streams in the auditory cortex: Nonhuman primates illuminate human speech processing. *Nature Neuroscience* 12 (6), 718–724. doi:10.1038/nn.2331.
Rijntjes, Michel, Weiller, Cornelius, Bormann, Tobias & Musso, Mariacristina (2012): The dual loop model: Its relation to language and other modalities. *Frontiers in Evolutionary Neuroscience* 4 (JUL). doi:10.3389/fnevo.2012.00009.
Rivera, S. M., Reiss, A. L., Eckert, M. A. & Menon, V. (2005): Developmental changes in mental arithmetic: Evidence for increased functional specialization in the left inferior parietal cortex. *Cerebral Cortex* 15 (11), 1779–1790. doi:10.1093/cercor/bhi055.
Rosenberg-Lee, Miriam, Iuculano, Teresa, Bae, Se Ri, Richardson, Jennifer, Qin, Shaozheng, Jolles, Dietsje & Menon, Vinod (2018): Short-term cognitive training recapitulates hippocampal functional changes associated with one year of longitudinal skill development. *Trends in Neuroscience and Education*. Elsevier GmbH 10 (June 2017). 19–29. doi:10.1016/j.tine.2017.12.001. https://doi.org/10.1016/j.tine.2017.12.001.
Rotzer, Stephanie, Loenneker, Thomas, Kucian, Karin, Martin, Ernst, Klaver, Peter & Michael, von Aster (2009): Dysfunctional neural network of spatial working memory contributes to developmental dyscalculia. *Neuropsychologia* 47 (13), 2859–2865. doi:10.1016/j.neuropsychologia.2009.06.009.
Rousselle, Laurence & Marie-Pascale, Noël (2007): Basic numerical skills in children with mathematics learning disabilities: A comparison of symbolic vs non-symbolic number magnitude processing. *Cognition*. Elsevier 102 (3), 361–395.
Rubinsten, Orly & Henik, Avishai (2005): Automatic activation of internal magnitudes: A study of developmental dyscalculia. *Neuropsychology*. American Psychological Association 19 (5), 641.
Saura, Dorothee, Kreher, Björn W., Schnell, Susanne, Kümmerera, Dorothee, Kellmeyera, Philipp, Vrya, Magnus Sebastian, Umarova, Roza et al. (2008): Ventral and dorsal pathways for language. *Proceedings of the National Academy of Sciences of the United States of America* 105 (46), 18035–18040, doi:10.1073/pnas.0805234105.
Sergent, Justine (1988): Some theoretical and methodological issues in neuropsychologial research. In Boller, F., Grafman, J. (eds): *Handbook of Neuropsychology*. Amsterdam: Elsevier.
Shalev, Ruth S & Gross-Tsur, Varda (2001): Developmental dyscalculia. *Pediatric Neurology*. Elsevier 24 (5), 337–342.
Shallice, Tim. (1988): *From Neuropsychology to Mental Structure*. Cambridge: Cambridge University Press.
Shaywitz, Sally E & Shaywitz, Bennett A (2012): Dyslexia and reading disorders. In *Psychopathology of Childhood and Adolescence: A Neuropsychological Approach*. Springer Publishing Company, 127.
Skeide, Michael A & Friederici, Angela D (2016): The ontogeny of the cortical language network. *Nature Reviews Neuroscience*. Nature Publishing Group 17 (5), 323.
Snowling, Margaret J. (2000): *Dyslexia*. Oxford: Blackwell publishing.
Stein, John & Walsh, Vincent (1997): To see but not to read; the magnocellular theory of dyslexia. *Trends in Neurosciences*. Elsevier 20 (4), 147–152.

Sumner, Emma, Connelly, Vincent & Barnett, Anna L (2013): Children with dyslexia are slow writers because they pause more often and not because they are slow at handwriting execution. *Reading and Writing*. Springer 26 (6), 991–1008.
Tallal, Paula (1980): Auditory temporal perception, phonics, and reading disabilities in children. *Brain and Language*. Elsevier 9 (2), 182–198.
Temple, Christine M. (1991): Procedural dyscalculia and number fact dyscalculia: Double dissociation in developmental dyscalculia. *Cognitive Neuropsychology*. Taylor & Francis 8 (2), 155–176.
Toll, Sylke W M, Van der Ven, Sanne H G, Kroesbergen, Evelyn H & Van Luit, Johannes E H (2011): Executive functions as predictors of math learning disabilities. *Journal of Learning Disabilities*. Sage Publications Sage CA: Los Angeles, CA 44 (6), 521–532.
Vandermosten, Maaike, Boets, Bart, Wouters, Jan & Pol, Ghesquière (2012): A qualitative and quantitative review of diffusion tensor imaging studies in reading and dyslexia. *Neuroscience and Biobehavioral Reviews*. Elsevier 36 (6), 1532–1552.
Warrington, Elizabeth K. (1982): The fractionation of arithmetical skills: A single case study. *The Quarterly Journal of Experimental Psychology Section A*. SAGE Publications Sage UK: London, England 34 (1), 31–51.
Weiller, Cornelius, Bormann, Tobias, Saur, Dorothee, Musso, Mariachristina & Rijntjes, Michel (2011): How the ventral pathway got lost – And what its recovery might mean. *Brain and Language* 118 (1–2), 29–39. doi:10.1016/j.bandl.2011.01.005.
Wernicke, Carl (1874): *Der aphasische Symptomencomplex: Eine psychologische Studie auf anatomischer Basis*. Breslau: Max Cohn & Weigert.
Wiese, Heike (2007): The co-evolution of number concepts and counting words. *Lingua* 117 (5), 758–772. doi:10.1016/j.lingua.2006.03.001.
Willmes, Klaus (1990): Statistical methods for a single-case study approach to aphasia therapy research. *Aphasiology* 4 (4), 415–436.
Willmes, Klaus (2008): Acalculia. In: Goldenberg, G, Miller BL (Hrsg). Handbook of clinical neurology, Bd 88 (3rd series). Neuropsychology and behavioral neurology. Elsevier: London. 339–358.
Willmes, Klaus, Moeller, Korbinian & Klein, Elise (2014): Where numbers meet words: A common ventral network for semantic classification. *Scandinavian Journal of Psychology* 55 (3), 202–211. doi:10.1111/sjop.12098.
Wilson, Anna J, Andrewes, Stuart G, Struthers, Helena, Rowe, Victoria M, Bogdanovic, Rajna & Waldie, Karen E (2015): Dyscalculia and dyslexia in adults: Cognitive bases of comorbidity. *Learning and Individual Differences*. Elsevier 37, 118–132.
Zaunmuller, Luisa, Domahs, Frank, Dressel, Katharina, Lonnemann, Jan, Klein, Elise, Ischebeck, Anja & Willmes, Klaus (2009): Rehabilitation of arithmetic fact retrieval via extensive practice: A combined fMRI and behavioural case-study. *Neuropsychological Rehabilitation* 19 (3), 422–443. doi:10.1080/09602010802296378.

Ann Dowker

Culture and language: How do these influence arithmetic?

International comparisons such as those carried out by TIMSS and PISA (e.g., Mullis et al., 2016a, b; OECD, 2016) tend to show considerably better arithmetical performance by children in some countries than in others. The position of countries can vary over time, but one consistent finding is that children from countries in the Far East, such as Japan, Korea, Singapore, and China, tend to perform better in arithmetic than do children in most parts of Europe and America.

Stevenson et al. (1993) looked at performance in different subjects. They found that Japanese and Korean children outperformed American children to a greater extent in mathematics than in reading. This may be in part because of specific difficulties with regard to reading that are posed by East Asian writing systems; but it is also likely that the results reflect a special focus on mathematics in East Asian countries.

1 Why do some countries seem to perform better than others?

There are a number of reasons why some countries may perform better than others. These include (1) suitability and comparability of the tests for different national or cultural groups; (2) the social and economic situation of the countries; (3) cultural attitudes toward mathematics; (4) teaching methods, and the emphasis placed on mathematics in school; (5) mathematical experience in out-of-school contexts; and (6) influences of language and in particular the counting system.

(1) *Suitability and comparability of tests*
Considering the emphasis that has been placed on international comparisons, there has been relatively little study of how suitable particular tests may be in different international contexts. Yet this may well be quite important. First of all, children in some countries and cultural groups are in general more accustomed than others to being tested and more familiar with the conventions of testing (e.g., the concept of being questioned in order to find out what they know, rather than to find out the answer). While those countries that participate

https://doi.org/10.1515/9783110661941-004

in international comparisons are unlikely to have many children who are totally inexperienced with regard to testing, there are certainly likely to be differences in the extent of testing experience that they have.

There is also the issue of context. Children are likely to perform better in a context that is familiar to them, as will be discussed in section (5); and contextual familiarity of any problem may vary between countries, locations within countries, and cultural groups.

More generally, it is difficult to standardize a test reliably for all countries in which it may be used, which creates problems for any international comparison. Kreiner and Christenson (2014) pointed out that the PISA results are not totally reliable, and can fluctuate significantly according to which test questions are used. The Rasch measurement model used by PISA is valid only if the questions are equally difficult for each country. This appears not always to be true: especially for reading, but also for mathematics.

(2) *The economic and political situation of a country*

The very poorest countries, where a high proportion of people have little or no secondary school education, are rarely included in international comparisons. However, even among those countries that are included in such comparisons, social and economic deprivation and political violence and insecurity are likely to be associated with reduced performance. The lowest-achieving countries for Grade 8 Mathematics in the TIMSS 2011 and 2015 studies (Mullis et al., 2016a, b) were mostly countries in the Middle East that are listed on the ISI Register of Developing Countries, and which were in many cases also experiencing political turmoil at the time when the testing took place. The lowest-achieving country of all was Ghana, which was also the only country in southern or western Africa to be included, and, while it is one of the more stable and prosperous countries in Africa, it still shares some of the economic problems of the region.

It should be noted at this point that although the use of international comparisons may give the impression that countries are largely homogeneous, there can be much variation within a country, which may have links to regional economic differences. For example, in the TIMSS comparisons, Massachusetts was very much above the average in mathematics achievement, while Alabama was somewhat below it. This makes a general rating for the United States unreliable. The difference is likely to reflect the economic situation of the two states: Massachusetts comes first or second and Alabama 45th in recent rankings of the 50 states in terms of GDP per capita. In China, it is well known that rural areas are much poorer economically than urban areas, but almost all testing for international comparisons has been in urban areas, and thus may not reflect the whole country.

(3) *Cultural attitudes toward mathematics*
Some cultures seem to value mathematics more highly than others, which may affect performance. There appears to be a tendency for East Asian cultures to place a higher value on academic performance in general, and mathematics and related subjects in particular, than do many other countries (Askew et al., 2010; Stevenson et al., 2000). This may lead to greater attention to, and practice in, mathematics.

The relationships between emotions and attitudes toward mathematics, mathematical performance, and national achievement in mathematics appear to be complex, though *within* any country positive attitudes toward mathematics are linked to better performance, and in particular mathematics anxiety is negatively related to performance (Foley et al., 2017; Lee, 2009). Children in high-achieving Asian countries, such as Korea and Japan, tended to demonstrate high mathematics anxiety; while those in high-achieving Western European countries, such as Finland, the Netherlands, Liechtenstein, and Switzerland, tended to demonstrate low mathematics anxiety. This may be because the high importance given to mathematics in East Asian countries makes failure more threatening. It may also reflect some cultural differences in the social acceptability of expressing anxiety about academic subjects.

(4) *Teaching methods, and amount of time devoted to arithmetic in school*
Since international comparisons have come into extensive use, there have been many proposals that countries with medium or low positions in the international league tables for school mathematics performance should emulate the teaching methods of higher scorers, such as the Pacific Rim countries. While this sounds plausible, it is important to exercise caution (Jerrim, 2011; Sturman, 2015). For one thing as stated above, factors such as the suitability of the measures used, and the economic situation of a country, could be what influences performance, as could the level of emphasis put on mathematics within a culture. For another, even as regards teaching methods there are usually many differences between mathematics instruction in different countries, and it may be difficult to isolate which factors are causing differences in mathematics performance.

For example, there has been considerable emphasis recently on the "Mathematics Mastery" approach of Singapore and other East Asian countries, and some UK schools have begun to introduce aspects of this approach. It can be difficult, however, to tease apart the numerous aspects of this approach that might contribute to better mathematical performance (Jerrim & Vignoles, 2015). For example, the Mathematics Mastery approach breaks different parts of the mathematics curriculum into units with clearly defined goals. It has a narrower, but deeper, focus than some other primary curricula such as that of the UK, aiming to teach a smaller number of topics within arithmetic in depth rather than a larger number more superficially. It also aims to ensure that all pupils have mastered each unit before going on to another one. While the approach is sometimes misinterpreted as involving whole-class teaching, with all children expected to succeed so that no allowances or individualized interventions are given to weaker pupils, in fact teachers are expected to look at the children's work, and to intervene immediately with individuals' misconceptions before moving on. This, of course, places high demands on the teacher.

Here we may possibly find another difference between East Asian countries and many others: the status of and requirements for the teaching profession in many East Asian cities may influence performance as much as any specific aspects of the curriculum. Teaching tends to be regarded as a high-status profession, which requires high academic qualifications for entry, and to involve extensive continued professional development (Ma, 1999).

Another reason for international differences in arithmetic may be the sheer amount of time devoted to it in different countries. In the UK, and certainly in England, primary school children study a wider variety of subjects than children in some other countries, resulting in less time devoted specifically to mathematics. Within mathematics, children study a wide variety of topics: not only arithmetic, but shape and space; measurement; recording data; applying mathematical knowledge to real-world problems, and so on. This could lead to English children being less good at arithmetic, but better at some other aspects of mathematics, than those in some other countries. One international comparison did indicate that English children were worse at arithmetical calculation, but better at applying mathematics to real-world problems, than those in most other European countries.

Children in Pacific Rim countries, at least the urban children who are most commonly included in international comparisons, spend more time in academic pursuits, both in school and in homework, than those in many other countries. Mathematics comprises a higher proportion of that time than it does in many other countries. Thus, the sheer amount of time devoted to mathematics may explain at least part of the superior performance in mathematics by children in these countries.

(5) Children and indeed adults may learn mathematical techniques, strategies, and concepts within a particular context, and may not transfer them to other contexts. People may not always transfer what they learn in school mathematics to real-world non-school contexts and vice versa.

Carraher et al. (1985) studied Brazilian child street traders aged between 9 and 15 years. All attended school, though many attended somewhat irregularly. They were given the same arithmetic problems in three different contexts: (1) a "street" context, where the researchers approached them as customers and asked them about prices and change; (2) a "word problem" context where they were given school-type word problems dealing with prices and change in hypothetical vending situations; and (3) a numerical context, where they were given the problems in the form of written sums. The children performed much better in the street context than on word problems and much better on word problems than on written sums. They solved almost all – 98% – of the problems correctly in the street context. Seventy-four percent of the same problems were solved correctly when presented in the form of word problems; but only 37% were solved correctly when presented in the form of written sums. By contrast, middle-class children, who attended school regularly but had no street market experience, performed better in a numerical context than a market-type context.

Other studies of the effects of schooling versus street trading experience were carried out by Saxe (1985, 1990; Saxe & Esmonde, 2005). Saxe studied the arithmetical strategies of Oksapmin children in Papua New Guinea. Some were street vendors with little or no schooling; some attended school but had no vending experience; and some had both types of experience. They were all given word problems based on the prices and profits for selling sweets. Those with more schooling relied more on written numbers and place value notation. Those with little or no schooling referred more to the specific features of the currency.

Among children with equal amounts of schooling, children with vending experience used more derived fact strategies. Those without vending experience relied more on well-learned, school-taught algorithms. Those with more schooling relied more on written numbers and place value notation. Those with little or no schooling referred more to the specific features of the currency.

Even apart from school-taught arithmetic, different cultures may have different preoccupations. For instance, in many cultures, including the UK and white Australia, age is an important preoccupation, whereas it is much less important, for example, to Aboriginal Australians. On the other hand, navigation and the estimation of distance and direction are very important in Aboriginal Australian culture: far more than they are for most white Australians. Presumably for these reasons, Kearins (1991) found that Australian Aboriginal children were better than non-Aboriginal children at estimating direction, which was traditionally very important in this group. On the other hand, non-Aboriginal Australian children were better than Aboriginal children at estimating age, which is very important in Western culture but much less so in Aboriginal culture.

Posner (1982) found that the Dioula, a mercantile group of people on the Ivory Coast, learned to use rather complex calculation strategies for trading and selling purposes. Even those merchants who had never been to school were adept at calculation. Baoule people in the same region, who were farmers rather than merchants, did not demonstrate such high-level calculation abilities. These findings indicate that groups that require sophisticated calculation strategies are likely to develop them, with or without schooling.

The extent and nature of use of technology may also influence arithmetical performance. Ever since calculators became widely used, there have been concerns that they may interfere with children's learning to calculate and/or result in an unthinking approach. Obviously much will depend on *how* calculators are used; but on the whole, studies of the effects of calculator use have suggested that these are surprisingly weak: the use of a calculator as such does not have a large effect on the development of arithmetical calculation or reasoning.

Technological aids to arithmetic did not begin with calculators. In Pacific Rim countries, many people still use the abacus, which involves the use of beads on strings, where the strings represent place value. Experienced calculators can become very fast and accurate, and even use a "mental abacus": visualizing operations on an abacus, even when there is none present (Hatano et al., 1977; Stigler, 1984). However, even highly expert abacus calculators do not always transfer their abacus skills to other arithmetical contexts. The use of the abacus alone is unlikely to explain the superior arithmetical skills of people in Pacific Rim countries.

(6) For centuries, there has been much debate as to the role of number words and numerals in arithmetic. Could we do arithmetic without words? Locke (1690) argued that small numbers can be represented without words by showing numbers of fingers, but words are needed to keep track of larger numbers. According to this theory, speakers of languages without number words would be restricted to the understanding of numbers that can be represented through fingers.

Most languages have number words at least up to 10. However, a few Native American and Native Australian languages (e.g., Warlpiri in Australia) have only words corresponding to "one, two, three, many." Some languages with somewhat more extensive counting systems have limits on how far one can count; for example, some of the languages of Papua New Guinea count by pointing to body parts and use the names of these body parts for their counts (Butterworth, 1999; Lancy, 1983). In the Kewa language "1" is represented by the right little finger, and "34" by the nose. The upper limit of the Kewa counting system is 68, while that of the Oksapmin system is 19. It is arguable that there is an upper limit on the counting sequence in a language; then this may interfere with arithmetic and quantity representation beyond that number. It may also limit the ability to understand the key mathematical concept of infinity:

Pica et al. (2004) studied 55 Munduruku-speaking participants. Munduruku is a language spoken by approximately 7,000 indigenous people in the state of

Para, Brazil. It only includes the words for the numbers one to five. The first test was to name the numbers for sets including from 1 to 15 points. The second test involved showing participants two clouds of dots and asking them to judge which of the two sets was more numerous. The third test involved approximate computation. Participants were shown short video clips illustrating simple operations. For example, approximately 20 seeds fall into a box, and then approximately 30 more were added. The participants were asked whether the total was more or less than another set (e.g., of approximately 40 seeds). The fourth test involved exact computation. The participants were again shown video clips, and were asked to give the result of a precise mathematical operation, for example, 6 seeds minus 4 seeds.

Results showed that participants could not carry out arithmetic operations with quantities above 5. For example, they could not calculate 6–4 or 7–7 accurately. However, they could do approximate arithmetic just as well as French controls! The researchers concluded that numerical approximation ability is a basic cognitive ability that is common to all human beings, and which may be independent of language.

It is possible that even these findings underestimate the mathematical abilities of speakers of languages with limited counting systems. In this study, most of the exact number tasks involved subtraction, which is usually found to be more difficult than addition.

Another study was carried out by Butterworth et al. (2008), involving studied child speakers of two Aboriginal Australian languages, Anindilyakwa and Warlpiri, and compared them with English speakers. These languages do not have number words beyond three. Nonetheless, they showed some capacity for *exact* nonverbal arithmetic.

In this study, participants were given four tasks:

(1) Memory for sets of counters. Children had to reproduce sets comprising two, three, four, five, six, eight, or nine randomly placed counters.
(2) Cross-modal matching. Children had to match numbers of counters with numbers of times that a block was tapped (numerosities ranged from 1 to 7).
(3) Nonverbal addition. Children watched an experimenter put one or more counters under a cover onto a mat; and then add more counters. They were asked to "make your mat like hers." Sums included 2 + 1, 3 + 1, 4 + 1, 1 + 2, 1 + 3, 1 + 4, 3 + 3, 4 + 2, and 5 + 3.
(4) Sharing. Children shared sets of play-dough disks among three toy bears. The trials comprised 6, 9, 7, and 10.

There were effects on performance of both age (6- to 7-year-olds performed better than 4- to 5-year-olds) and of set size (children performed better on problems involving smaller numbers than larger numbers). However, there was little or no effect of language. Speakers of the languages with the limited counting systems performed as well on these tasks as English-speaking Australian controls.

2 Transparency of counting systems

As early as 1798, Edgeworth and Edgeworth (1798) pointed out that English speakers may be at a disadvantage compared with speakers of some other languages due to the relatively irregular English counting system.

Transparency of a counting system involves two major components: (i) The regularity of the spoken number system: the degree to which it gives a clear and consistent representation of the base system (usually base 10) used in the language; and (ii) The degree and consistency of conformity between the spoken and the written number system. In practice, these usually amount to the same thing, as most languages use the same (Arabic) written counting system.

East Asian counting systems are more transparent than most others. Instead of "eleven, twelve, thirteen" and so on, such counting systems use the equivalent of "ten-one, ten-two, ten-three" and so on. Instead of "twenty, twenty-one, . . . thirty, thirty-one . . .," they use the equivalent of "two-tens, two-tens-one, . . . three-tens, three-tens-one . . .."

It is sometimes suggested that the relative transparency of Asian counting systems is a major contributory factor to the superior performance of Pacific Rim children in most aspects of arithmetic. Learning number names may be easier in systems where new numbers may be inferred rather than having to be learned by rote. One might also expect that the concept of place value would be easier to comprehend and use in a regular counting system.

There are indeed results that suggest that users of regular counting systems find it easier to count, even before they start formal schooling. Miller et al. (1995) studied counting in Chinese and American four- and five-year-olds. The two groups performed similarly in learning to count up to 12, but the Chinese children were about a year ahead of the American children in the further development and counting of higher numbers.

There is also evidence that primary school children who use transparent counting systems find it easier to represent two-digit numbers than children who use less transparent counting systems. Miura, Okamoto, and colleagues studied six-year-old children of different nationalities (Miura et al., 1988; Okamoto, 2015). Three groups used regular counting systems; Japanese, Korean, and Chinese. Three groups used less regular counting systems – American, French, and

Swedish. The children were given tasks involving representation of two-digit numbers with base ten blocks (unit blocks and tens blocks; the latter being blocks with ten segments shown on them).

The users of transparent counting systems were far more likely to represent the tens and units by means of the blocks: typically representing 42 by four tens blocks and two unit blocks. The American, French, and Swedish children tended to attempt to represent the numbers as collections of units, such as by representing the number 42 as 42 unit blocks. The researchers concluded that the users of transparent counting systems find it easier to represent multi-digit numbers and that this leads to better arithmetical performance.

Several studies have supported this view, but some have not, and in general it seems likely that the effects of using a transparent counting system are specific to some aspects of arithmetic, rather than affecting all aspects. Some of the studies have involved representing numbers on empty number lines. Siegler and Mu (2008) found that Chinese kindergarten children performed better than American children on mental number line estimation tasks involving a number line spanning from 1 to 100. Laski and Yu (2014) found that Chinese children performed better on such tasks than Chinese-American children, who in turn performed better than monolingual English-speaking American children. This could indicate either that the *extent* to which children use a transparent counting system has a significant effect on their arithmetic (for Chinese-American children, the effect may be diluted by their use of English as well as Chinese) or, perhaps more likely, that both linguistic and educational factors are important to children's number representation.

On the other hand, Muldoon et al. (2011) did not find any such differences in number line performance between Chinese and Scottish four- and five-year-olds, despite the fact that that the Chinese children performed better than the Scottish children on a standardized arithmetic test.

Mark and Dowker (2015) studied children in Chinese and English medium schools in Hong Kong. They found that those in the Chinese medium school did perform somewhat better at a standardized arithmetic test, and at backward and forward counting, but only younger children (6 to 7), and not older children (8 to 9), showed group differences in reading and comparing two-digit numbers.

However, it is difficult to draw firm conclusions about the implications of these results, because there are so many other cultural and educational differences between Asian and Western children (Towse & Saxton, 1998).

The Welsh language can offer important insights here. The main counting system used for school mathematics is, like the counting systems used in Pacific Rim countries, completely regular (Roberts, 2000). The number words are easily constructed by knowing the numbers 1 to 10 and the rule for combining them.

For example, eleven in Welsh is "un deg un" (one ten one), twelve is "un deg dau" (one ten two), and twenty two is "dau ddeg dau" (two tens two).

Wales provides an unusual opportunity for research on linguistic influences on mathematics, since it is a region in which languages with both transparent and non-transparent counting systems are used in schools. In Wales, children receive either English or Welsh medium schooling within the same country, educational system, curriculum, and cultural environment. About 80% of children in Wales, like those elsewhere in the UK, receive their school education in English, but 20% attend Welsh medium schools, where they study in Welsh. Children whose parental language is English may still receive their education from age 4 entirely in Welsh. This makes it possible to compare children, who are following the same National Curriculum, and where the *only* educational difference is in the language used.

Maclean and Whitburn (1996) studied children in their first year of school, and found that those in Welsh medium schools performed better than those in English medium schools on certain numerical measures. In particular, they could count higher. However, comprehension and use of multi-digit numbers were hard to assess in their study, as most of the children were six years old or under, and had not been much exposed to oral and written representations of tens and units.

Dowker et al. (2008) carried out a study investigating the performance of numerical tasks by Welsh children who had just begun dealing with such representations (6-year-olds) and those who had greater experience (8-year-olds).

They investigated the performance of numerical tasks by 30 Welsh children who had just begun dealing with such representations (6-year-olds) and 30 who had greater experience (8-year-olds). One third of the children in each age group spoke Welsh both at home at school; one third spoke English at home but were educated at a Welsh medium school; and one third spoke English both at home and at school.

The children were given three standardized tests: the British Abilities Scales (BAS) Basic Number Skills test (2nd edition), which measures written calculation; the WISC Arithmetic subtest, which measures mental arithmetical reasoning, especially word problem solving; and the WISC Block Design subtest, which measures nonverbal reasoning.

They were also given a Number Comparison task, based on that used by Donlan and Gourlay (1999). In the Number Comparison task, 24 pairs of two-digit numbers were presented to children in a flip booklet. All participants were required to read each pair of numbers aloud before pointing to which was the bigger.

The groups did not differ in Block Design scores. They also did not differ in terms of overall arithmetical reasoning or calculation ability. A two-factor analysis of variance with School and Age as factors was applied to the WISC Arithmetic and

BAS Number Skills scores. No statistically significant differences were found between schools or age groups on the scaled score on either test. This suggests that the counting system on its own does not appear to have an impact on global arithmetical ability in otherwise culturally and educationally similar groups. But might the nature of the counting system have an effect on some more specific aspects of arithmetic?

There were indeed group differences in more specific areas of arithmetical ability – notably in the ability to read and judge number pairs, as shown by the Number Comparison Task. The composite Comparison Error score was found to show highly significant differences in a two-way analysis of variance between schools and between age groups. Not surprisingly, eight-year-olds performed better than six-year-olds. Children who spoke Welsh both at home and at school performed better than those who spoke Welsh only at school, who in turn performed better than those who spoke English both at home and at school. These results suggest that the transparency of the counting system does not have a global effect on arithmetical performance when other aspects of education and culture are kept constant, and is therefore unlikely to be the sole or main reason for superior performance by children in Pacific Rim countries. However, it does appear to have *specific* effects on performance in particular aspects of arithmetic.

This appears to be supported by other studies of Welsh children. Dowker and Roberts (2015) studied children in English and Welsh medium schools in Wales. The study found a trend for children in Welsh medium schools to be more accurate and quicker on number line tasks, but the difference did not quite reach significance. However, the Welsh medium pupils did show significantly lower standard deviations than the English medium pupils, indicating more consistency and lower variability in performance.

Some languages have *less* transparent counting systems than English: in particular those such as German, Dutch, and Arabic, where the tens and units in multi-digit numbers are inverted in speech. For example, in German the written number "24" is spoken as "vier und zwanzig" (four and twenty). While this does not seem to have broad negative effects on arithmetic as a whole (Germany and the Netherlands usually do relatively well in international comparisons), it does seem to affect specific aspects of numerical abilities. Children who use such counting systems are less accurate in placing numbers on empty number lines children who use counting systems with little or no inversion (e.g., Dowker & Nuerk, 2016; Bahnmueller et al., 2018; Göbel et al., 2011; Helmreich et al., 2011; Klein et al., 2013; Lonneman et al., 2016; Moeller et al., 2015). Krinzinger et al. (2011) compared German, Austrian, French, Flemish, and Walloon second-grade children on several numerical tasks. The first two groups had such inversion effects in their language; the others did not. Results showed that inversion had a clear effect on writing Arabic numerals to dictation, but not on reading and recognizing them, or on calculations. Once again, we see specific but not global effects of the level of transparency of the counting system.

3 Conclusions

There are numerous ways in which culture may affect arithmetic: ranging from the effects of learning in different contexts, to the effects of counting in different languages. Many of these affect specific aspects of arithmetic but do not affect mathematical ability globally, or lead to strikingly different levels of performance in different contexts.

In education, it is important to remember that a child's apparent weakness in mathematics in one context does not necessarily mean that they will not be able to carry out apparently similar mathematical tasks in another context. An ideal would be to find ways of helping children to transfer knowledge and skills from one context to others, but that is often remarkably difficult.

It is important to bear in mind, when teaching children mathematics, that the language that they speak and the counting system that they use may influence how easy or difficult they find it to acquire and understand certain aspects of place value. However, the effects of the counting system do not apply to every aspect of mathematics; and even speakers of languages with very limited counting systems can acquire many number concepts and skills.

The many cultural variations that we find should not obscure the fact there is a universal potential for arithmetical reasoning: indeed the variations themselves demonstrate this potential. Arithmetical reasoning can develop in a very wide variety of contexts, not only in a conventional school context, but at home; in school; in games; in shopping, budgeting, and other financial contexts; in jobs ranging from street trading to carpentry; in cooking, sewing, and other domestic activities; and in dealing with measurement in many situations.

Bibliography

Ann, Dowker & Nuerk, Hans-Christoph (2016): Linguistic influences on mathematics. *Frontiers in Psychology* 7, 1035.

Askew, Mike, Hodgen, Jeremy, Hossain, Sarmin & Bretscher, Nicola (2010): *Values and Variables: Mathematics Education in High-Performing Countries*. London: Nuffield Foundation.

Bahnmueller, Julia, Nuerk, Hans-Christoph & Moeller, Korbinian (2018): A taxonomy proposal for types of interactions of language and place-value processing in multi-digit numbers. *Frontiers in Psychology* 9, 1024.

Butterworth, Brian (1999): *The Mathematical Brain*. London: Macmillan.

Butterworth, Brian, Reeve, Robert, Reynolds, Fiona & Lloyd, Delyth (2008): Numerical thought with and without words. *Proceedings of the National Academy of Sciences of the United States of America* 105, 13179–13184.

Carraher, Terezinha Nunes, Carraher, David William & Schliemann, Analucia Dias (1985): Mathematics in the streets and in the schools. *British Journal of Developmental Psychology* 3, 21–29.

Donlan, Chris & Gourlay, Sarah (1999): The importance of nonverbal skills in the acquisition of place-value knowledge: Evidence from normally-developing and language-impaired children. *British Journal of Developmental Psychology* 17, 1–19.

Dowker, Ann & Roberts, Manon (2015): Does the transparency of the counting system affect children's numerical abilities? *Frontiers in Psychology* 6, 945.

Dowker, Ann, D., Bala, Sheila & Lloyd, Delyth (2008): Linguistic influences on mathematical development: How important is the transparency of the counting system. *Philosophical Psychology* 21, 523–538.

Edgeworth, Maria & Edgeworth, Richard Lovell (1798): *Practical Education*. London: Johnson and Johnson.

Foley, Alana E., Herts, Julianne B., Borgonovi, Francesca, Guerriero, Sonia, Levine, Susan C. & Beilock, Sian L. (2017): The math anxiety-performance link: A global phenomenon. *Current Directions in Psychological Science* 26, 52–58.

Göbel, Silke M., Shaki, Samuel & Fischer, Martin H. (2011): The cultural number line: A review of cultural and linguistic influences on the development of number processing. *Journal of Cross-Cultural Psychology* 42, 543–565.

Hatano, Giyoo, Miyake, Yoshio & Binks, Martin G. (1977): Performance of expert abacus operators. *Cognition* 5, 47–55.

Helmreich, Iris, Zuber, Julia, Pixner, Silvia, Kaufmann, Liane, Nuerk, Hans-Christoph & Moeller, Korbinian (2011): Language effects on children's mental number line: How cross-cultural differences in number word systems affect spatial mappings of numbers in a non- verbal task. *Journal of Cross Cultural Psychology* 42, 598–613.

Jerrim, John (2011). England's "plummeting" PISA test scores between 2000 and 2009: Is the performance of our secondary school pupils really in relative decline? DoQSS Working Paper No. 11–09. London: University of London Institute of Education.

Jerrim, John & Vignoles, Anna (2015). The causal effect of East Asian 'mastery' teaching methods on English children's mathematics skills. DoQSS Working Paper No. 15–05. London: University College London Institute of Education.

Kearins, Judith (1991): Number experience and performance in Australian Aboriginal and Western children. In Durkin, Kevin, Shire, Beatrice (eds.): *Language in Mathematical Education: Research and Practice*. Milton Keynes: Open University Press, 247–255.

Klein, Elise, Bahnmueller, Julia, Mann, Anne, Pixner, Silvia, Kaufmann, Liane, Hans-Christoph., Nuerk et al. (2013): Language influences on numerical development – inversion effects on multi-digit number processing. *Frontiers in Psychology* 4, 480.

Kreiner, Svend & Christensen, Karl Bang (2014): Analyses of model fit and robustness. A new look at the PISA scaling model underlying ranking of countries according to reading literacy. *Psychometrika* 79, 210–231.

Krinzinger, Helga, Gregoire, Jacques, Desoete, Annemie, Kaufmann, Liane, Nuerk, Hans-Christoph & Willmes, Klaus (2011): Differential language effects on numerical skills in second grade. *Journal of Cross-Cultural Psychology* 42, 614–629.

Lancy, David (1983): *Cross-Cultural Studies in Cognition and Mathematics*. New York: Academic Press.

Laski, Elida V. & Yu, Quingyi Y. (2014): Number line estimation and mental addition: Examining the potential roles of language and education. *Journal of Experimental Child Psychology* 117, 29–44.

Lee, Jihyun (2009): Universals and specifics of math self-concept, math self-efficacy, and math anxiety across 41 PISA 2003 participating countries. *Learning and Individual Differences* 19, 355–365.

Locke, John (1690): *An Essay Concerning Human Understanding*. London: Printed for Tho. Basset. Sold by Edw. Mory.

Lonneman, Jan, Linkersdorfer, Janosch, Hasselhorn, Marcus & Lindberg, Sven (2016): Differences in arithmetical performance between Chinese and German children are accompanied by difference in processing of symbolic numerical magnitude. *Frontiers in Psychology* 7, 1337.

Ma, Liping (1999): *Knowing and Teaching Elementary Mathematics: Teachers' Understanding of Mathematics in China and the United States*. Mahwah, NJ: Erlbaum.

Maclean, Morag & Whitburn, Julia (1996). Number name systems and children's early number knowledge: A comparison of Welsh and English speakers. Paper presented at XVIth Biennial ISSBD Conference, Quebec City, August 1996.

Mark, Winifred & Dowker, Ann (2015): Linguistic influence on mathematical development is specific rather than pervasive: Revisiting the Chinese number advantage in Chinese and English children. *Frontiers in Psychology* 6, 203.

Miller, Kevin F., Smith, Catherine M., Zhu, Jianjun & Zhang, Houcan (1995): Preschool origins of cross-national differences in mathematical competence: The role of number-naming systems. *Psychological Science* 6, 56–60.

Miura, Irene, Kim, Chungsoon, Chang, Chih-Mei & Okamoto, Yukari (1988): Effects of language characteristics on children's cognitive representations of number: Cross-cultural comparisons. *Child Development* 59, 1445–1450.

Moeller, Korbinian, Shaki, Samuel, Goebel, Silke M. & Nuerk, Hans-Christoph (2015): Language influences number processing – a quadrilingual study. *Cognition* 136, 150–155.

Muldoon, Kevin, Simms, Victoria, Towse, John, Menzies, Victoria & Yue, Guoan (2011): Cross-cultural comparisons of 5-year-olds' estimating and mathematical ability. *Journal of Cross-Cultural Psychology* 42 669–681.

Mullis, Ina V.S., Martin, Michael O., Foy, Pierre & Hooper, Martin (2016a): *TIMSS 2015 International Results in Mathematics*. Boston, USA: International TIMSS and PIRLS Study Centre.

Mullis, Ina V.S., Martin, Michael O. & Loveless, Tom (2016b): *20 Years of TIMSS: International Trends in Mathematics and Science Achievement, Curriculum, and Instruction*. Boston, USA: International TIMSS and PIRLS Study Centre.

OECD. (2016): Organisation for Economic Co-operation and Development. (2016a). *PISA 2015: Results in Focus*. Paris: OECD.

Okamoto, Yukari (2015): Mathematics learning in the USA and Japan: Influences of language. In Kadosh, Roi Cohen, Dowker, Ann (eds.): *Oxford Handbook of Mathematical Cognition*. Oxford: Oxford University Press, 412–429.

Pica, Pierre, Lemer, Cathy, Izard, Veronique & Dehaene, Stanislas (2004): Exact and approximate arithmetic in an Amazonian indigene group. *Science* 306 (5695), 499–503.

Posner, Jill K. (1982): The development of mathematical knowledge in two West African societies. *Child Development* 53, 200–208.

Roberts, Gareth (2000): Bilingualism and number in Wales. *International Journal of Bilingual Education and Bilingualism* 3, 44–56.

Saxe, Geoffrey B. (1985): Effects of schooling on arithmetical understandings: Studies with Oksapmin children in Papua New Guinea. *Journal of Educational Psychology* 77, 503–513.

Saxe, Geoffrey B. (1990): The interplay between children's learning in school and out-of-school contexts. In Gardner, Marjorie, Greeno, James G. (eds.): *Toward a Scientific Practice of Science Education*. Hillsdale, N.J.: Erlbaum.

Saxe, Geoffrey B. & Esmonde, Indigo (2005): Studying cognition in flux: A historical treatment of fu in the shifting structure of Oksapmin mathematics. *Mind, Culture and Activity* 12, 171–225.

Siegler, Robert S. & Mu, Yan (2008): Chinese children excel on novel mathematical problems even before elementary school. *Psychological Science* 19, 759–763.

Stevenson, Harold W., Chen, Chuansheng & Lee, Shin-Ying (1993): Mathematics achievement of Chinese, Japanese and American children: Ten years later. *Science* 259, 53–58.

Stevenson, Harold W., Hofer, Barbara K. & Randel, Bruce (2000): Mathematics achievement and attitudes about mathematics in China and the West. *Journal of Psychology in Chinese Societies* 1, 1–16.

Stigler, James (1984): "Mental abacus": The effect of abacus training on Chinese children's mental calculation. *Cognitive Psychology* 16, 145–176.

Sturman, Linda (2015): What is there to learn from international surveys of mathematics achievement? In Kadosh, R. Cohen, Dowker, A. (eds.): *Oxford Handbook of Numerical Co Oxford University Press*. Oxford: Oxford University Press, 993–1017.

Towse, John & Saxton, Matthew (1998): Mathematics across national boundaries. In Donlan, Chris (ed.): *The Development of Mathematical Skills*. Hove: Psychology Press, 129–150.

Susanne Prediger and Ángela Uribe

Exploiting the epistemic role of multilingual resources in superdiverse mathematics classrooms: Design principles and insights into students' learning processes

1 Introduction

Many language policy documents worldwide have called for treating students' multiple languages as resources in subject matter classrooms (Beacco et al., 2015). Many qualitative observation studies (Adler, 2001; Barwell et al., 2016; Planas, 2018) and some quantitative intervention studies in mathematics education (Schüler-Meyer et al., 2019) have also shown that teaching approaches that activate students' multilingual resources can be beneficial.

However, most of these studies refer to classrooms with shared bilingualism, which means teachers and students share at least two languages. In contrast, Meyer et al. (2016) describe European schools as superdiverse language contexts in which more than five non-shared languages can be present in a classroom with only the language of instruction being shared. This is exemplified in the German-language context: In German schools, 30% of all students are multilingual, with most of them being second- to fourth-generation children of immigrant families. Typical classes in urban areas have five to seven languages, with only German being shared by all students and English learned as a foreign language. Teachers might speak one immigrant language, but not all. Usually, using home languages is allowed in small group work, but not encouraged and built upon.

The prevalence of non-shared multilingualism raises the question of how teaching approaches for activating multilingual resources can be transferred from classrooms with shared bilingualism to superdiverse classrooms with non-shared multilingualism.

In this paper, we argue that this transfer is possible when the focus is not only on the communicative role of multiple languages (i.e., so that students can

Acknowledgment: This paper was developed in the MuM-Multi 2: Fostering Language in Multilingual Mathematics Classrooms project, funded from 2017–2020 by the German Federal Ministry of Education and Research (BMBF; grant 01JM1703A to S. Prediger and A. Redder). We thank our colleagues for the interdisciplinary collaboration.

https://doi.org/10.1515/9783110661941-005

engage more easily in discussions about mathematics), but also on the *epistemic role of multiple languages* (Prediger et al., 2019). Section 2 of this paper presents the theoretical background of this argumentation. In Section 3, we introduce design principles and the design elements used to realize those. Sections 4 and 5 present a case study from an ongoing design research project in Grade 7, in which we explore how to adapt the design principles of multilingual mathematics classrooms for the mathematical topic of covariation. Thus, we pursue the following research question for a topic-specific case study on covariation: *What are the benefits and limitations of activating non-shared multilingual resources in a superdiverse mathematics classroom?*

The current paper is an expanded version of an unpublished conference paper on this topic (Uribe & Prediger, 2020). Whereas the empirical part is mainly preserved, the theoretical part is significantly enhanced.

2 Theoretical background

2.1 Existing research on building upon students' multilingual resources and its epistemic role

As the majority of the societies in the world are multilingual, many students are obliged to learn mathematics in languages of instruction that do not correspond to their home languages. The call for instructional approaches that build upon students' home languages as resources for mathematics learning (Barwell, 2009; Planas, 2018) has its sources in early language policy discourses (Ruíz, 1984), and it is now widely acknowledged in many educational policies (e.g., for the Council of Europe: Beacco et al., (2015)).

Many qualitative and quantitative studies have identified the benefits of activating multilingual students' home languages as resources for mathematics learning (Barwell et al., 2016; Planas 2018). These benefits include (1) higher engagement in classrooms discussions, (2) strengthened identities, (3) better connections to students' everyday school experiences, (4) better support in literacy development, and, ideally, (5) higher mathematical achievements (quantitative empirical evidence for efficacy has been provided in only few controlled trials, e.g., in Schüler-Meyer et al., 2019).

Many researchers in these studies have traced back the possible effects of using multiple languages in code-switching practices, with a strong focus on the *communicative role* of students' home languages. The hypothesized relationship was that being encouraged to use *all* languages strengthened their ability to participate in communication (1), which then results in the other possible benefits (2–5).

In recent years, researchers have also increasingly emphasized a second role of multilingual resources: the *epistemic role of languages as tools for thinking and knowledge construction*, especially for students' meaning-making processes (Barwell, 2018; Planas, 2018; Prediger et al., 2019). This epistemic role goes far beyond early ideas of simplified communication by code-switching: It has value whereby it emphasizes that connecting languages unconsciously or deliberately can reveal epistemic and didactic potentials for students' processes of meaning-making. For example, Prediger et al. (2019) show how discussing different conceptualizations of fractions in German and Turkish could enrich multilingual students' multifaceted understanding of the part–whole relationship. In these cases, multiperspectivity on mathematics concepts is strengthened by connecting a variety of language-specific nuances (Planas, 2018; Prediger et al., 2019). These benefits cannot be explained by code-switching in complementary bilingual modes (languages do not only complement each other in moments of insufficiency), but require a *connective bilingual mode* in which the individuals deliberatively or unconsciously connect their languages mentally or in external communication. Hence, the *connective bilingual mode* has been presented to realize ideas of translanguaging (Li, 2011) for mathematics learning. The relevance of the *connective bilingual mode* with its added epistemic value for knowledge construction is also the reason, why we speak of activating *multilingual* resources, in other words, the full linguistic repertoire rather than just activating *home languages* (which might be restricted to a complementary bilingual mode rather than a connective mode).

The distinction between the *communicative* and *epistemic role of languages* is crucial for superdiverse classrooms: The communicative role of multilingual resources can mainly be exploited in classrooms with *shared bilingualism*, when teachers and students share at least two languages to be used for communication. In contrast, the epistemic role of multilingual resources can also be exploited in *superdiverse language contexts* with non-shared multilingualism, even for monolingual students in these classrooms, as we will show in this paper.

2.2 The epistemic role of multilingual resources in the literacy engagement framework

The possible connection of the five potential benefits identified in Section 2.1 can be explained through Cummin's (2015) literacy engagement framework, especially in its form in Fig. 1, which was adapted to mathematics learning by including the grey boxes (Uribe & Prediger, 2020).

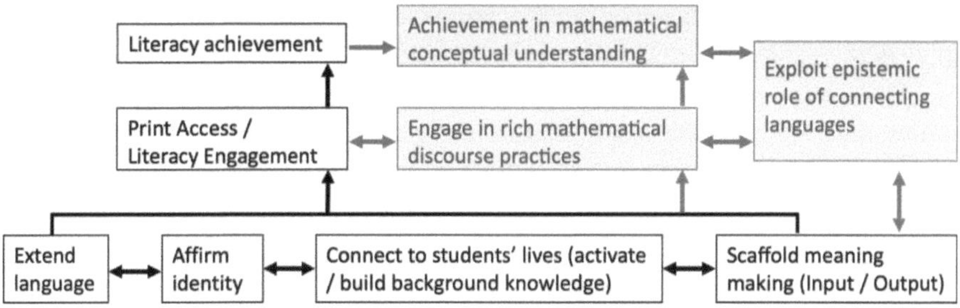

Fig. 1: Literacy engagement framework (Cummins 2015, p. 240), adapted for mathematics by adding grey boxes.

When we *extend* the students' *language* resources by allowing the use of all languages, registers, and representations, we are then not only able to strengthen students' *identities* and *connect* to students' everyday *experiences*, but we also enhance their *meaning-making* processes. This opens the base for extending students' language repertoires with more elaborated language, which can contribute to increased participation in language activities (referred to as print access and *literacy engagement* by Cummins, 2015) and students' language learning in the long run (referred to as *literacy achievement*).

Scaffolded meaning-making can contribute to *exploiting the epistemic role* of connecting languages (Schüler-Meyer et al., 2019; Barwell, 2018), which can strengthen students' access to joint language activities and, in particular, *engage students in rich mathematical discourse practices*. This can lead to *higher achievement in mathematical conceptual understanding* (Erath et al., 2018).

Exploiting epistemic potentials of language comparisons for enhancing students' conceptual understanding is of particular importance when students are in the process of learning to understand highly compacted concepts (such as functional relationships; see Prediger & Zindel, 2017) by unfolding them in joint discourse practices. This involves de-composing the condensed language (e.g., de-composing nominalizations or condensed adjectives into verbal forms) and connecting it again to the more compacted language. Planas (2019: 27) provides an example of translanguaging practices from a classroom with shared bilingualism in which the connection of nominalized expressions in one language with de-composed expressions in the second language supported students' meaning-making processes when learning about equations.

Although these possible impacts have been shown to *potentially exist*, *realizing* them deliberately in teaching-learning situations is nevertheless a challenging task for instructional designers and teachers. Realization is easier in classrooms with shared bilingualism, but the following section will present design principles and design elements for realizations in mathematics classrooms with multiple languages where all (but one) languages are not shared.

Since most of the existing studies stem from language contexts of shared multi- and bilingualism (e.g., California, Ireland, South Africa), Meyer et al. (2016) called for more research to explore how existing teaching approaches for activating multilingual resources can be transferred to superdiverse classrooms with non-shared multilingualism. As the empirical part of the paper will show, the bilingual connective mode is crucial as it can also be activated internally when only one speaker of a certain home language is present.

3 Design principles and design elements for exploiting multilingual resources epistemically

Exploiting the epistemic role of multilingual resources can best take place in teaching-learning arrangements that are based on the most relevant design principles for language-responsive teaching-learning arrangements (which will eventually be applied in monolingual contexts; see Erath et al., 2021 for an overview on the research base and Schüler-Meyer et al., 2019 for their adaptation to bi- and multilingual classrooms):

DP1 Engaging students in rich discourse practices
DP2 Macro-scaffolding along two coordinated learning trajectories
DP3 Connecting multiple representations, languages, and registers
DP4 Variation for initiating reflection

DP1 means that teachers should demand and support discourse practices so that *all* students can participate in explaining meanings, arguing, and discussing divergent ideas. DP2 means that the language learning opportunities should be sequenced in parallel to the mathematical learning opportunities starting from students' everyday resources, establishing a common meaning-related language before introducing formal, technical language. Both design principles are crucial and unchanged in both multilingual and monolingual language-responsive teaching-learning arrangements.

In a multilingual setting, DP3, *connecting multiple representations and registers*, is enriched by also connecting languages (Prediger et al., 2016), preferably in the connective bilingual mode of each individual, but perhaps only with inner speech if no partner with the same languages is present. To initiate this connection, we use the following design elements as examples:

- *Recurrent explicit and implicit prompts to use two languages* (DE3.1) are required in contexts that are usually monolingual before students dare to activate multilingual resources (Meyer et al., 2016). Also, students from monolingual German families are encouraged to use their foreign language, English, in order to treat all students equally as multilinguals. Explicit prompts are required to overcome the usually established monolingual norms (at least in most German schools), but always respect the principle of voluntary language use. Once new multilingual norms are established, this design element is no longer necessary.
- *Graphical representations* (DE3.2), such as the double number line, facilitate students' explanations and descriptions of mathematical relationships by activating the whole range of multimodal resources: Gestures, deictic means, and embodied language are more important meaning-making resources than well-elaborated technical phrases in the home language (Barwell, 2018).
- *Activities to make sense of crucial meaning-related phrases* (DE3.3). Explaining (not only translating) crucial meaning-related phrases in two languages is a key activity to mobilize the epistemic function of multiple languages. This activity usually elicits slightly different nuances of meaning and by this consolidates the conceptual understanding (Prediger et al., 2019).

The last design principle, the variation principle, DP4, means that language-related or mathematical reflection can be initiated by slightly varying some language pieces (form, function, etc.) and comparing them explicitly (Erath et al., 2021). In the multilingual setting, this specifically involves the comparison of expressions and conceptualizations for the same mathematical concept in different languages. It can be realized, for example, by the following design elements:

- *Activities for comparing phrase variations with different meanings* (DE4.1). Concept cartoons or varied word problems can initiate the students' comparison of phrases that look very similar but have different meanings. This can raise students' language awareness of small details (Dröse & Prediger, 2020).
- *Activities for comparing multiple phrases with same meaning* (DE4.2). When students provide synonymous phrases for the same key meaning-related phrase (in the language of instruction and other languages; see DE3.3), the teacher does not necessarily need to be able to translate it: Sometimes it is also interesting to compare the syntactic form. Multilingualism becomes an occasion to reflect on language and on the different words in use in terms

of conceptual relevance: When students translate German lexical means into different languages in their language repertoires, they can reflect on the resulting variations that emerge through the translation process.

In the following case studies, we illustrate how the design elements DE3.3 and DE4.2 initiate a rich mathematical reflection and moments of conceptual insight.

4 Methodological framework of the case study

4.1 Research context of the case study

4.1.1 Embedding in larger design research project

The case study presented here belongs to the larger project MuM-Multi 2 being carried out in a German-language context. The investigation of transfer possibilities between contexts of shared and non-shared multilingualism has been conducted using a design research methodology that iteratively combines the design of teaching-learning arrangements with the qualitative study of initiated teaching-learning processes (Gravemeijer & Cobb, 2006). The design elements and design principles in Section 3 were a first outcome of the design research project on the practical side. Here, we provide some empirical insights into the initiated learning processes.

4.1.2 Mathematical topic in view: Covariational reasoning

The mathematical topic in view is covariation, a core idea for functional relationships that is fundamental to students' mathematical learning: Using functional relationships, we can investigate how one quantity varies with another one as a dynamic phenomenon (Thompson & Carlson, 2017). Students first encounter covariation in the coupling of two quantities that vary simultaneously in proportional reasoning: "The more apples I buy, the higher the total price." In the first step of the learning trajectory, qualitative ideas about covariation are to be developed on the double number line, which proved to be a powerful visualization to activate students' intuitive resources (van Galen et al., 2008) and a tool that supports verbalizing and imagining quantities' values that vary smoothly or continuously. In a later step of the learning trajectory, qualitative covariation is quantified using fixed scalars, and the double number line is then used to engage students in reasoning flexibly up and down (van Galen et al., 2008).

With respect to the epistemic function of languages, covariation is an interesting exemplary topic, as it can be expressed in different degrees of compaction and reification (see Fig. 2).

Increasingly compacted and reified ↓	**E0** "The apples get more. The total price gets more." **E1** "If I buy more apples, then I also have to pay more." **E2** "The more apples, the more expensive they are" **E3** "The more the number of apples increases, the more the price increases" **E4** "The higher the number of apples, the higher the total price" **E5** "The price depends on the number of apples"	↑ Must be unfolded with de-composed language

Fig. 2: Context-embedded expressions for covariation with different degrees of compaction.

For our case study, we hypothesize that different expressions for covariation can also provide multiple approaches for understanding this compacted concept. We explore whether this kind of epistemic support can also be provided in non-shared multilingualism.

4.1.3 Teaching learning arrangement in view

The language-responsive teaching-learning arrangement on covariational reasoning was designed based on the design principles and design elements in Section 3. The learning trajectory starts from students' intuitive mathematical resources and language resource (DP2) on informal qualitative covariation. By engaging the students in rich discourse practices of explaining meanings (DP1), the aim is to develop proportional reasoning across different contexts. The double number line, introduced as the continuously used graphical representation, is used to elicit students' multiple multimodal resources (DE3.2).

Fig. 3 shows two of the initial tasks of the learning trajectory, which first are treated in language-homogeneous small groups and then discussed in the whole class. Task 1 introduces the double number line by eliciting its intuitive use. The estimation prompt leads the students into their first thinking about covariation. Task 2 asks students to express their own ideas in two languages (DE3.3); Tarkan's model shows that this is expected to result in establishing a new norm (DE3.1). Students from monolingual German families are encouraged to use their foreign language, English, in order to treat all students equally as multilinguals (DE3.1). Whereas the pre-formulated answer articulates the correspondence approach, the later whole-class discussion of estimation strategies addresses covariation ideas (DE4.1). The whole-class discussion is designed to compare multiple phrases with same meanings in different languages (DE4.2).

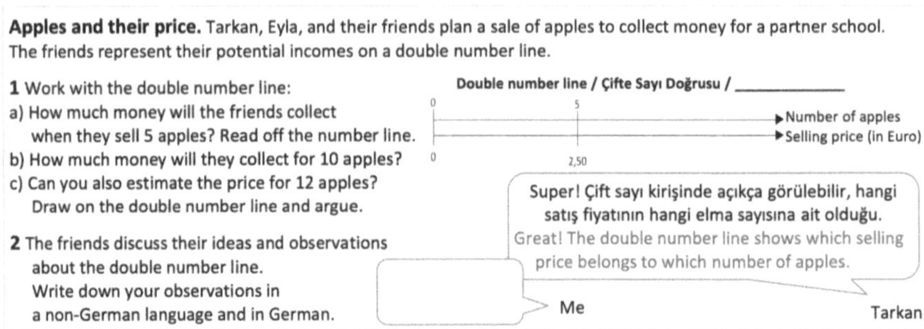

Fig. 3: Initial tasks for activating multilingual resources on covariation.

4.2 Methods of data gathering

The video data of the case study presented here stems from the first lesson of the design experiment Cycle 3 (encompassing four lessons in total), in which the intervention in a regular Grade 7 classroom was video-recorded and partly transcribed.

All students speak German (and English as their common foreign language learned in school); additional home languages are Turkish (spoken by 12 students), Polish (4 students), Arabic (2 students), Albanian (1 student), and Kurdish (1 student). The episode analyzed for this paper stems from a whole-class discussion collecting students' work on Task 2 (shown in Fig. 3).

4.3 Methods of data analysis

The transcript was qualitatively analyzed with respect to students' multiple language use, their conceptions of covariance, and the processes of compacting and unfolding using more or less condensed language. The analytic scheme was deductively derived from Prediger and Zindel (2017) and then inductively adapted to the other learning content. For the final coding scheme the codes E0–E5 were assigned for the expressions (see Fig. 2) used by the students and the teacher in order to identify communicative and epistemic benefits and limitations of work in a non-shared multilingual context. The codes were assigned based on the following criteria:

E0: Variation of quantities is described separately without expressing the relationship between them.
E1: Covariation of both quantities is expressed in complete sub-clauses with verbs connected by the conjunctions "if . . . then . . . ."
E2: Covariation of both quantities is expressed in compacted adverbs, connected, for instance, by "the more apples . . . the more . . ."
E3: Quantities are addressed explicitly by nominalizations (e.g., "the value," "the amount"), and the covariation is expressed by verbs (e.g., "increase") connected by "if . . . then . . ." or "the . . . the . . . ."
E4: Quantities are addressed explicitly using nominalizations (e.g., "the value," "the amount"), and the covariation is compacted into adjectives "the . . . the higher the price," connected by "the . . . the . . . ."
E5: Quantities are addressed explicitly by nominalizations (e.g., "the value," "the amount"), and the covariation is not addressed explicitly but compacted into a phrase such as "depends on."

5 Empirical insights into benefits and limitations of exploiting the epistemic role of multilingual resources

5.1 Qualitative analysis of three sequences

We analyzed a transcript of the whole-class discussion on students' multilingual writings about covariation on the double number line (Task 2 in Fig. 3) in order to show how the students and the teacher made sense of covariation by connecting multiple phrases (DE3.3).

Sequence 1: Separate variations in two languages

The whole-class discussion starts with a conversation on Baydar's German-Turkish writing product (Fig. 4) which is projected onto the whiteboard.

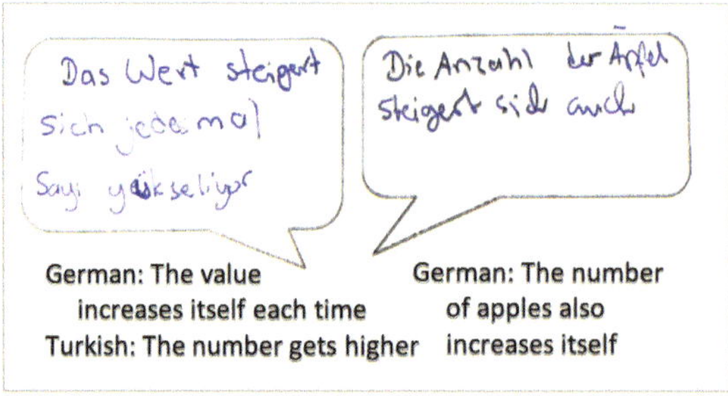

Fig. 4: Baydar's writing product (translated).

89	Teacher	Baydar has written something very nice: that the value increases each time. What is written here? [*points to Baydar's Turkish sentence*]
90	Fatma	[*translates*] That the number gets higher.
91	Teacher	That the number gets higher. And then he says here that the number of apples also increases itself.
92	Fatma	That is the same.
93	Teacher	Is it the same?
94	Shenay	No, that is "number" and that is "value."
95	Fatma	I see! Yes.
97	Shenay	The one means that the value [meaning the price] increases itself each time and, yes, that the value increases. And the other one, that the apples, though the number increases.
98	Teacher	Thus, what does he mean by the value?
99	Shenay	Eh, the selling price.
100	Teacher	The selling price. Ok, he sees that the selling price changes. And what else does he observe?
101	Thilo	That the selling price changes and that the number changes. So, that five apples, eh, ten apples, cost 10 €.

The teacher chooses what Baydar wrote because it describes the increase of both quantities separately in two speech bubbles (included as expression **E0** in the analytic graph in Fig. 5). She asks other Turkish-speaking students to translate the Turkish part for the class (Turn 89), which affirms their identity as competent multilinguals (Fig. 1). Fatma translates (Turn 90) and states that both speech bubbles describe the same thing (Turn 92), which is not true for the Turkish sentence.

As the teacher cannot read the Turkish, she misses this subtlety and continues working with the German text. Shenay distinguishes the value (the selling price) from the number of apples (Turn 94). She uses re-voicing to repair the language slip, changing "increase itself" to "increase" without further comments (a productive practice).

As the teacher's focus is first on striving for precision by turning "value" into "selling price" (Turn 98–101), the students continue articulating two separate statements of variation, without combining them into a statement of covariation until Turn 101. When Thilo tries to explain the two separate variations in Turn 101, he expresses the correspondence, but not yet the covariation approach (still coded as **E0, Corr**).

Sequence 1 shows how students' interest can be attracted to understanding students' bilingual writing, even those students who do not share the language. It also shows a limitation of not sharing a language with the students, as the teacher missed the opportunity to compare the Turkish "number" with the German "value."

Sequence 2: German expressions for covariation

In the next turns, the two separate variations are condensed into a covariation, and the direction of the dependency is questioned:

103	Dennis	That's normal, though! If the value increases itself, then more apples are there. Otherwise, it makes no sense.	E1/E3
105	Teacher	But what depends on what?	
106	Dennis	The, well, the price depends at [*sic*] the number of apples.	E5
107	Teacher	The price depends on the number of apples.	
108	Dennis	Yes.	
109	Teacher	That is more precise, isn't it?	
110	Dennis	Somewhat.	
111	Teacher	How would you articulate the sentence?	
112	Shenay	Em, the more number of apples, the more the price increases.	E3
113	Teacher	The more apples, the higher the number of apples, the . . .	E4
114	Lale	. . . more expensive	
115	Shenay	The higher increases the price.	
116	Lale	Eh, I would say, the more apples, the more expensive they are.	E2
117	Teacher	The more apples, the more expensive.	

Dennis articulates an if-then sentence that first combines the two variations in a covariation (coded as **E1** in Turn 103). Since he uses condensed verbs for expressing the change, the utterance is also coded as **E3**.

The teacher reacts to his inverse direction of dependency by asking a question, but this question is phrased in a much more condensed structure, coded as **E5** (Turn 105). Dennis corrects himself by using the highly condensed expression **E5** that he picks up from the teacher (Turn 106). In order to activate the design principle DP4, variation for initiating reflection, the teacher calls for further ways of articulating the covariance (Turn 111). The collection of utterances (Turns 111–117) reveals four different ways of expressing the covariance (**E5, E3,**

E4, and **E2**). These four phrases successively de-compose the highly compacted phrase of the teacher (**E5**).

This sequence shows nicely how the call for further phrases can add multiple modes of expression to the covariation concept in view.

Sequence 3: Multilingual expressions for covariation

So far, three students contributed utterances in German only. In order to activate *all* students to recapitulate the discussion, the teacher calls for finding further expressions in *other* languages:

121	Teacher	Further possibilities, a further idea how to construct the sentence? Who has an idea in Turkish, Polish, or Arabic, how could you construct the sentence in these languages? [. . .] Think about it for a minute. And also think [*directed to the students of monolingual German-speaking families*], how would you express that in English?	
122	Lale	We have one!	
123	Teacher	Yes, you have one. We will allow some more time for the others. Write it into the speech bubbles. [*6-second break*]. You can also use dictionaries but only for selected words, no direct translation.	
124	Teacher	[*After a 5-minute break*] Now, do you want to read the sentence in Turkish and what it means?	
125	Fatma	"Elmalarin sayisi yükselirse fiyati da yükselir." [Translated from Turkish: If the number of apples raised, then the price would also raise.]	E1
126	Teacher	Have you also written "the . . . the" or have you rephrased it?	
127	Lale	We have simply – directly written it.	
128	Teacher	What would it mean exactly, what you have written?	
129	Lale	If the number of apples gets higher, then the price also gets higher.	E1/
130	Teacher	Super, that is also creative. How is it for you?	E3
131	Thilo	We have only written "the more apples" [English in original] and then we wanted to write, eh, "more," eh, the more? [*Student searches for "the . . . the" – "je . . . desto" in German*]	E2
132	Teacher	"Expensive"	
133	Thilo	Yes	
134	Teacher	Well, well, well. And you [*to Dennis*]?	
135	Dennis	We don't have it completely in Polish, but I can try.	
136	Teacher	Yes, so that we can see how Polish . . .	

137	Dennis	"Ile masz jabłko, tak będzie" [Translated from Polish: So many apples as you have, so . . . it will be] and the rest I do not know, there is a word missing.	
138	Teacher	Which word is missing?	
139	Dennis	Eh, expensive	
140	Shenay	Shall I google that quickly?	Corr
141	Dennis	Yeah, do it.	
. . .			
149	Shenay	"Kostownie" [Translated from Polish: valuable]	
150	Dennis	"Ile masz jabłko, tak będzie kostownie." [So many apples as you have, so valuable it will be.]	
151	Teacher	[. . .] How would this sentence be in German?	E4
152	Dennis	This would be, eh. I have so many apples. The more apples I have, the more evaluable it'll be.	
153	Shenay	Yes, you cannot translate everything directly.	
. . .			
155	Qaiss	"Lil- mazīd minat-tuffāḥ kallama kān ʿalayka ʾan tadfaʿa ʾakṯar" [Translated from Arabic: For more apples, all you have to do is pay more]	E1
. . .			
157	Teacher	When you buy more apples, then you have to pay more. Thus, when the number increases, then you have to spend more money.	E1/ E3

In this sequence, the classroom transforms into a *translanguaging space* in which all students feel free to deal with multiple languages for meaning-making in mathematics, regardless of their proficiency level. Even if the conversation takes place mainly in their shared language, German, they individually make use of their multilingual repertoires in an epistemic function that supports a consolidated meaning making of covariation. In Turn 125, Fatma brings in an important sentence variation in Turkish (**E1**). After Lale first declares it to be a literal translation (Turn 127, **E1/E3**), the teachers' repeated questions result in a more accurate translation (Turn 129) that reveals the variation from "the . . . the" to "if . . . then" and a denominalization, both supporting the unpacking of the condensed concept.

The group of foreign English-language speakers searches for the translation of "je . . . desto" in German ("the . . . the" in English) in Turn 131 (**E2**). The teacher provides vocabulary support for "valuable" without noticing this obstacle. (Later, the students use a more de-composed version: As they still do not know "the . . . the," they construct a longer sentence.)

Dennis and three other Polish-speaking students are also unfamiliar with the use of Polish within a mathematical discourse and feel uncomfortable using it (Turn 135). The teacher encourages them several times, and Dennis reads his incomplete sentence (Turn 137). With dictionary support of the non-Polish-speaking peer Shenay (Turns 141–147), he completes their Polish sentence, showing the students' engagement and identity strengthening. Dennis' Polish sentence reveals an interesting perspective, the correspondence approach, assigning one price for each number of apples (Turn 150). Again, the students' limitations in a language serve as catalysts for multiple perspectives and for enriching the learning situation. However, Dennis' translation back to German (Turn 152, **E4**) does not articulate the correspondence approach anymore, so the teacher cannot exploit it. This shows again the limitations of exploiting non-shared languages. A similar situation occurs with the Arabic sentence in Turn 155, which is only partially exploited (**E1**).

5.2 Overview of three sequences

A summary of the interpretations in the analytic graph in Fig. 5 shows the course of expressions brought in by students and the teacher in Sequences 1–3. During the 70 lines of transcript, the students articulate all six expressions (**E0–E5** as shown in Fig. 2) several times and connect them by moving forward and backward. Sequence 1 started with a misconception **E0** and the teacher funneled the most compacted expression **E1**. Many class discussions stop at this point.

The created translanguaging space, however, allowed the students to connect the highly compacted **E5** to four other versions, which allowed them a multi-perspective meaning-making. This was possible for the students even in languages that are not shared in the whole group.

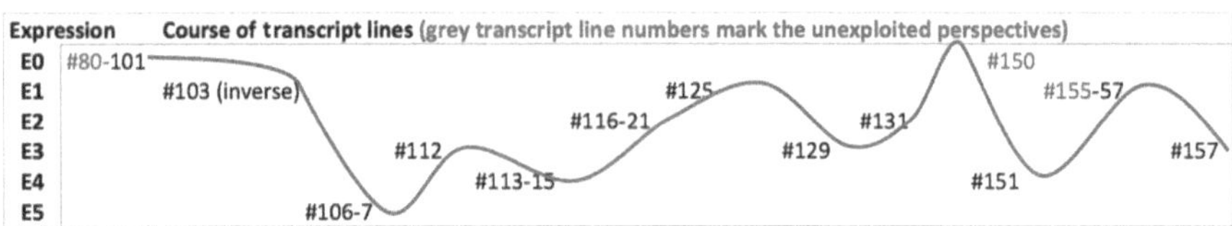

Fig. 5: Analytic graph of the process of connecting multilingual expressions in Sequences 1–3.

6 Discussion and outlook

Instructional strategies for activating multilingual resources for mathematics learning have been shown to be beneficial not only for overcoming communicative obstacles, but also for epistemic purposes, especially by providing multiple perspectives for meaning-making (Cummins, 2015; Planas, 2019; Prediger et al., 2019). The current case study investigates whether these results from classrooms with shared bilingualism can also be transferred to classrooms with non-shared multilingualism.

Qualitative analysis shows that teachers can also activate students' home and foreign languages and conduct productive discussions about multiple languages (design principles DP3 and DP4 when the conversation itself is in German; by striving for preciseness in different languages, DE4.1), and by comparing different expressions in German and other languages (DE3.3 and DE4.2), the students in the case study were able to realize that the covariation statement "the more *x*, the more *y*" can be expressed in different degrees of compaction. They benefited mathematically by unfolding a highly compacted mathematical concept. Although the macro-scaffolding principle suggests starting with less compacted expressions and connecting them to more compacted ones, the learning is also deepened when different degrees of compaction are mutually connected and explicitly compared. Thus, mobilizing students' multiple language not only increases their agency by being addressed as competent multilingual speakers, but also strengthens the students' mathematical reflection.

The case study also shows that it is not always necessary to share the languages: Help for Polish expressions was provided by a non-Polish speaker offering to google a missing word. The analysis shows how the "bilingual connective mode" can also be established by individual mental connections that are only partially accessible to the non-shared space.

However, the case study also reveals various examples of limitations where the teacher was unable to react or include some students' statements, especially when she cannot exploit subtle differences because they are not covered by the students' translations back to the shared language of instruction.

In future research, the methodological limitations of this case study (being bound to a specific classroom and languages, with a limited scope and small sample) should be overcome by extending the investigations to further cases, mathematical topics, and language contexts.

References

Adler, Jill (2001): *Teaching Mathematics in Multilingual Classrooms*. Dordrecht: Kluwer.

Barwell, Richard (2009): Multilingualism in mathematics classrooms: An introductory discussion. In Barwell, R. (ed.): *Multilingualism in Mathematics Classrooms*. Bristol, Buffalo, Toronto: Multilingual Matters, 1–13.

Barwell, Richard (2018): From language as a resource to sources of meaning in multilingual mathematics classrooms. *Journal of Mathematical Behavior* 50, 155–168.

Barwell, Richard, Clarkson, Philip, Halai, Anjum, Kazima, Mercy, Moschkovich, Judit, Planas, Nuria & Villavicencio, Mamokgethi (eds.) (2016): *Mathematics Education and Language Diversity: The 21st ICMI Study*. Dordrecht: Springer.

Beacco, Jean-Claude., Byram, Michael, Cavalli, Marisa, Coste, Daniel, Cuenat, Mirjam E., Goullier, Francis & Panthier, Johanna (2015): *Guide for the Development And Implementation of Curricula for Plurilingual and Intercultural Education*. Strasbourg: Council of Europe.

Cummins, Jim (2015): Language differences that influence reading development: Instructional implications of alternative interpretations of the research evidence. In Afflerbach, Peter (ed.): *Handbook of Individual Differences in Reading Reader, Text, and Context*. London: Routledge, 223–244.

Dröse, Jennifer, Prediger, Susanne (2020). Enhancing Fifth Graders' Awareness of Syntactic Features in Mathematical Word Problems: A Design Research Study on the Variation Principle. *Journal für Mathematik-Didaktik* 41(2), 391–422. https://doi.org/10.1007/s13138-019-00153-z.

Erath, Kirstin, Ingram, Jenni, Moschkovich, Judith & Prediger, Susanne (2021, online first): Designing and enacting teaching that enhances language in mathematics classrooms – A review of the state of the art. *ZDM Mathematics Education* 53(3), https://doi:org/10.1007/s11858-020-01213-2.

Erath, Kirstin, Prediger, Susanne, Uta, Quasthoff & Heller, Vivien (2018): Discourse competence as important part of academic language proficiency in mathematics classrooms. The case of explaining to learn and learning to explain. *Educational Studies in Mathematics* 99 (2). 161–179.

Gravemeijer, Koeno & Cobb, Paul (2006): Design research from a learning design perspective. In Akker, Jan v. d., Gravemeijer, Koeno, McKenney, Susan, Nieveen, Nienke (eds.): *Educational Design Research: The Design, Development and Evaluation of Programs, Processes And Products*. London: Routledge, 17–51.

Li, Wei (2011): Moment analysis and translanguaging space: Discursive construction of identities by multilingual Chinese youth in Britain. *Journal of Pragmatics* 43 (5). 1222–1235.

Meyer, Michael, Prediger, Susanne, César, Margarida & Norén, Eva (2016): Making use of multiple (non-shared) first languages: State of and need for research and development in the european language context. In Barwell, Richard et al. (eds.): *Mathematics Education and Language Diversity*. Dordrecht: Springer, 47–66.

Planas, Nuria (2018): Language as resource: A key notion for understanding the complexity of mathematics learning. *Educational Studies in Mathematics* 98 (3). 215–229.

Planas, Nuria (2019): Transition zones in mathematics education research for the development of language as resource. In Venkat, Hamsa, Graven, Mellony, Essien, Anthony, Vale, Pamela (eds.): *Proceedings of 43th Annual Meeting of the International Group for the Psychology of Mathematics Education -PME 43*. Vol. 1, Pretoria: PME, 17–31.

Prediger, Susanne, Clarkson, Philip & Bose, Arindam (2016): Purposefully relating multilingual registers: Building theory and teaching strategies for bilingual learners based on an integration of three traditions. In Barwell, Richard, Clarkson, Philip, Halai, Anjum, Kazima, Mercy, Moschkovich, Judit, Planas, Nuria, Setati-Phakeng, Mamokgethi, Valero, Paola, Ubillús, Martha Villavicencio (eds.): *Mathematics Education and Language Diversity*. Dordrecht: Springer, 193–215.

Prediger, Susanne, Kuzu, Taha, Schüler-Meyer, Alexander & Wagner, Jonas (2019): One mind, two languages – separate conceptualizations? *Research in Mathematics Education* 21 (2). 188–207.

Prediger, Susanne & Zindel, Carina (2017): School academic language demands for understanding functional relationships. *EURASIA Journal of Mathematics, Science & Technology Education* 13 (7b). 4157–4188.

Ruíz, Richard (1984): Orientations in language planning. *NABE – Journal for the National Association for Bilingual Education* 8 (2). 15–34.

Schüler-Meyer, Alexander, Prediger, Susanne, Kuzu, Taha, Wessel, Lena & Redder, Angelika (2019): Is formal language proficiency in the home language required to profit from a bilingual teaching intervention in mathematics? *International Journal of Science and Mathematics Education* 17 (2), 317–399. doi:10.1007/s10763-017-9857-8.

Thompson, Patrick W. & Carlson, Marilyn P. (2017): Variation, covariation, and functions. In Cai, Jinfa (ed.): *Compendium for Research in Mathematics Education*. Reston: NCTM, 421–456.

Uribe, Ángela & Prediger, Susanne (2020). Activating multilingual resources in a superdiverse covariation classroom – a design research study. Unpublished conference paper presented at ICME 14 in Shanghai.

Van Galen, Frans, Feijs, Els, Figueiredo, Nisa, Gravemeijer, Koeno, Van Herpen, Els & Keijzer, Ronald (2008): *Fractions, Percentages, Decimals, and Proportions: A Learning-Teaching Trajectory*. Rotterdam: Sense.

II Language learning and mathematics development

Julia Bahnmueller, Júlia Beatriz Lopes-Silva,
Vitor Geraldi Haase, Ricardo Moura, and Korbinian Moeller

Ties of math and language: A cognitive developmental perspective

Because numerical and mathematical competencies play an important role in our everyday life (e.g., Butterworth et al., 2011), it is crucial to understand underlying cognitive processes and factors influencing the acquisition of these numerical and mathematical competences. In particular, a better understanding at the level of cognitive processes may help to develop targeted interventions, inform, and enhance the quality of mathematics teaching, which may raise student attainment (cf. The Royal Society & The British Academy, 2018).

To be able to deal with numbers and mathematical content in a competent and efficient way, a set of concepts, procedures, and (math) facts need to be acquired starting even before (formal) education in kindergarten, preschool, and (elementary) school years. Crucially, and probably more so than it is the case in many other school subjects, mathematics education is largely hierarchical in nature (e.g., Clements & Sarama, 2021). As such it is important and necessary to be able to draw on previously acquired competences and knowledge, because new numerical and mathematical content usually builds on these previously acquired competences, concepts, and procedures.

Besides considerable developmental variability on the individual level, international studies evaluating scholastic abilities have consistently reported large cross-cultural differences in mathematical achievement (e.g., OECD, 2019a). In addition to differences in schooling and cultural valuation (e.g., OECD, 2019b), it has been argued that influences of domain-general factors such as language also need to be considered as a potential source of but also resource for overcoming difficulties in the acquisition of numerical and mathematical competences. In particular, language may refer to a range of different linguistic aspects and/or specific aspects of language skills, each of which might interact with specific steps in the acquisition of numerical and mathematical competences.

So far, a wide range of studies investigated various language aspects critical for the acquisition of numerical and mathematical concepts. And indeed, findings of many of these studies are in line with a weak Whorfian hypothesis suggesting that different aspects of language seem to influence the way we acquire, think about, perceive, represent, and apply numerical and mathematical concepts, procedures, and (math) facts. In an attempt to classify and structure previously observed associations of language and mathematics as well as influences

https://doi.org/10.1515/9783110661941-006

of language on mathematics, Dowker and Nuerk (2016) proposed a taxonomy differentiating linguistic categories that have previously been identified to influence numerical and mathematical processing in various ways (see also Bahnmueller et al., 2018; Berch et al., 2018). In particular, Dowker and Nuerk (2016) specified six categories: (1) lexical, (2) syntactic, (3) phonological, (4) visuo-spatial orthographic, (5) semantic, and (6) conceptual influences of language. Structuring associations of language and math from a linguistic point of view may thereby foster a broader, more theoretically guided approach to the investigation of how language and math are intertwined throughout development.

Drawing from this proposed taxonomy, this chapter will give an overview of a subset of specific linguistic categories covering those aspects we deem most influential with respect to the development of (early) numerical and mathematical competences. In particular, after a brief description of selected linguistic categories (i.e., lexical, syntactic, phonological, and semantic), we will discuss associations of language and numerical cognition along three consecutive content strands: (i) early numerical competences: number words, counting, and cardinality understanding; (ii) processing of multi-digit numbers; and (iii) basic arithmetic operations. Afterward, a summarizing paragraph will highlight differences, commonalities, and implications of the reported associations of language and numerical and mathematical development.

1 Linguistic influences in numerical/ mathematical development

Linguistics is the objective study of natural languages addressing characteristics of language concerning the lexicon (e.g., words, morphemes, compound words), knowledge about language structures (phonology, morphology, syntax) as well as the creation and understanding of meaning of words and sentences in different contexts (semantics, pragmatics; Pickett et al., 2018). Drawing from linguistic categories, Dowker and Nuerk (2016) proposed above-mentioned taxonomy of different linguistic categories that were shown to influence numerical and mathematical processing. In the following, key aspects of (1) lexical, (2) syntactic, (3) phonological, and (4) semantic linguistic influences will be outlined briefly as they seem particularly relevant in the context of associations of language and mathematics from a developmental context.

Within the proposed taxonomy, *lexical* influences reflect the degree to which number words vary to obscure or emphasize features of a number system such as the most widely used Arabic number system. In this context, the transparency (i.e., the consistent reflection of the Arabic number system in a language's number word system) or rather the lack thereof poses a specific hurdle for children that needs to be overcome to master more sophisticated numerical and mathematical competences. For example, one of the most widely investigated intransparencies is the so-called inversion property of number words with respect to the digital-Arabic notation. Number word inversion reflects that in some languages (German, Dutch, Maltese, etc.) the unit digit is named first in two-digit number words which is inverted with respect to the order of tens and units in the digital-Arabic notation (e.g., the number word for 24 is "vierundzwanzig" – literally four and twenty in German; for an overview, see e.g., Klein et al., 2013). Overall, the *lexical* category seems the most widely investigated one in contexts of multi-digit number processing. There is now accumulating evidence suggesting that a lack of transparency has detrimental effects on different aspects of numerical processing (e.g., number transcoding: Imbo et al., 2014; number magnitude comparison: Pixner et al., 2011; addition: Göbel et al., 2014) as well as numerical and mathematical development (e.g., Moeller et al., 2011 for longitudinal influences of inversion-related difficulties on later arithmetic performance).

Syntactic influences usually result from (language-specific) grammatical rules and thus do not reflect influences on the word level but rather on the sentence level. For example, effects of grammatical number fall within this category. Effects of grammatical number on the early acquisition of cardinality knowledge result from differences in singular, dual, and plural marking between certain languages (e.g., Almoammer et al., 2013; Sarnecka et al., 2007; for an overview, see Sarnecka, 2014). In this context, Sarnecka and colleagues (2007) report, for instance, that children speaking Japanese (a language with hardly any marking of singular/plural) learned the meaning of the number word "one" later than English- as well as Russian-speaking children (with English and Russian having explicit plural marking). Thus, grammatical number seems to foster the very early acquisition of the meaning of small numbers.

Another important linguistic category reflects *phonological* influences, which cover effects of phonological language processes as well as effects related to verbal working memory.[1] As regards the former, one subcomponent of phonological processing, namely, phonemic awareness (i.e., the ability to perceive and manipulate phonemes that constitute words, Wagner & Torgesen, 1987), is of particular interest. Phonemic awareness has been argued to be associated with, for example, the early acquisition of number words (e.g., Koponen et al., 2013;

1 Dowker & Nuerk (2016) actually consider influences of verbal working memory in a 7th category ("other language-related skills").

Krajewski & Schneider, 2009; Soto-Calvo et al., 2015) as well as performance in multi-digit number transcoding (e.g., Lopes-Silva et al., 2014; see also Chapter 15 of this volume by Haase et al.) and arithmetic fact retrieval (e.g., De Smedt et al., 2010). As regards verbal working memory, it has been found that the ability to temporarily store and manipulate verbal information influences a multitude of different numerical and mathematical tasks (among other working memory components; for reviews, see e.g., Friso-Van den Bos et al., 2013; Peng et al., 2016) and was suggested to represent an integral part in numerical and mathematical development.

Besides Arabic numbers and number words, there are other words and symbols conveying numerical or mathematical meaning by means of their semantics (e.g., more, less, some many, buy, sell). As such, the proposed category of *semantic* influences shows considerable overlap with conceptualizations and investigations of domain-specific mathematical language (e.g., knowledge of terms such as more, less, near, and far; e.g., Purpura et al., 2017; Purpura & Reid, 2016). For example, results of a study by Purpura and Reid (2016) suggest that mathematical language might be a more important predictor of early numerical competences as compared to more general language-related predictors such as vocabulary. Furthermore, research on, for example, text problem solving nicely illustrates the context-dependency of certain (numerical/mathematical) words. For example, it was suggested that words such as "more," "buy," and "get" facilitate the processing of text problems requiring additions whereas words like "less" and "sell" interfere with solving addition problems (e.g., Verschaffel et al., 1992; see Daroczy et al., 2015 for a review on text problems).

Taken together, a variety of different linguistic influences seem to affect the acquisition of numerical and mathematical competencies. Crucially, some linguistic aspects seem to affect specific numerical and/or mathematical competences and concepts early on while others only follow later and might be critical for different competences and concepts. In the following, we will elaborate on selected linguistic influences (i.e., lexical, syntactic, phonological, and semantic) on three different consecutive content strands: (i) early numerical competencies including the acquisition of number words, counting principles, and cardinality understanding; (ii) multi-digit number processing; and (iii) basic arithmetic operations (see Fig. 1 for an overview). Please note, however, that providing a comprehensive and exhausting overview goes beyond the scope of this chapter. Therefore, and because not all linguistic influences seem to be of equal importance for each of the three content strands, we will give an overview of selected linguistic influences on above-named content strands of early numerical and mathematical development from a cognitive developmental perspective.

LINGUISTIC LEVEL

CONTENT STRAND	LEXICAL	SYNTACTIC	PHONOLOGICAL	SEMANTIC
NUMBER WORDS, COUNTING, CARDINALITY	lack of transparency of teen numbers (e.g., Miller et al., 1995)	grammatical number & cardinality knowledge (e.g., Sarnecka et al., 2007)	phonemic awareness & counting sequence (e.g., Krajewski & Schneider, 2009)	mathematical language (e.g., Purpura & Reid, 2016)
MULTI-DIGIT NUMBERS	lack of transparency of numbers >20 & transcoding/number comparison (e.g., Pixner et al., 2011 a,b)		verbal working memory & transcoding (e.g., Imbo et al., 2014)	
BASIC ARITHMETIC OPERATIONS	lack of transparency & arithmetic (e.g., Göbel et al., 2014; Van Rinsveld et al., 2015)		verbal working memory & arithmetic, phonemic awareness & fact retrieval (e.g., Friso-Van den Bos et al., 2013;)	consistency effect in word problems (e.g., van der Schoot et al., 2009)

Fig. 1: Examples for associations of selected linguistic categories (i.e., lexical, syntactic, phonological, and semantic; cf. Dowker & Nuerk, 2016) with three consecutive content strands discussed in the current chapter. Please note that empty cells do not necessarily indicate that there is no association but rather that these specific aspects are not covered in the current chapter.

2 Early numerical competences: Number words, counting, and cardinality understanding

In this first content strand, we will elaborate on the role of above-described language-related aspects in the context of early numerical competencies that build the foundation for further, more advanced numerical and arithmetic competences. In particular, we will address language-related benefits but also pitfalls that were observed to influence the acquisition of number words and the number word sequence as well as the cardinality of small numbers. Moreover, we will discuss the role of mathematical language for early numerical development.

The development of early numerical competencies is a complex process that begins well before formal mathematics instruction starts. In this context, the acquisition of number words as well as the counting sequence alongside specific counting principles (i.e., one-to-one principle, stable order, cardinality-principle; e.g., Gelman & Gallistel, 1978) represents an early milestone in numerical learning. As regards *phonological* influences, it has been demonstrated that, beyond the prominent relation of phonemic awareness with reading and writing skills (for a review, see e.g., Melby-Lervåg et al., 2012), phonemic awareness also seems to be associated with and predictive of the acquisition of number words and the number word sequence (e.g., Koponen et al., 2013; Krajewski & Schneider, 2009). For instance, in a longitudinal study with 5-year-old kindergarten children (at T1), Krajewski and Schneider (2009) investigated the association of phonemic awareness with future mastery of the counting sequence (e.g., counting forward and backward, identifying the successor and predecessor of a number). Results showed a substantial association of phonological awareness and mastery of the counting sequence. Thus, previous studies seem to be in line with the idea that phonemic awareness fosters the acquisition of number words by supporting the construction of sound-based representations in the same way as it facilitates the acquisition of other word categories (e.g., Gathercole, 2006).

Considering that number words are often embedded in sentences, it is not surprising to see language-specific grammatical rules to also shape the learning of the *semantic* meaning of (small) number words (i.e., number words "one," "two," and "three"). Substantiating this idea, several studies investigated influences of grammatical number on the acquisition of early cardinality knowledge typically assessed via the Give-N task (i.e., asking children to produce a set of a given size; e.g., Almoammer et al., 2013; Barner et al., 2009; Li et al., 2013; Sarnecka et al., 2007; for a review, see Sarnecka, 2014). The term "grammatical number" refers to singular, dual, and plural markings of certain languages like, for example, the morpheme "s" used as a suffix for plural marking in English. Research suggests that trajectories of number word learning and with this the acquisition of cardinality understanding are influenced by the presence or absence of explicit plural/dual markings in a certain language. As mentioned above, Sarnecka and colleagues (2007) report that English- as well as Russian-speaking children aged between 2 and 3 (with English and Russian having explicit singular/plural markings) learned the meaning of the number word "one" earlier than Japanese-speaking children (with Japanese being a language with hardly any marking of singular/plural). Moreover, 2- to 4-year-old children speaking a language with explicit dual marking such as Slovenian or Arabic (i.e., a specific form referring to exactly two things), appear to learn the meaning of "two" earlier than English-speaking children (Almoammer et al., 2013). Thus, not only the frequency of exposure to number words and the counting sequence but also numerical information expressed and made explicit through grammatical structures seems to influence learning the meaning of (small) number words (see also Barner et al., 2009 for syntactical aspects in quantifiers).

Next to grammatical structures, various studies suggest that the way number words are formed determines number word learning trajectories. As mentioned above, number word systems vary considerably with respect to their transparency by which the place-value structure of the Arabic number system is reflected in number word formation. From a *lexical* point of view, differences between language groups in number word acquisition should be comparably small for numbers up to ten as in most languages there are exactly ten arbitrary but ordered words (eleven when including zero) that need to be learned and mapped to the respective numerical magnitude. And indeed, cross-cultural studies investigating counting skills (i.e., correctly reciting the counting sequence) in children aged between 3 and 6 suggest that for numbers up to 10 average performance is fairly similar across different language groups (e.g., LeFevre et al., 2002; Miller et al., 1995; Miller & Stigler, 1987).

However, within the number range up to 20 (i.e., numbers "eleven" to "nineteen"; see Section 3 ("Processing multi-digit numbers") for effects for numbers >20), the transparency by which the place-value structure of the Arabic number system is reflected in number words starts to vary considerably between languages. Importantly, the teen number range seems to be special in that a lack of transparency for teen number words can be found in quite many languages (e.g., Arabic, English, Hindi, Italian, Polish, Russian, Spanish, Swedish) – even though number words for numbers larger than 20 are often quite transparent again in these languages. In particular, languages show a variety of peculiarities for teen number words such as, for example, (i) exceptional number words not indicating the teen range at all (e.g., English: "eleven" and "twelve"); (ii) inverted number words [e.g., English: “fourteen” instead of ten four; Polish: “jedenaście” (oneteen); German: “dreizehn” (three-ten)], or (iii) inconsistent construction of teen number words within a language [e.g., Italian: “undici” (one ten) but “diciotto” (ten eight); cf. Lewis et al. (2020)]. This may represent a source of considerable difficulty for children because rather than simply applying a consistent rule for the first two-digit number words children are confronted with, they have to deal with irregularities that may not facilitate the acquisition of numerical and place-value and concepts more broadly.

And indeed several studies investigated the acquisition of teen numbers in different languages and reported a delay in number word acquisition for numbers larger than 10 for languages with less transparent number words (e.g., Aunio et al., 2008; Cankaya et al., 2014; LeFevre et al., 2002; Lonnemann et al., 2019; Miller et al., 1995; Miller & Stiegler, 1987). For instance, Miller and colleagues (1995) compared early counting skills in three- to five-year-old English- and Mandarin-speaking preschoolers. Compared to perfectly transparent Mandarin number words [“shí yī” (ten one), “shí èr” (ten two), “shí sān” (ten three), etc.], English number words in the teens range are fairly in-transparent (“twelve” instead of ten-two, “fourteen” instead of ten-four, etc.). As mentioned before, this study found no differences in counting skills between language groups for numbers up to 10. However, the authors found significant language differences favoring Mandarin-speaking children starting in the teen range. When looking at individual teen numbers more closely, language differences in counting were most pronounced for numbers above 12 (Miller et al., 1995). Importantly, while approximately 75% of Mandarin-speaking children were able to correctly count up to 20, only 50% of English-speaking children were able to do so by the age of five. Based on these findings, the authors concluded that English-speaking children need more time to master number names for teen numbers and beyond because of the lack of transparency between their number word system and the place-value structure of the Arabic number system.

Notably, however, others have questioned specificities in number word systems as sole contributing factor to observed cross-cultural differences and argue that differences in, for instance, approaches to teaching and learning (e.g., Aunio et al., 2008) as well as differences in home experiences (e.g., LeFevre et al., 2002) need to be considered as plausible additional or even alternative explanations for the observed differences in counting. In this context, evaluating a counting intervention in three- and four-year-old Turkish- and English-speaking children, Cankaya and colleagues (2014) came to the conclusion that both the transparency of the number word system and prior experience with numeracy-related activities were crucial for the acquisition of counting skills. As in Mandarin, Turkish teen number words are very transparent [e.g., on üç (13) translates to ten three]. The authors report that in their intervention learning gains in counting were higher and more consistent for Turkish- compared to English-speaking children. The authors attributed this finding to the more transparent Turkish number words compared to the English ones. However, despite the transparent number word system, Turkish-speaking children showed poorer counting skills and poorer general numerical performance overall before the intervention. Thus, while the transparency of the number word system does seem to matter, additional culture-specific variables likely contribute to differences in developmental trajectories when learning to count.

Next to *lexical* influences concerning the (lack of) transparency of many number word systems (especially in the teen number range), both before and during formal schooling children need to learn further specific numerical and mathematical language. Such mathematical words may, for example, convey less precise numerical meaning than number words (e.g., many, fewer, less than) or may have a different meaning in a mathematical context (e.g., quarter, break apart; e.g., Powell et al., 2017; Powell & Nelson, 2017). Generally, Harmon et al. (2005) suggest that language used in numerical and mathematical contexts is highly content-specific and, thus, may require more explicit teaching of the meaning of specific words at times. Accumulating evidence suggests that mathematical language proficiency (e.g., Powell et al., 2017; Powell & Nelson, 2017; Purpura et al., 2017; Purpura & Reid, 2016; Schleppegrell, 2007; Toll & Van Luit, 2014a, 2014b) but also parent and teacher usage of this specific language (e.g., Boonen et al., 2011; Gunderson & Levine, 2011; see also Chapter 7 in this volume by Desoete et al.) is associated with and predicts future development of numerical and mathematical competences.

For young children, knowledge of two types of mathematical language terms seems to be critical: quantitative (e.g., more than, many, fewer; cf. quantifier knowledge, e.g., Hurewitz et al., 2006) and spatial terms (e.g., before, close to; e.g., Mix & Cheng, 2012; Pruden et al., 2011). In this early numerical context, results of a study by Purpura and Reid (2016) in 3- to 5-year-old preschool children suggest that mathematical language might be a more important predictor of early numerical competences (e.g., counting or relational knowledge) when compared to more general language-related skills (e.g., vocabulary, phonemic awareness). The relevance of mathematical language for early numerical skills was further substantiated by an intervention study conducted by Purpura et al. (2017) focusing on quantitative and spatial mathematical language in 3- to 5-year-old children. After the eight-week intervention, children in the intervention group not only outperformed the business-as-usual control group with respect to their knowledge of mathematical language terms but also with respect to early numerical skills (e.g., one-to-one correspondence, number order, set and numeral comparison, numeral identification, among others; Purpura et al., 2017).

Taken together, this evidence is in line with the idea that different aspects of language seem to shape the development of early numerical competences way before formal education starts. As such, very specific language aspects such as phonemic awareness and grammatical number but also formation of number words and use of unspecific quantifiers seem to influence the typical development of children's early numerical development to a considerable degree - even without formal instruction but by simply living in the respective language environment. Importantly, however, while linguistic influences seem to reflect one important factor during the development of early numerical competences, other cultural and environmental factors (e.g., approaches to learning and teaching, home environment) ought not to be neglected when trying to evaluate the specific contribution of language to arrive at a comprehensive understanding of the driving factors in early numerical development.

3 Processing multi-digit numbers

One key concept of the Arabic, base-10 number system is its place-value structure. The place-value structure defines that overall magnitude of multi-digit numbers is represented by powers of ten increasing from right to left combined by specific multiplicative and additive composition rules (e.g., $342 = \{3\} \times 10^2 + \{4\} \times 10^1 + \{2\} \times 10^0$; McCloskey et al., 1985). In particular, deriving the magnitude of a specific number requires understanding that any digit in a multi-digit number informs about both the size (through the digit's face value) and the power of ten it represents (through the digit's position within the digit string). As such, mastery of the place-value structure of the Arabic number system is critical for understanding multi-digit numbers.

In this context, mastery of the place-value structure of the Arabic number system was indeed observed to be associated with current but also predictive of later arithmetic performance (e.g., Chan et al., 2014; Moeller et al., 2011). Moreover, deficient place-value knowledge has been discussed as a predictor or source of dyscalculia and mathematical learning difficulties (e.g., Cawley et al., 2007; Chan & Ho, 2010; Haase et al., 2014). For instance, Chan and Ho (2010) assessed 8- as well as 10-year-old children with and without mathematical difficulties and demonstrated that conceptual understanding of the place-value structure differentiated reliably between children with and without mathematical difficulties (see also Lambert & Moeller, 2019). Thus, mastering the place-value structure of the

Arabic number system represents an important milestone in numerical development (see also Herzog et al., 2017, 2019 for a developmental model of conceptual place-value understanding).

However, children often experience specific difficulties when learning the place-value structure and, thus, with many tasks involving multi-digit numbers. Moreover, as mentioned in the context of teen numbers before, number words do not always reflect the place-value structure properly, which further complicates the learning process. Such *lexical* influences are the most widely investigated linguistic aspects in the context of multi-digit numbers processing. Thus, this section will focus on linguistic influences on place-value processing resulting from specificities in the formation of number words.

Regarding multi-digit number processing, *lexical* influences cover both the (lack of) transparency of power (e.g., in Mandarin, power is expressed explicitly in both number symbols and words: 42 = 四十二 = sì shí èr = 4-10-2) and the (lack of) transparency of order (e.g., the inversion of number words in, e.g., German: the number word for 23 is "dreiundzwanzig," literally three-and-twenty). Although many cultures share the Arabic number system, number word systems clearly vary with respect to the degree of transparency in which power and order are conveyed (see above for the case of teen numbers). While for many cultures using Arabic digits and for most numbers, power is expressed by different words for the same symbol depending on the position in the digit string (e.g., in English the number word for 4 is "four" and the number word for 42 is "*forty*-two") or by adding a multiplier indicating the power (the number word for 342 is "three *hundred* forty-two"), many exceptions are found for specific number ranges in different number word systems.

In French number words, for example, most two-digit numbers are transparently composed of two words each reflecting the power of a digit in accordance with the place-value structure (e.g., the French number word for 42 is "quarante-deux," literally forty-two). However, number words larger than 60 are constructed quite irregularly by drawing on a vigesimal system (i.e., a base-20 system), which is inconsistent with the base-10 structure of the Arabic number system (e.g., the number word for 72 is "soixante-douze," literally sixty-twelve). Finally, the French number word for 80 is "quatre-vingt" (literally four-twenty) and larger numbers are constructed accordingly (e.g., the number word for 96 is "quatre-vingt seize," literally four-twenty sixteen), which adds even more construction principles to the already complex number word system.

Lack of transparency with respect to order can also be found in English number words. As mentioned before, English number words for teens from 13 to 19 are inverted with respect to the digital-Arabic notation (e.g., 19 = "nineteen"), although English number words for two-digit numbers are otherwise fairly transparent (42 = "forty-two"). While in modern English the phenomenon of inversion is restricted to teen numbers, in old English and many other modern languages (e.g., Arabic, Danish, Dutch, German, Flemish, Malagasy, Maltese, and also partly in Czech and Norwegian; Comrie, 2005) number words of wider number ranges are inverted. For example, in German, all two-digit numbers are inverted (e.g., the number word for 42 is "zweiundvierzig," literally two and forty). Moreover, although hundreds and thousands are not inverted (e.g., the number word for 2342 is "zweitausenddreihundertzweiunsvierzig," literally two thousand three hundred two and forty), for thousands and ten thousands (e.g., powers 10^3 and 10^4) the inversion of numbers words occurs again (the word for 42,342 is "zweiundvierzigtausenddreihundertzweiundvierzig," literally two and forty thousand three hundred two and forty).

This exemplary illustration of some aspects of lack of transparency shows that there are number word systems that do not reflect basic principles of the Arabic number system such as its base-10 structure or/and the place coding scheme correctly (i.e., that value increases from right to left). Noteworthy, a long list of studies showed that lack of transparency of number word systems with respect to the place-value structure of the Arabic number system influences and, crucially, complicates multi-digit number processing (for an overview, see e.g., Klein et al., 2013).

As mentioned before in the context of teen numbers, in addition to, for instance, specific approaches to teaching and learning, Asian children seem to benefit from their highly transparent number word systems (i.e., power and order are transparently reflected in the number words themselves; e.g., in Mandarin the number word for 42 is sì shí èr (四十二), literally four-ten-two). When investigating the understanding of the place-value structure of the Arabic number system in children from various Asian and Western countries, several early studies demonstrated better place-value understanding in Asian compared to Western preschoolers and 1st graders (e.g., Miura et al., 1988; Miura & Okamoto, 2003; Miura et al., 1993). In particular, while Asian children preferred representing multi-digit numbers by using ten and one blocks (i.e., matching the place-value structure of multi-digit numbers), Western children preferred using collections of one blocks for longer suggesting delayed understanding of the place-value structure of the Arabic number system in Western children with less transparent number systems (e.g., Miura, et al., 1994; Towse & Saxton, 1998). Importantly, differences were already observed before the concept of the place-value system was explicitly taught (e.g., in school) questioning influences of differences in the teaching approaches as the sole determining factor (see Vasilyeva et al., 2015 for an opposing view).

Detrimental effects of the lack of transparency of certain number word systems were demonstrated in different numerical tasks. The probably most obvious effects can be observed in number transcoding (i.e., writing down numbers to dictation) in elementary school children. Typically, errors children commit suggest that they "simply write down what they hear". More formally speaking, errors in transcoding often reflect insufficient knowledge of additive (e.g., "three hundred and forty-two" is written down as 30042) and/or multiplicative composition rules (e.g., "three hundred" is written down as 3100) or further language-specific (in)transparencies.

As regards the latter, it was shown, for example, that children speaking languages with inverted (e.g., German, Dutch), as compared to children speaking languages with non-inverted, number words (e.g., French, Italian) do not only commit more transcoding errors overall (e.g., Krinzinger et al., 2011; Pixner et al., 2011b; but see Imbo et al., 2014), but also commit up to 50% inversion-related errors (e.g., "vierundzwanzig" (24) – literally "four and twenty" – is written down as 42; Imbo et al., 2014; Krinzinger et al., 2011; see Pixner et al., 2011b for a within-culture approach in Czech).

Another example for the language specificity of transcoding errors is described in the study by Van Rinsveld and Schilz (2016) who investigated effects of the vigesimal structure in French number words larger than 60 (e.g., the number word for 72 is "soixante-douze," literally sixty twelve; see also Seron & Fayol, 1994) in two computerized transcoding tasks (i.e., choosing the Arabic number with auditory verbal presentation and reading out loud the Arabic number presented on the screen). Results in both tasks indicated that performance in English-speaking fifth graders (aged 10) was comparable to French-speaking fifth graders for numbers up to 60. However, for numbers larger than 60 and, thus, the number range where number words in French follow the vigesimal and number words in English follow the decimal structure, English-speaking children were faster in both tasks and made fewer errors in the recognition task than French-speaking children. The fact that these results were observed in fifth graders is of particular interest because it illustrates that although the impact of specificities in number word formation might get smaller with age and experience, traces of in-transparent number word formation can still be detected in children way beyond the age of early numerical development.

The latter also applies to the processing of multi-digit number magnitude. Here, influences of lack of transparency in number word formation were shown for the unit-decade compatibility effect (Nuerk et al., 2001) in two-digit number magnitude comparison. When children (or adults) are asked to indicate the larger of two two-digit numbers, they usually respond faster to unit-decade compatible number pairs for which separate comparisons of tens and units bias the same decision (e.g., 32_57, $3 < 5$ and $2 < 7$). In contrast, in decade-unit incompatible number pairs comparing tens and units separately leads to opposing decision biases (37_62, $3 < 6$ but $7 > 2$), resulting in comparably slower responses due to the interference between comparisons of tens and units. The unit-decade compatibility effect (i.e., the performance difference between compatible and incompatible number pairs) was replicated for both adults and children (e.g., adults: Bahnmueller et al., 2019; Ganor-Stern et al., 2007; Macizo & Herrera, 2011; Nuerk et al., 2005; children: Landerl & Kölle, 2009; Pixner et al., 2011a; Van Rinsveld et al., 2016).

Importantly, the unit-decade compatibility effect is found in number pairs for which the comparison of the unit digit is actually completely irrelevant because identification of the larger number can be based solely on the comparison of the tens digits. This suggests two things: first, the unit digit is processed automatically although it is irrelevant for the task at hand, and second, that magnitudes of tens and units are processed in a decomposed way but in accordance with the place-value structure of the Arabic number system (i.e., tens are compared with tens and units are compared with units; see Wood et al. (2005) for expansions on this).

Although the compatibility effect is not a language-specific phenomenon (i.e., the effect was observed in many different languages), the effect was found to be modulated by language, and more specifically by the inversion property of number words. For example, Pixner, Moeller, and colleagues (2011) investigated the unit-decade compatibility effect in German-, Italian-, and Czech-speaking first graders. While German number words are inverted and Italian number words are not, in Czech there are two number word systems – one inverted and the other one not. Clear differences in the compatibility effect were observed with German-speaking children showing a significantly larger compatibility effect than the other two language groups and, for reaction times, Czech-speaking children showed a compatibility effect falling in between the German and the Italian group. This pattern of results suggests that number word formation influences the processing of two-digit number magnitude in an entirely symbolic number magnitude comparison task as neither the input nor the output in this task was verbal. In particular, in languages with inverted number words such as German, the unit digit is named first (e.g., the number word for 23 is "dreiundzwanzig," literally three and twenty) and might, thus, lead to increased unit-based interference in incompatible trials, which in turn would increase the unit-decade compatibility effect for inverted languages. Thus, these results suggest that verbal number word information influences number processing even when it is not present in or necessary for the task at hand.

Research regarding transcoding competencies (i.e., writing numbers to dictation) in elementary school children further investigated possible *phonological* influences relating to verbal working memory capacities in multi-digit number processing. In this context, several studies in different language groups with and without inverted number words evaluated the idea that transcoding might be influenced by working memory capacity because incoming number word information needs to be manipulated and mapped onto the digital-Arabic notation. These studies observed that better working memory was associated with better transcoding performance in general (Imbo et al., 2014; Pixner et al., 2011b; Simmons et al., 2012; Zuber et al., 2009) but with a lower number of inversion-related errors in German-speaking first graders in particular (Zuber et al., 2009). However, while there is broad agreement that working memory is important for transcoding, findings are so far inconsistent with respect to specific working memory components (cf. Baddeley, 2000; Baddeley & Hitch, 1974). For instance, while some studies primarily reported associations of transcoding performance with verbal working memory capacities (e.g., Imbo et al., 2014), others highlight the relevance of visual-spatial working memory capacities (e.g., Simmons et al., 2012; van der Ven et al., 2017), or the central executive (Pixner et al., 2011b; Zuber et al., 2009). So far, findings suggest that for transcoding numbers from the verbal number word to the digital notation conveying the correct order of digits seems more relevant than solely being able to temporarily store verbal number word information – at least in children that are busy learning the place-value structure of the Arabic number system (cf. van der Ven et al., 2017).

Taken together, the presented studies provide further strong evidence that cognitive representations of multi-digit numbers are shaped by and differ between languages as indicated by significant linguistic influences in a variety of different tasks ranging from transcoding between the verbal and the digital-Arabic notation to tasks requiring the explicit processing of number magnitude information. Importantly, these observed language-related influences do not only foster our understanding of underlying principles of multi-digit number processing, but they may also be of diagnostic value. For instance, Moeller and colleagues (2011) showed that the number of inversion-related errors in transcoding as well as the size of the compatibility effect in the first grade predicted arithmetic performance in the third grade (including mathematics grades). Thus, better understanding language-specific aspects of place-value processing might help identifying children that may develop mathematical difficulties early on.

4 Basic arithmetic operations

Building on previously described numerical competencies, mastery of basic arithmetic operations – addition, subtraction, multiplication, and division – represents a further cornerstone in numerical and mathematical development. Not only is mastery of basic arithmetic operations a pervasive requirement in everyday life, it also represents a crucial basis for more advanced mathematical competencies (e.g., Geary & Hoard, 2005). Children use a variety of different strategies to solve arithmetic problems that may vary with the type of operation they are presented with. Moreover, strategies used by children become more efficient and adaptive with age and experience (Ashcraft, 1982; Carpenter & Moser, 1984; Geary & Hoard, 2005; Geary et al., 2004; Siegler, 1996; Siegler & Shrager, 1984). Usually, two major types of strategies are distinguished separating (i) procedural strategies including counting (cf. Fuson, 1982), mental computations and/or transformations, or keeping track of intermediate solutions; and (ii) retrieval strategies that allow direct retrieval of previously learned arithmetic facts from memory (Ashcraft, 1982). Strategy choices were reported to depend on a range of factors, including problem size (i.e., the numerical magnitude of the components of an arithmetic problem; e.g., De Smedt et al., 2010) and the respective arithmetic operation (e.g., Imbo & Vandierendock, 2007), as well as the presentation format or context in which a problem is presented (e.g., digital-Arabic format vs. embedded in word problem). Because heterogeneous solving strategies are involved in arithmetic problem solving, language may affect arithmetic processing differently depending on the strategy that is used when solving a particular problem. In this final section, we will therefore elaborate on *phonological* influences on arithmetic processing with respect to the use of both procedural and retrieval-based strategies. Moreover, we will describe *lexical* influences on multi-digit arithmetic problem solving as well as *semantic* influences in the context of word problems.

Regarding *phonological* influences on the development of arithmetic competence, a considerable body of research is concerned with the relation of working memory resources and arithmetic performance in both adults and children (for reviews, see DeStefano & LeFevre, 2004; Friso-van den Bos et al., 2013; Peng et al., 2016; Raghubar et al., 2010). While researchers seem to generally agree that working memory is crucial for arithmetic processing and learning, inconsistent findings also suggest that the relation between a specific working memory component (e.g., verbal and visual-spatial working memory, central executive, e.g., Baddely, 2000; Baddeley & Hitch, 1974) and arithmetic performance likely depends on several factors (age, mathematical outcome variable, working memory task, etc.; e.g., Raghubar et al., 2010). Concerning arithmetic processing in primary school, Friso-van den Bos and colleagues (2013) suggested verbal working memory to show the most pronounced association with arithmetic competencies.

Generally, working memory has been suggested to be of specific importance when procedural strategies including maintaining and manipulating intermediate results during calculation have to be used to solve a problem (e.g., DeStefano & LeFevre, 2004). This is, for example, the case for more complex problems with larger problem sizes (e.g., Barrouillet, Mignot & Thevenot, 2008; Imbo & Vandierendock, 2008), for addition problems requiring a carry procedure (e.g., Ashcraft & Kirk, 2001; Fürst & Hitch, 2000), or – more generally – for problems for which solutions cannot (yet) be retrieved from memory.

Critically, empirical evidence suggests that the respective contribution of verbal and visual-spatial working memory components changes with age (e.g., De Smedt et al., 2009; Rasmussen & Bisanz, 2005; Van de Weijer-bergsma et al., 2015). For example, van de Weijer-Bergsma and colleagues (2015) observed that while the importance of verbal working memory for all four arithmetic operations was shown to increase from the second to sixth grades, visual-spatial working memory influences decreased. A similar conclusion was drawn by McKenzie et al. (2003), who investigated influences of verbal and visual-spatial working memory on simple arithmetic competencies in 6- to 7- and 8- to 9-year-old children experimentally by using a dual task paradigm. Children were asked to solve simple, auditorily presented addition problems (e.g., 9 + 4, 4 + 3 + 7) in three conditions: a baseline condition without added interference, a verbal interference condition in which children heard an audiotaped story while solving the addition problems, and a visual-spatial interference condition in which children solved addition problems and at the same time saw a matrix of black and white squares that randomly changed on the screen. Results indicated that while performance of children in both age groups was affected by visual-spatial interference, verbal interference only decreased performance in the older group of children. Thus, this study seems to substantiate that younger children may rely more on visual-spatial working memory when acquiring arithmetic competences, whereas older children seem to draw from both verbal and visual-spatial working memory resources when solving arithmetic problems.

Studies that specifically address the involvement of verbal working memory resources when retrieving arithmetic facts provided somewhat mixed results (for a review, see DeStefano & LeFevre, 2004). For instance, some studies suggest that the retrieval of multiplication facts is interrupted by concurrent verbal processing (e.g., Lee & Kang, 2002; Lemaire et al., 1996); however, in other studies fact retrieval remained largely unaffected under verbal load (De Rammelaere et al., 2001; Seitz & Schumann-Hengsteler, 2000). Thus, the degree to which verbal working memory influences arithmetic problem solving seems to depend on the respective strategies available and used to solve a particular problem.

Interestingly, further studies have addressed an additional *phonological* language aspect in that they focused on the influence of phonemic awareness on arithmetic performance assessed by standardized tests (e.g., Fuchs et al., 2006; Hecht et al., 2001; Krajewski & Schneider, 2009; Leather & Henry, 1994; Rasmussen & Bisanz, 2005; Simmons et al., 2008) but also more specifically on arithmetic fact retrieval (De Smedt & Boets, 2010; De Smedt et al., 2010). While many studies provided quite substantial evidence for an association of phonemic awareness with general arithmetic skills, the precise mechanism driving this association seems less well understood. Following up on this, De Smedt et al. (2010) suggested that one mechanism driving the association of phonemic awareness with general arithmetic skills might lay in its functional role for the retrieval of arithmetic facts. To investigate this claim, 9- to 11-year-old children were asked to solve addition, subtraction, and multiplication problems of both small (<25) and large problem size. The idea was that problems with a small problem size are more likely to be solved via retrieval-based strategies and should, thus, show a more pronounced association with phonemic awareness than problems with a large problem size. And, indeed, results showed a significant association of phonemic awareness with performance on problems with a small but not with a large problem size. Interestingly, this was observed independent of the respective operation. Thereby, the results of De Smedt and colleagues (2010; see also De Smedt & Boets, 2010 for additional evidence in dyslexics) support the idea that phonemic awareness may play a critical role for the acquisition of arithmetic facts.

Beyond *phonological* influences and similar to previously reported tasks involving multi-digit numbers, *lexical* influences related to the lack of transparency of certain number word systems were also observed for basic arithmetic. Investigating inversion-related effects, Göbel and colleagues (2014) evaluated performance differences in mental addition between German- and Italian-speaking second graders (with German having inverted and Italian having non-inverted number words). The authors specifically focused on the so-called carry effect which describes the observation that it takes considerably longer and more errors are committed in addition problems that require a carry procedure compared to problems that do not contain a carry (e.g., Deschuyteneer et al., 2005; Fürst & Hitch, 2000; Imbo et al., 2007; Klein et al., 2010). For example, a carry procedure is needed for 18 + 35 = 53 because the units add up to a sum larger than 9 (i.e., 8 + 5 = 13) and, thus, the tens digit of the unit sum has to be carried to the sum of the tens digits. Göbel and colleagues (2014) observed a regular carry-effect for both language groups; however, the effect was more pronounced in

German-speaking as compared to the Italian-speaking children (for similar results in adults, see Lonnemann & Yan, 2015). In carry problems it is crucial to keep track of place-value information because a successful carry operation requires to carry the tens digit of the unit sum to the tens position of the result. The more pronounced carry effect in children speaking German – a language with inverted number words – was attributed to increased demands on the manipulation and the mapping of the digital-Arabic notation and number words due to the inversion-related lack of transparency in the German number word formation.

Next to inversion-related language effects, there are also effects of number word systems (partially) following vigesimal (i.e., base-20) structuring (e.g., in, e.g., French or Basque the number word for 35 literally means to twenty-fifteen). For instance, Van Rinsveld et al. (2015) investigated performance in addition problems in German-French bilinguals across grades 7 to 10. Results indicated that when problems had to be solved in French, it took participants longer and they made more errors for problems with sums larger than 70 as compared to when the same problems had to be solved in German. Similarly, Colomé, Laka and Sebastián-Gallés (2010) manipulated addition problems so that problems either did not (e.g., 25 + 10 =) or did match with a vigesimal number word structure (e.g., 20 + 15 =). While performance between conditions did not differ for Italian and Catalan speakers, Basque speakers were specifically faster when addends followed the same vigesimal structure as Basque number words.

A last important aspect in the context of *semantic* influences on basic arithmetic abilities concerns the fact that throughout formal education arithmetic (and other) problems are regularly presented as word problems. The difficulty of arithmetic word problems is influenced by many factors related to both linguistic and numerical aspects (e.g., single- vs. multi-digit numbers, type of operation; see Daroczy et al., 2015 for an overview). On the one hand, linguistic aspects of arithmetic word problems such as sentence structure and length (e.g., Abedi & Lord, 2001; Spanos et al., 1988) but also the presence or absence of additional irrelevant information (e.g., Muth, 1992) certainly affect arithmetic word problem difficulty. On the other hand, the role of mathematical language and, in particular, the role of explicit verbal cues has also been investigated (e.g., Boonen et al., 2016; Hegarty et al., 1992; Van der Schoot et al., 2009; Verschaffel et al., 1992). Explicit verbal cues include words and phrases whose semantic usually directly hints at a respective operation that needs to be performed to arrive at the solution of the problem (e.g., subtraction: "Henry has 9 books. He *sells* 4 books at the flea market. How many books does he have *left*?"; multiplication: "Henry has 5 friends that he will meet in the park later today. He wants to bring 3 gummy bears for *each* of his friends. How many gummy bears does he have to bring?").

Unfortunately, verbal cues must not be used blindly because in some instances they are misleading. For example, the relational term "less" in the compare word problem "At the supermarket, a chocolate bar costs £1. This is 30 pence *less* than at the kiosk. How much do you have to you pay at the kiosk?" is inconsistent with the required operations (i.e., less would suggest a subtraction problem, however, to solve the problem correctly an addition needs to be performed). In this context, the consistency effect describes the finding that such inconsistent arithmetic word problems are more prone to errors compared to consistent problems (i.e., in which the term "less" indeed requires a subtraction; e.g., Hegarty et al., 1992; van der Schoot et al., 2009). Thus, while it is important to learn the semantic meaning of verbal cues and their associated arithmetic operations, it is also crucial to emphasize the integration of additional information across sentences to derive a proper mental model of the problem and with this a first step to a successful solution.

Taken together, as mentioned above, mathematics education is largely hierarchical in nature and, therefore, it is necessary to be able to draw on previously acquired competences, because new numerical and mathematical content usually builds on these previously acquired competences. Regarding some of the linguistic influences (i.e., *lexical, phonological, semantic*) there appears to be a similar pattern: some aspects that have been observed to already influence early numerical competences (i.e., counting and cardinality understanding) seem to persist or even increase their impact on more complex mathematical content strands such as arithmetic problem solving. This means that one may not assume linguistic influences on basic numerical competences to be overcome entirely with time. Instead, it seems that they exhibit a lasting influence on human numerical cognition.

5 Conclusion

In this chapter, we discussed (i) *lexical*, (ii) *syntactic*, (iii) *phonological*, and (iv) *semantic* aspects of language that seem to influence numerical and mathematical development and illustrated their relevance for selected numerical and mathematical content strands of (i) counting and cardinality understanding, (ii) multi-digit number processing, and (iii) basic arithmetic operations. In this last section we aim at discussing differences between but also commonalities across linguistic influences and content strands, before we elaborate on potential implications of the reported linguistic influences that arise for numerical and mathematical development.

First, it needs to be mentioned that linguistic influences seem to be most obvious, relevant, and detectable during specific time windows of numerical and mathematical development. On the one hand, some linguistic influences begin to affect numerical and mathematical development very early on even before formal education starts (e.g., effects of grammatical number, effects of phonemic awareness on the acquisition of the counting sequence). Others start to show their effect later when more advanced numerical and mathematical competences are acquired (e.g., lexical effects regarding the transparency of number words on multi-digit number processing and mental arithmetic). On the other hand, some linguistic influences seem to fade out rather quickly and, thus, can be observed only in a comparably small time window (e.g., effects of grammatical number), whereas others keep being relevant or become even more relevant throughout elementary school years when more and more complex mathematical competences are acquired (e.g., influences of verbal working memory, semantic influences regarding mathematical language). From this pattern of effects, it seems that linguistic influences occur in waves that peak for and when new numerical or mathematical concepts or procedures are learned. It seems that at these times of high external demands due to new to-be-learned content the cognitive system is more susceptible to influences of internal biases of numerical representations reflecting influences of *lexical*, *syntactic*, *phonological*, and *semantic* linguistic specificities of the respective language.

Second, because we aimed at summarizing linguistic influences on numerical and mathematical processing in children, we did not specifically consider evidence on adolescents or adults throughout this chapter. It is worth mentioning though that most of the reported linguistic influences can still be observed in highly skilled adults. For example, in a cross-cultural study, Moeller et al. (2015) realized a natural 2 by 2 design for the variables number word inversion (inverted vs. non-inverted) and reading direction (left-to-right vs. right-to-left) in a quadrilingual study with German-, English-, Hebrew-, and Arabic-speaking adults. Results were comparable to those observed for children by Pixner et al. (2011a) indicating *lexical* influences. In particular, Moeller et al. (2015) observed that unit-decade compatibility effects were larger when reading direction and order of tens and units as named in number words were in conflict (i.e., for German, left-to-right reading but units named before tens, and Hebrew, right-to-left reading but tens named before units) than for English- and Arabic-speaking participants for which reading direction and the order in which tens and units are named in number words match. Thus, even though linguistic effects might be more pronounced in children, traces of linguistic influences can also be found in adults.

Nevertheless, it needs to be noted that effects in adults are usually quantitatively smaller (a few dozen milliseconds) and, thus, are only detectable using more sensitive measures (e.g., reaction time measures). However, the consistent observation of linguistic influences in adults suggests that they are not a purely transient phenomenon but shape how we process numbers for good. Yet, studying linguistic influences in adults seems more relevant from a theoretical cognitive perspective aiming at understanding the underlying principles of numerical and mathematical cognition. Implications for numerical and mathematical learning or even educational practice may be limited because effects and differences in the millisecond range may not reflect practically relevant differences in numerical and mathematical competence in everyday life.

Finally, however, for a teaching practitioner, knowing that certain language aspects influence typical numerical and mathematical development in a certain time window might help identifying children that struggle or might struggle in the future. As mentioned earlier, Moeller and colleagues (2011) showed, for instance, that the number of inversion-related errors in transcoding as well as the size of the unit-decade compatibility effect in the first grade predicted arithmetic performance in the third grade. Thus, better understanding language-specific aspects of place-value processing might help identifying children that may develop mathematical difficulties later on. Moreover, while considering linguistic influences on numerical and mathematical development when developing interventional strategies is certainly necessary, it is also important to know that not all linguistic influences seem to cause lasting disadvantages for a particular language group or mathematical task.

In turn, this allows for a reconciliatory ending of this chapter. Although we presented explicit effects of linguistic aspects on numerical and mathematical development, it does not seem to be the case that any of the discussed linguistic aspects (alone) is a necessary predictor of numerical and mathematical development in an all or nothing manner. Instead, specific linguistic aspects may be detrimental to some aspects of numerical cognition while others may even facilitate numerical and mathematical learning (e.g., explicit plural markings or a transparent number word system). As such, it is important to be aware of the width of linguistic influences to be able to adapt teaching and learning approaches accordingly. These adaptations may then allow to compensate for disadvantageous influences and to foster beneficial linguistic aspects to help children to successfully develop sufficient numerical and mathematical competences to master everyday demands and needs.

References

Abedi, Jamal & Lord, Carol (2001): The language factor in mathematics tests. *Applied Measurement in Education* 14, 219–234. doi:https://doi.org/10.1207/S15324818AME1403_2.

Almoammer, Alhanouf, Sullivan, Jessica, Donlan, Chris, Marušič, Franc, Zaucer, Roc, O'Donnell, Timothy & Barner, David (2013): Grammatical morphology as a source of early number word meanings. *Proceedings of the National Academy of Sciences* 110 (46), 18448–18453. doi:https://doi.org/10.1073/pnas.1313652110.

Ashcraft, Mark H. (1982): The development of mental arithmetic: A chronometric approach. *Developmental Review* 2 (3), 213–236. doi:https://doi.org/10.1016/0273-2297(82)90012-0.

Ashcraft, Mark H. & Kirk, Elizabeth P. (2001): The relationships among working memory, math anxiety, and performance. *Journal of Experimental Psychology: General* 130 (2), 224–237. doi:https://doi.org/10.1037/0096-3445.130.2.224.

Aunio, Pirjo, Aubrey, Carol, Godfrey, Ray, Pan, Yuejuan & Liu, Yan (2008): Children's early numeracy in England, Finland and People's Republic of China. *International Journal of Early Years Education* 16 (3), 203–221. doi:https://doi.org/10.1080/09669760802343881.

Baddeley, Alan D. (2000). The episodic buffer: a new component of working memory?. *Trends in Cognitive Sciences*, 4(11), 417–423. doi: https://doi.org/10.1016/S1364-6613(00)01538-2

Baddeley, Alan D. & Hitch, Graham (1974): Working memory. In Bower, G. (ed.): *Psychology of Learning and Motivation*. 8, New York City Academic press, 47–89. doi: https://doi.org/10.1016/S0079-7421(08)60452-1

Bahnmueller, Julia, Maier, Carolin A., Göbel, Silke M. & Moeller, Korbinian (2019): Direct evidence for linguistic influences in two-digit number processing. *Journal of Experimental Psychology: Learning, Memory, and Cognition* 45 (6), 1142–1150. doi:https://doi.org/10.1037/xlm0000642.

Bahnmueller, Julia, Nuerk, Hans-Christoph & Moeller, Korbinian (2018): A taxonomy proposal for types of interactions of language and place-value processing in multi-digit numbers. *Frontiers in Psychology* 9, 1024. doi:https://doi.org/10.3389/fpsyg.2018.01024.

Barner, David, Libenson, Amanda, Cheung, Pierina & Takasaki, Mayu (2009): Cross-linguistic relations between quantifiers and numerals in language acquisition: Evidence from Japanese. *Journal of Experimental Child Psychology* 103 (4), 421–440. doi:https://doi.org/10.1016/j.jecp.2008.12.001.

Barrouillet, Pierre, Mignon, Mathilde & Thevenot, Catherine (2008): Strategies in subtraction problem solving in children. *Journal of Experimental Child Psychology* 99 (4), 233–251. doi:https://doi.org/10.1016/j.jecp.2007.12.001.

Berch, Daniel B., Geary, David C. & Koepke, Kathleen Mann (2018): Introduction: Language and culture in mathematical cognitive development. In Berch, Daniel B., Geary, David C., Koepke, Kathleen Mann (eds.): *Language and Culture in Mathematical Cognition*. Academic Press, 1–29.

Boonen, Antin J., de Koning, Björn B., Jolles, Jelle & van der Schoot, Menno (2016): Word problem solving in contemporary math education: A plea for reading comprehension skills training. *Frontiers in Psychology* 7, 191. doi:https://doi.org/10.3389/fpsyg.2016.00191.

Boonen, Anton J., Kolkman, Meijke E. & Kroesbergen, Evelyn H. (2011): The relation between teachers' math talk and the acquisition of number sense within kindergarten classrooms. *Journal of School Psychology* 49 (3), 281–299. doi:https://doi.org/10.1016/j.jsp.2011.03.002.

Butterworth, Brian, Varma, Sashank & Laurillard, Diana (2011): Dyscalculia: from brain to education. *Science* 332 (6033), 1049–1053. doi:https://doi.org/10.1126/science.1201536.

Cankaya, Ozlem, LeFevre, Jo-Anne & Dunbar, Kristina (2014): The role of number naming systems and numeracy experiences in children's rote counting: Evidence from Turkish and Canadian children. *Learning and Individual Differences* 32, 238–245. doi:https://doi.org/10.1016/j.lindif.2014.03.016.

Carpenter, Thomas P. & Moser, James M. (1984): The acquisition of addition and subtraction concepts in grades one through three. *Journal for Research in Mathematics Education* 179–202. doi:https://doi.org/10.2307/748348.

Cawley, John F., Parmar, Rene S., Lucas-Fusco, Lynn M., Kilian, Joy D. & Foley, Teresa E. (2007): Place value and mathematics for students with mild disabilities: Data and suggested practices. *Learning Disabilities: A Contemporary Journal* 5 (1), 21–39.

Chan, Becky Mee-yin & Ho, Connie Suk-han (2010): The cognitive profile of Chinese children with mathematics difficulties. *Journal of Experimental Child Psychology* 107 (3), 260–279. doi:https://doi.org/10.1016/j.jecp.2010.04.016.

Chan, Winnie Wai Lan, Au, Terry K. & Tang, Joey (2014): Strategic counting: A novel assessment of place-value understanding. *Learning and Instruction* 29, 78–94. https://doi.org/10.1016/j.learninstruc.2013.09.001.

Cheng, Yi-Ling & Mix, Kelly S. (2012): The relation between space and math: Developmental and educational implications. In Benson, J. B. (ed.): *Advances in Child Development and Behavior*. 42, JAI, Atlanta, 197–243. doi:https://doi.org/10.1016/B978-0-12-394388-0.00006-X

Clements, Douglas H. & Sarama, Julie (2021): *Learning and teaching early math – The learning trajectories approach*. New York: Routledge.

Colome, Angels, Laka, Itziar & Sebastián-Gallés, Núria (2010): Language effects in addition: How you say it counts. *The Quarterly Journal of Experimental Psychology* 63 (5), 965–983. doi:http://doi.org/10.1080/17470210903134377.

Comrie, Bernard (2005): Endangered numeral systems. In Wohlgemuth, Jan, Dirksmeyer, Tyko (eds.): *Bedrohte Vielfalt: Aspekte des Sprach(en)tods [Endangered Diversity: Aspects of Language Death]*. Berlin, Germany: Weißensee Verlag, 203–230.

Daroczy, Gabriella, Wolska, Magdalena, Meurers, Walt Detmar & Nuerk, Hans-Christoph (2015): Word problems: a review of linguistic and numerical factors contributing to their difficulty. *Frontiers in Psychology* 6, 348. doi:https://doi.org/10.3389/fpsyg.2015.00348.

De Rammelaere, Stijn, Stuyven, Els & Vandierendonck, André (2001): Verifying simple arithmetic sums and products: Are the phonological loop and the central executive involved? *Memory & Cognition* 29 (2), 267–273. doi:https://doi.org/10.3758/BF03194920.

De Smedt, Bert & Boets, Bart (2010): Phonological processing and arithmetic fact retrieval: Evidence from developmental dyslexia. *Neuropsychologia* 48 (14), 3973–3981. doi: https://doi.org/10.1016/j.neuropsychologia.2010.10.018.

De Smedt, Bert, Janssen, Rianne, Bouwens, Kelly, Verschaffel, Lieven, Boets, Bart & Ghesquière, Pol (2009): Working memory and individual differences in mathematics achievement: A longitudinal study from first grade to second grade. *Journal of Experimental Child Psychology* 103 (2), 186–201. doi:https://doi.org/10.1016/j.jecp.2009.01.004.

De Smedt, Bert, Taylor, Jessica, Archibald, Lisa & Ansari, Daniel (2010): How is phonological processing related to individual differences in children's arithmetic skills? *Developmental Science* 13 (3), 508–520. doi:https://doi.org/10.1111/j.1467-7687.2009.00897.x.

Denise, Muth, K. (1992): Extraneous information and extra steps in arithmetic word problems. *Contemporary Educational Psychology* 17, 278–285. doi:https://doi.org/10.1016/0361-476X(92)90066-90068.
Deschuyteneer, Maud, De Rammelaere, Stijn & Fias, Wim (2005): The addition of two-digit numbers: Exploring carry versus no-carry problems. *Psychology Science* 47 (1), 74–83.
DeStefano, Diana & LeFevre, Jo-Anne (2004): The role of working memory in mental arithmetic. *European Journal of Cognitive Psychology* 16 (3), 353–386. doi:https://doi.org/10.1080/09541440244000328.
Dowker, Ann & Nuerk, Hans-Christoph (2016): Linguistic influences on mathematics. *Frontiers in Psychology* 7, 1035. doi:https://doi.org/10.3389/fpsyg.2016.01035.
Friso-van den Bos, Illona, Van der Ven, Sanne, Kroesbergen, Evelyn & Van Luit, Johannes E (2013): Working memory and mathematics in primary school children: A meta-analysis. *Educational Research Review* 10, 29–44. https://doi.org/10.1016/j.edurev.2013.05.003.
Fuchs, Lynn S., Fuchs, Douglas, Compton, L., Powell, Sarah R., Seethaler, Pamela M., Capizzi, A. M. & Fletcher, Jack M. (2006): The cognitive correlates of third-grade skill in arithmetic, algorithmic computation, and arithmetic word problems. *Journal of Educational Psychology* 98 (1), 29. doi:https://doi.org/10.1037/0022-0663.98.1.29.
Fürst, Ansgar J. & Hitch, Graham J. (2000): Separate roles for executive and phonological components of working memory in mental arithmetic. *Memory & Cognition* 28 (5), 774–782. doi:https://doi.org/10.3758/BF03198412.
Fuson, Karen C. (1982): An analysis of the counting-on solution procedure in addition. In Carpenter, Thomas P., Moser, James M., Romberg, Thomas A. (eds.): *Addition and Subtraction: A Cognitive Perspective*. Hillsdale, NJ: Erlbaum, 67–81.
Ganor-Stern, Dana, Tzelgov, Joseph & Ellenbogen, Ravid (2007): Automaticity and two-digit numbers. *Journal of Experimental Psychology: Human Perception and Performance* 33 (2), 483–496. doi:http://doi.org/10.1037/0096-1523.33.2.483.
Gathercole, Susan E. (2006): Nonword repetition and word learning: The nature of the relationship. *Applied Psycholinguistics* 27 (4), 513. doi:https://doi.org/10.1017.S0142716406060383.
Geary, David C., Hoard, Mary K., Byrd-Craven, Jennifer & Catherine, DeSoto, M. (2004): Strategy choices in simple and complex addition: Contributions of working memory and counting knowledge for children with mathematical disability. *Journal of Experimental Child Psychology* 88 (2), 121–151. doi:https://doi.org/10.1016/j.jecp.2004.03.002.
Geary, David C. & Hoard, Mary K. (2005). Learning disabilities in arithmetic and mathematics. In Campbell, I. D. (ed.): *Handbook of Mathematical Cognition*. Psychology Press, 253–268.
Gelman, Rochel & Gallistel, Charles R. (1978): *The child's Concept of Number*. Cambridge, MA: Harvard.
Göbel, Silke M., Moeller, Korbinian, Pixner, Silvia, Kaufmann, Liane & Nuerk, Hans-Christoph (2014): Language affects symbolic arithmetic in children: The case of number word inversion. *Journal of Experimental Child Psychology* 119, 17–25. doi:https://doi.org/10.1016/j.jecp.2013.10.001.
Gunderson, Elizabeth A. & Levine, Susan C. (2011): Some types of parent number talk count more than others: Relations between parents' input and children's cardinal-number knowledge. *Developmental Science* 14 (5), 1021–1032. doi:https://doi.org/10.1111/j.1467-7687.2011.01050.x.
Haase, Vitor G., Júlio-Costa, Annelise, Lopes-Silva, Julia B., Starling-Alves, Isabella, Antunes, Andressa, Pinheiro-Chagas, Pedro & Wood, Guilherme (2014): Contributions from specific

and general factors to unique deficits: Two cases of mathematics learning difficulties. *Frontiers in Psychology* 5, 102. doi:https://doi.org/10.3389/fpsyg.2014.00102.

Harmon, Janis M., Hedrick, Wanda B. & Wood, Karen D. (2005): Research on vocabulary instruction in the content areas: Implications for struggling readers. *Reading & Writing Quarterly* 21 (3), 261–280. doi:https://doi.org/10.1080/10573560590949377.

Hecht, Steve A., Torgesen, Joseph K., Wagner, Richard K. & Rashotte, Carol A. (2001): The relations between phonological processing abilities and emerging individual differences in mathematical computation skills: A longitudinal study from second to fifth grades. *Journal of Experimental Child Psychology* 79 (2), 192–227. doi:https://doi.org/10.1006/jecp.2000.2586.

Hegarty, Mary, Mayer, Richard E. & Green, Charles E. (1992): Comprehension of arithmetic word problems: Evidence from students' eye fixations. *Journal of Educational Psychology* 84 (1), 76–84. doi:https://doi.org/10.1037/0022-0663.84.1.76.

Herzog, Moritz, Ehlert, Antje & Fritz, Annemarie (2017): A competency model of place value understanding in South African primary school pupils. *African Journal of Research in Mathematics, Science and Technology Education* 21 (1), 37–48. doi:https://doi.org/10.1080/18117295.2017.1279453.

Herzog, Moritz, Ehlert, Antje & Fritz, Annemarie (2019): Development of a sustainable place value understanding. In Fritz, A., Haase, V. G., Räsänen, P (eds.): *International Handbook of Mathematical Learning Difficulties*. Springer, New York, 561–579.

Hurewitz, Felicia, Papafragou, Anna, Gleitman, Lila & Gelman, Rochel (2006): Asymmetries in the acquisition of numbers and quantifiers. *Language Learning and Development* 2 (2), 77–96. doi:https://doi.org/10.1207/s15473341lld0202_1.

Imbo, Ineke, Vanden Bulcke, Charlotte, De Brauwer, Jolien & Fias, Wim (2014): Sixty-four or four-and-sixty? The influence of language and working memory on children's number transcoding. *Frontiers in Psychology* 5, 313. doi:https://doi.org/10.3389/fpsyg.2014.00313.

Imbo, Ineke & Vandierendonck, Andre (2007): The development of strategy use in elementary school children: Working memory and individual differences. *Journal of Experimental Child Psychology* 96 (4), 284–309. doi:https://doi.org/10.1016/j.jecp.2006.09.001.

Imbo, Ineke & Vandierendonck, Andre (2008): Effects of problem size, operation, and working-memory span on simple-arithmetic strategies: Differences between children and adults? *Psychological Research* 72 (3), 331–346. doi:https://doi.org/10.1007/s00426-007-0112-8.

Klein, Elise, Bahnmueller, Julia, Mann, Anne, Pixner, S., Kaufmann, Liane, Hans-Christoph., Nuerk & Moeller, Korbinian (2013): Language influences on numerical development – Inversion effects on multi-digit number processing. *Frontiers in Psychology* 4, 480. doi: https://doi.org/10.3389/fpsyg.2013.00480.

Klein, Elise, Moeller, Korbinian, Dressel, Katharina, Domahs, Frank, Wood, Guilherme, Willmes, Klaus & Nuerk, Hans-Christoph (2010): To carry or not to carry – Is this the question? Disentangling the carry effect in multi-digit addition. *Acta Psychologica* 135 (1), 67–76. doi:http://doi.org/10.1016/j.actpsy.2010.06.002.

Koponen, Tuine, Salmi, Paula, Eklund, Kenneth & Aro, Tuija (2013): Counting and RAN: Predictors of arithmetic calculation and reading fluency. *Journal of Educational Psychology* 105 (1), 162. doi:https://doi.org/10.1037/a0029285.

Krajewski, Kristin & Schneider, Wolfgang (2009): Exploring the impact of phonological awareness, visual–spatial working memory, and preschool quantity–number competencies on mathematics achievement in elementary school: Findings from a 3-year longitudinal study. *Journal of Experimental Child Psychology* 103 (4), 516–531. doi: https://doi.org/10.1016/j.jecp.2009.03.009.

Krinzinger, Helga, Gregoire, Jacques, Desoete, Annemie, Kaufmann, Liane, Nuerk, Hans-Christoph & Willmes, Klaus (2011): Differential language effects on numerical skills in second grade. *Journal of Cross-Cultural Psychology* 42 (4), 614–629. doi:http://doi.org/10.1177/0022022111406252.

Lambert, Katharina & Moeller, Korbinian (2019): Place-value computation in children with mathematics difficulties. *Journal of Experimental Child Psychology* 178, 214–225. doi: https://doi.org/10.1016/j.jecp.2018.09.008.

Landerl, Karin & Kölle, Christina (2009): Typical and atypical development of basic numerical skills in elementary school. *Journal of Experimental Child Psychology* 103 (4), 546–565. doi:http://doi.org/10.1016/j.jecp.2008.12.006.

Leather, Cathy V. & Henry, Lucy A. (1994): Working memory span and phonological awareness tasks as predictors of early reading ability. *Journal of Experimental Child Psychology* 58 (1), 88–111. doi:https://doi.org/10.1006/jecp.1994.1027.

Lee, Kyoung-Min & Kang, So-Young (2002): Arithmetic operation and working memory: Differential suppression in dual tasks. *Cognition* 83 (3), B63–B68. doi:http://doi.org/10.1016/S0010-0277(02)00010-0.

LeFevre, Jo-Anne, Clarke, Tamara & Stringer, Alex P. (2002): Influences of language and parental involvement on the development of counting skills: Comparisons of French-and English-speaking Canadian children. *Early Child Development and Care* 172 (3), 283–300. doi:https://doi.org/10.1080/03004430212127.

Lemaire, Patrick, Abdi, Herve & Fayol, Michel (1996): Working memory and cognitive arithmetic: Evidence from the disruption of the associative confusion effect. *European Journal of Cognitive Psychology* 8 (1), 73–103.

Lewis, Carolin Annette., Bahnmueller, Julia, Wesierska, Marta, Moeller, Korbinian & Göbel, Silke Melanie (2020): Inversion effects on mental arithmetic in English-and Polish-speaking adults. *Quarterly Journal of Experimental Psychology* 73 (1), 91–103. doi:https://doi.org/10.1177/1747021819881983.

Li, Xia, Sun, Ye, Baroody, Arthur J. & Purpura, David (2013): The effect of language on Chinese and American 2-and 3-year olds' small number identification. *European Journal of Psychology of Education* 28 (4), 1525–1542. doi:https://doi.org/10.1007/s10212-013-0180-7.

Lonnemann, Jan, Su, Li, Zhao, Pri, Linkersdörfer, Janosch, Lindberg, Sven, Hasselhorn, Marcus & Yan, Song (2019): Differences in counting skills between Chinese and German children are accompanied by differences in processing of approximate numerical magnitude information. *Frontiers in Psychology* 9, 2656. https://doi.org/10.3389/fpsyg.2018.02656.

Lonnemann, Jan & Yan, Song (2015): Does number word inversion affect arithmetic processes in adults? *Trends in Neuroscience and Education* 4 (1–2), 1–5. doi:http://doi.org/10.1016/j.tine.2015.01.002.

Lopes-Silva, Julia B., Moura, Ricardo, Júlio-Costa, Aneeliese, Haase, Vitor G. & Wood, Guilherme (2014): Phonemic awareness as a pathway to number transcoding. *Frontiers in Psychology* 5 (13), doi:https://doi.org/10.3389/fpsyg.2014.00013.

Macizo, Pedro & Herrera, Amparo (2011): Cognitive control in number processing: Evidence from the unit–decade compatibility effect. *Acta Psychologica* 136 (1), 112–118. doi:http://doi.org/10.1016/j.actpsy.2010.10.008.

McCloskey, Michael, Caramazza, Alfonso & Basili, Annamaria (1985): Cognitive mechanisms in number processing and calculation: Evidence from dyscalculia. *Brain and Cognition* 4 (2), 171–196. doi:https://doi.org/10.1016/0278-2626(85)90069-7.

McKenzie, Bruce, Bull, Rebecca & Gray, Colin (2003): The effects of phonological and visual-spatial interference on children's arithmetical performance. *Educational and Child Psychology* 20 (3), 93–108.

Melby-Lervåg, Monica, Lyster, Solveig-Alma H. & Hulme, Charles (2012): Phonological skills and their role in learning to read: A meta-analytic review. *Psychological Bulletin* 138 (2), 322. doi:https://doi.org/10.1037/a0026744.

Miller, Kevin F., Smith, Catherine M., Zhu, Jianjun & Zhang, Houcan (1995): Preschool origins of cross-national differences in mathematical competence: The role of number-naming systems. *Psychological Science* 6 (1), 56–60. doi:https://doi.org/10.1111/j.1467-9280.1995.tb00305.x.

Miller, Kevin F. & Stigler, James W. (1987): Counting in Chinese: Cultural variation in a basic cognitive skill. *Cognitive Development* 2 (3), 279–305. doi:https://doi.org/10.1016/S0885-2014(87)90091-8.

Miura, Irene T., Kim, Chungsoon C., Chang, Chih Mei & Okamoto, Yukari (1988): Effects of language characteristics on children's cognitive representation of number: Cross-national comparisons. *Child Development* 59, 1445–1450. doi:https://doi.org/10.2307/1130659.

Miura, Irene T., Okamoto, Y., Kim, Chungsoon C., Chih-Mei., Chang, Steere, Marcia & Fayol, Michel (1994): Comparisons of children's cognitive representation of number: China, France, Japan, Korea, Sweden, and the United States. *International Journal of Behavioral Development* 17 (3), 401–411. doi:https://doi.org/10.1177/016502549401700301.

Miura, Irene T., Okamoto, Y., Kim, Chungsoon C., Steere, Marcia & Fayol, Michel (1993): First graders' cognitive representation of number and understanding of place value: Cross-national comparisons: France, Japan, Korea, Sweden, and the United States. *Journal of Educational Psychology* 85 (1), 24–30. doi:https://doi.org/10.1037/0022-0663.85.1.24.

Miura, Irene T. & Okamoto, Yukari (2003): Language supports for mathematics understanding and performance. In Baroody, A. J., Dowker, A. (eds.): *Studies in Mathematical Thinking and Learning. The Development of Arithmetic Concepts and Skills: Constructing Adaptive Expertise*. Hillsdale, NJ: Lawrence Erlbaum Associates Publishers, 229–242.

Moeller, Korbinian, Pixner, Silvia, Zuber, Julia, Kaufmann, Liane & Nuerk, Hans-Christoph (2011): Early place-value understanding as a precursor for later arithmetic performance – A longitudinal study on numerical development. *Research in Developmental Disabilities* 32 (5), 1837–1851. doi:https://doi.org/10.1016/j.ridd.2011.03.012.

Moeller, Korbinian, Shaki, Samuel, Göbel, Silke M. & Nuerk, Hans-Christoph (2015): Language influences number processing – A quadrilingual study. *Cognition* 136, 150–155. doi: http://doi.org/10.1016/j.cognition.2014.11.003.

NCTM (National Council of Teachers of Mathematics) (2000): *Principles and Standards for School Mathematics*. Reston, VA: NCTM.

Nuerk, Hans-Christoph, Weger, Ulrich & Willmes, Klaus (2001): Decade breaks in the mental number line? Putting the tens and units back in different bins. *Cognition* 82 (1), B25–B33. doi:http://doi.org/10.1016/S0010-0277(01)00142-1.

Nuerk, Hans-Christoph, Weger, Ulrich & Willmes, Klaus (2005): Language effects in magnitude comparison: Small, but not irrelevant. *Brain and Language* 92 (3), 262–277. doi:http://doi.org/10.1016/j.bandl.2004.06.107.

OECD (2019a): *PISA 2018 Results (Volume I): What Students Know and Can do*. Paris: OECD Publishing. https://doi.org/10.1787/5f07c754-en.

OECD (2019b): *PISA 2018 Results (Volume III): What school Life Means for Students' Lives*. Paris: OECD Publishing. https://doi.org/10.1787/acd78851-en.

Peng, Peng, Namkung, Jessica, Barnes, Marcia & Sun, Congying (2016): A meta-analysis of mathematics and working memory: Moderating effects of working memory domain, type of mathematics skill, and sample characteristics. *Journal of Educational Psychology* 108 (4), 455. doi:https://doi.org/10.1037/edu0000079.

Pickett, Joseph P., Rickford, John R., Pinker, Steven, Watkins, Calvert & Huehnergard, John (2018): *The American Heritage Dictionary of the English Language*. Boston Houghton Mifflin Harcourt.

Pixner, Silvia, Moeller, Korbinian, Hermanova, V., Hans-Christoph., Nuerk & Kaufmann, Liane (2011a): Whorf reloaded: Language effects on nonverbal number processing in first grade – A trilingual study. *Journal of Experimental Child Psychology* 108 (2), 371–382. doi:https://doi.org/10.1016/j.jecp.2010.09.002.

Pixner, Silvia, Zuber, Julia, Heřmanová, V., Kaufmann, Liane, Nuerk, Hans-Christoph & Moeller, Korbinian (2011b): One language, two number-word systems and many problems: Numerical cognition in the Czech language. *Research in Developmental Disabilities* 32 (6), 2683–2689. doi:http://doi.org/10.1016/j.ridd.2011.06.004.

Powell, Sarah R., Driver, Melissa K., Roberts, Greg & Fall, Anna-Maria (2017): An analysis of the mathematics vocabulary knowledge of third- and fifth-grade students: Connections to general vocabulary and mathematics computation. *Learning and Individual Differences* 57, 22–32. doi:https://doi.org/10.1016/j.lindif.2017.05.011.

Powell, Sarah R. & Nelson, Gena (2017): An investigation of the mathematics-vocabulary knowledge of first-grade students *The Elementary School Journal* 117(4), 664–686. doi:https://doi.org/10.1086/691604.

Pruden, Shannon M., Levine, Susan C. & Huttenlocher, Janellen (2011): Children's spatial thinking: Does talk about the spatial world matter? *Developmental Science* 14 (6), 1417–1430. doi:https://doi.org/10.1111/j.1467-7687.2011.01088.x.

Purpura, David J., Napoli, Amy R., Wehrspann, Elizabeth A. & Gold, Zachary S. (2017): Causal connections between mathematical language and mathematical knowledge: A dialogic reading intervention. *Journal of Research on Educational Effectiveness* 10 (1), 116–137. doi:https://doi.org/10.1080/19345747.2016.1204639.

Purpura, David J. & Reid, Erin E. (2016): Mathematics and language: Individual and group differences in mathematical language skills in young children. *Early Childhood Research Quarterly* 36, 259–268. doi:https://doi.org/10.1016/j.ecresq.2015.12.020.

Raghubar, Kimberly P., Barnes, Marcia A. & Hecht, Steven A. (2010): Working memory and mathematics: A review of developmental, individual difference, and cognitive approaches. *Learning and Individual Differences* 20 (2), 110–122. doi:https://doi.org/10.1016/j.lindif.2009.10.005.

Rasmussen, Carmen & Bisanz, Jeffrey (2005): Representation and working memory in early arithmetic. *Journal of Experimental Child Psychology* 91 (2), 137–157. doi:https://doi.org/10.1016/j.jecp.2005.01.004.

Sarnecka, Barbara W. (2014): On the relation between grammatical number and cardinal numbers in development. *Frontiers in Psychology* 5, 1132. doi:https://doi.org/10.3389/fpsyg.2014.01132.

Sarnecka, Barbara W., Kamenskaya, Valentina G., Yamana, Yuko, Ogura, Tamiko & Yudovina, Yulia B. (2007): From grammatical number to exact numbers: Early meanings of 'one','two', and 'three'in English, Russian, and Japanese. *Cognitive Psychology* 55 (2), 136–168. doi:https://doi.org/10.1016/j.cogpsych.2006.09.001.
Schleppegrell, Mary J. (2007): The linguistic challenges of mathematics teaching and learning: A research review. *Reading & Writing Quarterly* 23 (2), 139–159. doi:https://doi.org/10.1080/10573560601158461.
Seitz, Katja & Schumann-Hengsteler, Ruth (2000): Mental multiplication and working memory. *European Journal of Cognitive Psychology* 12 (4), 552–570. doi:https://doi.org/10.1080/095414400750050231.
Seron, Xavier & Fayol, Michel (1994): Number transcoding in children: A functional analysis. *British Journal of Developmental Psychology* 12 (3), 281–300. doi:http://doi.org/10.1111/j.2044-835X.1994.tb00635.x.
Siegler, Robert S. (1996): *Emerging Minds: The Process of Change in Children's Thinking*. Oxford: Oxford University Press.
Siegler, Robert S. & Shrager, Jeff (1984): Strategy choice in addition and subtraction: How do children know what to do? In Sophian, C. (ed.): *Origins of Cognitive Skills*. Hillsdale, NJ: Erlbaum, 229–293.
Simmons, Fiona, Singleton, Chris & Horne, Joanna (2008): Brief report – Phonological awareness and visual-spatial sketchpad functioning predict early arithmetic attainment: Evidence from a longitudinal study. *European Journal of Cognitive Psychology* 20 (4), 711–722. doi:https://doi.org/10.1080/09541440701614922.
Simmons, Fiona R., Willis, Catherine & Adams, Anne-Marie (2012): Different components of working memory have different relationships with different mathematical skills. *Journal of Experimental Child Psychology* 111 (2), 139–155. doi:https://doi.org/10.1016/j.jecp.2011.08.011.
Society, The Royal & Academy, The British (2018). *Harnessing educational research*. https://royalsociety.org/topics-policy/projects/royal-society-british-academy-educational-research/
Soto-Calvo, Elena, Simmons, Fiona R., Willis, Catherine & Adams, Anne-Marie (2015): Identifying the cognitive predictors of early counting and calculation skills: Evidence from a longitudinal study. *Journal of Experimental Child Psychology* 140, 16–37. doi:https://doi.org/10.1016/j.jecp.2015.06.011.
Spanos, George, Rhodes, Nancy C. & Dale, Theresa Corasaniti (1988): Linguistic features of mathematical problem solving: Insights and applications. In Cocking, R. R., Mestre, J. P. (eds.): *Linguistic and Cultural Influences on Learning Mathematics*. Hillsdale, NJ: Lawrence Erlbaum, 221–240.
Toll, Sylke W. & Van Luit, Johannes E. (2014a): Explaining numeracy development in weak performing kindergartners. *Journal of Experimental Child Psychology* 124, 97–111. doi:https://doi.org/10.1016/j.jecp.2014.02.001.
Toll, Sylke W. & Van Luit, Johannes E. (2014b): The developmental relationship between language and low early numeracy skills throughout kindergarten. *Exceptional Children* 81 (1), 64–78. doi:https://doi.org/10.1177/0014402914532233.
Towse, John & Saxton, Matthew (1998): Mathematics across national boundaries: Cultural and linguistic perspectives on numerical competence. In Donlan, C. (ed.): *Studies in Developmental Psychology. The Development of Mathematical Skills*. Psychology Press/Taylor & Francis, 129–150.

Van de Weijer-bergsma, Eva, Kroesbergen, Evelyn H. & Van Luit, Johannes E. (2015): Verbal and visual-spatial working memory and mathematical ability in different domains throughout primary school. *Memory & Cognition* 43 (3), 367–378. doi:https://doi.org/10.3758/s13421-014-0480-4.
Van der Schoot, Menno, Arkema, Annemieke H. B., Horsley, Tako M. & van Lieshout, Ernest C. (2009): The consistency effect depends on markedness in less successful but not successful problem solvers: An eye movement study in primary school children. *Contemporary Educational Psychology* 34 (1), 58–66. doi:https://doi.org/10.1016/j.cedpsych.2008.07.002.
Van der Ven, Sanne H., Klaiber, Jonathan D. & van der Maas, Han L. (2017): Four and twenty blackbirds: How transcoding ability mediates the relationship between visuospatial working memory and math in a language with inversion. *Educational Psychology* 37 (4), 487–505. doi:https://doi.org/10.1080/01443410.2016.1150421.
Van Rinsveld, Amandine, Brunner, Martin, Landerl, Karin, Schiltz, Christine & Ugen, Sonja (2015): The relation between language and arithmetic in bilinguals: Insights from different stages of language acquisition. *Frontiers in Psychology* 6, 265. doi:https://doi.org/10.3389/fpsyg.2015.00265.
Van Rinsveld, Amandine & Schiltz, Christine (2016): Sixty-twelve = Seventy-two? A cross-linguistic comparison of children's number transcoding. *British Journal of Developmental Psychology* 34 (3), 461–468. doi:http://doi.org/10.1111/bjdp.12151.
Van Rinsveld, Amandine, Schiltz, Christine, Landerl, Karin, Brunner, Martin & Ugen, Sonja (2016): Speaking two languages with different number naming systems: What implications for magnitude judgments in bilinguals at different stages of language acquisition? *Cognitive Processing* 17 (3), 225–241. doi:http://doi.org/10.1007/s10339-016-0762-9.
Vasilyeva, Marina, Laski, Elida V., Ermakova, Anna, Lai, Weng-Feng, Jeong, Yoonkyung & Hachigian, Amy (2015): Reexamining the language account of cross-national differences in base-10 number representations. *Journal of Experimental Child Psychology* 129, 12–25. doi:https://doi.org/10.1016/j.jecp.2014.08.004.
Verschaffel, Lieven, De Corte, Erik & Pauwels, Ann (1992): Solving compare problems: An eye movement test of Lewis and Mayer's consistency hypothesis. *Journal of Educational Psychology* 84 (1), 85. doi:https://doi.org/10.1037/0022-0663.84.1.85.
Wagner, Richard K. & Torgesen, Joseph K. (1987): The nature of phonological processing and its causal role in the acquisition of reading skills. *Psychological Bulletin* 101 (2), 192. doi:https://doi.org/10.1037/0033-2909.101.2.192.
Wood, Guilherme, Mahr, Moritz & Nuerk, Hans-Christoph (2005): Deconstructing and reconstructing the base-10 structure of Arabic numbers. *Psychology Science* 47 (1), 84–95.
Zuber, Julia, Pixner, Silvia, Moeller, Korbinian & Nuerk, Hans-Christoph (2009): On the language specificity of basic number processing: Transcoding in a language with inversion and its relation to working memory capacity. *Journal of Experimental Child Psychology* 102 (1), 60–77. doi:http://doi.org/10.1016/j.jecp.2008.04.003.

Desoete Annemie, Ceulemans Annelies, Sofie Rousseau, and Mathieu Roelants

The relative importance of "parental talk" as a predictor of the diversity in mathematics learning in young children

Abstract: This study explored the importance of the amount of "parental talk" focusing on numerical cues as "opportunity factor" in the prediction of diversity in mathematics learning. Thirty-one children were followed up from toddlerhood (24 months of age) till kindergarten (48 months of age). Mathematics learning was tested with a number discrimination task at 24 months. At 48 months children's mathematics learning was examined with a procedural and conceptual counting task and a calculation task. The amount of parental talk was operationalized via a questionnaire and via a structured play Duplo or Lego building session. The study confirmed a substantial amount of diversity in the frequency of parental talk with the results of the questionnaire and the observation positively related to each other. A positive concurrent association was found between the amount of observed parental talk and children's calculation skills in kindergarten. Parental talk with toddlers was also positively predicting children's mathematics learning in kindergarten. There was a trend of positive association between the amount of parental talk with toddlers and children's conceptual counting abilities in kindergarten. There was a positive quadratic predictive contribution of parental talk in toddlers for "calculation" in kindergarten. These results confirmed that mathematics learning might not be unitary even in young children and that parental talk should be considered as one of the opportunity factors to explain some of the diversities in mathematics learning.

Keywords: parental talk, opportunity-propensity model, toddlers, kindergartners, mathematics learning, number discrimination, procedural counting, conceptual counting, calculation

1 Introduction

1.1 Mathematics learning

Nelson and Powell (2018) revealed findings based on 35 longitudinal studies that mathematics learning in elementary school was one of the biggest predictors

https://doi.org/10.1515/9783110661941-007

for future academic achievement. Mathematics skills were stronger predictors than reading skills, even after controlling for intelligence and socioeconomic status. Deary et al. (2000) demonstrated diversity in mathematics learning from childhood to old age. A longitudinal study on 17.638 participants revealed that mathematics learning at the age of seven was positively associated with the socioeconomic status (SES) of individuals at the age of 42 years. This effect was significant even when controlling for intelligence and SES at birth on top of intelligence and SES at birth (Ritchie & Bates, 2013). Poor mathematics learning was revealed to have an impact on daily life, resulting in more employment in low paid professions and in negative consequences for (mental) health (Duncan & Magnuson, 2011; Geary, 2011; Wilson et al., 2015). These studies indicated the importance of mathematics and indicated the need to improve the understanding of mathematics learning, and in particular diversity in mathematics learning.

1.2 Mathematics learning in young children

Mathematics learning might not be unitary, but is rather made up of many different subcomponents, such as number discrimination, counting procedures, counting principles, and calculation with possible discrepancies among subcomponents (Dowker et al., 2019).

According to Clements and Sarama (2014) there is no age too young for mathematical thought. ***Number discrimination*** can be seen as an early marker of diversity in infants (Xu & Arriaga, 2007). Previous research has shown that number discrimination in toddlerhood even has some predictive value for mathematical learning in kindergarten (Ceulemans et al., 2015, 2017).

Previous studies also revealed that children’s counting proficiency played a role in the development of mathematics learning. A secondary analysis on 7,665 children indicated that counting was one of best predictors of school success (Claessens & Engel, 2013). The knowledge of counting procedures (or procedural knowledge) and the knowledge of counting principles (conceptual knowledge) can be seen as two distinctive aspects of counting. **Procedural counting knowledge** is needed to determine that there are five objects in an array. **Conceptual counting knowledge** reflects a child’s understanding of the essential counting principles: the stable order principle, the one-one-correspondence principle and the cardinality principle (Desoete & Roeyers, 2009; Stock et al., 2009).

Finally, mathematics learning also involves basic knowledge and skills to ***calculate*** accurately in order to solve mathematical tasks. In later years, not only calculation accuracy but also calculation fluency will be needed (LeFevre et al., 2009).

1.3 Opportunity (O)-Propensity (P) model to explain diversity

The Opportunity-Propensity (O-P) framework (Byrnes, 2020; Byrnes & Miller, 2016; Wang et al., 2013) aims to explain some of the diversity of mathematical learning, visualized in Fig. 1.

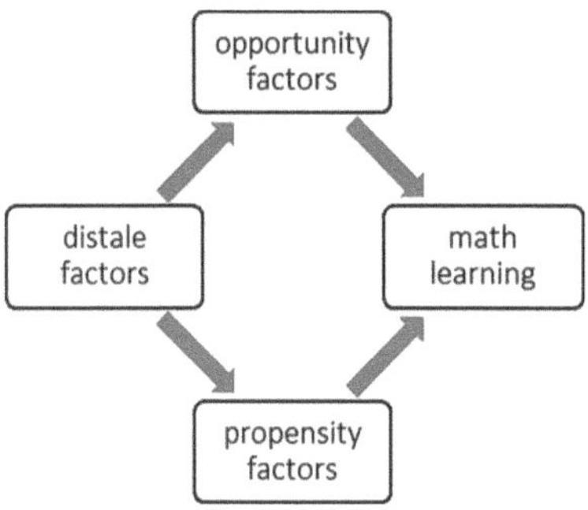

Fig. 1: The Opportunity-Propensity model.

Propensity factors (P) in the O-P model refer to the variables that make people able (e.g., intelligence) and/or willing (e.g., motivation) to learn mathematics. **Opportunity factors** (O) have been defined as contexts and variables that expose children to learning content (e.g., home and school environment, including parental talk). **Distal variables** (e.g., SES) were included in the model to explain why some people are exposed to richer opportunity contexts and have stronger propensities for learning than others.

The O-P model has been validated in large secondary data sets, including lower-income pre-kindergarten children, children followed up from kindergarten until primary school and secondary school pupils (Byrnes, 2020; Byrnes & Miller, 2016; Wang et al., 2013). These studies have shown that mathematics learning (as outcome variable) improved with more propensities (P-factor). In addition, some of the diversity in mathematical learning could be explained by the opportunities in the school and home environment (O-factors).

Previous studies informed us about the information on the influence of the school environment (Baten & Desoete, 2018; Byrnes & Miller, 2016) in the prediction of mathematics learning. The impact of these school related O-factors depended on the specific support factors (Byrnes & Wasik, 2007, 2009). However, other studies have shown that the home numeracy environment **also** mattered (as O-factor) for mathematics learning (Missall et al., 2014; Segers et al., 2015). Parent–child interactions that included experiences with numerical content in daily life have been positively associated with children's mathematics learning (e.g., Blevins-Knabe & Austin, 2016). Anders et al. (2012) demonstrated that the quality of the home environment at the beginning of kindergarten (mean age of 3 years) was strongly associated with mathematics learning in preschool, with this advantage maintained at the end of preschool (mean age 5 years). Their results underlined the differential impact of school and home environments on mathematics learning. Niklas et al. (2016) and Casey et al. (2018) demonstrated that offering more parental support to children resulted in better mathematics learning. Kindergartners who received rich early numerical home opportunities developed better mathematics skills compared to those with fewer learning opportunities (e.g. Clements & Sarama, 2014; Kleemans et al., 2012; LeFevre et al., 2009). However, Yildiz et al. (2018) found that home numeracy factors operationalized in a questionnaire (parent's reports) were positively related to children's calculation abilities, on the other hand, contrary to expectation, the observed parental talk was negatively related to children's calculation abilities. They concluded that questionnaires and observations might tap different aspects of home numeracy.

1.4 Parental talk as Opportunity (O) factor

Some component of the association between home environment and mathematics learning might be at least partially explained by parental involvement (Hong et al., 2010; Wilder, 2014) or by parental responsiveness (Dieterich et al., 2006). Parental **involvement** can be described as the overall quality of the interaction between parent and child (e.g., Melhuish & Phan, 2008; Sy et al., 2013). In addition, some studies revealed that also parental **responsiveness** was positively associated with later language and literacy development of children (e.g., Dieterich et al., 2006). On top of the constructs involvement and responsiveness, math-related parental talk is a component of home numeracy (e.g., Karrass & Braungart-Rieker, 2005).

Parental talk can be defined as the formal or direct numeracy talk (such as counting objects) and informal or indirect (such as mealtime) numeracy talk of parent–child dyads. Susperreguy and Davis-Kean (2016) revealed that all mothers involved their preschool child in a variety of math exchanges during mealtime, although there were differences in the amount of input that children received. Several studies have shown that parental talk often involved counting and labeling cardinal values of sets (Ramani et al., 2015; Zhou et al., 2006). Talking about large sets of objects was the strongest predictor of mathematics learning (Gunderson & Levine, 2011).

1.5 Diversity in parental talk and in subcomponents of mathematics learning

Dowker (2019) has shown that mathematics learning might not be unitary and may differ with age. A study in kindergarten, grades 1 and 2, revealed that basic numerical skills were positively associated with informal parental talk, whereas calculation fluency was related to both formal and informal parental talk (LeFevre et al., 2009). Yildiz et al. (2018) confirmed the positive relationship between parental talk and calculation in the last year of kindergartners (mean age 5.64 years), but only with the parental talk assessed via a questionnaire. Levine et al. (2011) demonstrated that in 14- to 30-month-olds the frequency of parental talk about numbers predicted the children's cardinal knowledge (e.g., knowing that the word "four" refers to sets with four items) at 46 months of age. The study of Skwarchuk et al. (2014) indicated that formal talk (such as practicing sums) predicted symbolic number system knowledge, whereas informal talk was related to children's abilities to non-symbolically represent and manipulate quantities in children starting in kindergarten (mean age 58 months). Benavides-Varela et al. (2016) provided evidence for the unexpected finding that the construct "home environment" was related to the exact, but not to the approximate, number representation in children with a mean age of 5 years 11 months. Casey et al. (2018) found that (observed) parental support (labeling sets of objects) at 36 months predicted mathematics learning at 4½ and 6–7 years.

To conclude, although there is evidence that parental talk is predictive for mathematics learning, the importance might depend on the age of children, on the subcomponent of mathematics learning that is studied, and even on the technique that is used to operationalize parental talk and mathematics learning.

1.6 Current study

In this study we use questionnaires and observations to assess parental talk and to study the association with mathematics learning in toddlers and in kindergarten. This results in the following specific Research Questions (RQ).

- RQ1. Is there diversity in the parental talk with young children? Are questionnaires and observations of parental talk positively associated?
- RQ2. Is there diversity in mathematics learning assessed at 24 months (with a number discrimination task) and at 48 months (with procedural and conceptual counting tasks and a calculation task)?
- RQ3. Does parental talk predict mathematics learning, controlling for parental involvement and sensitivity?

2 Method

2.1 Participants

Participants were part of a birth cohort, living in different Flemish districts in Belgium. They were recruited within the scope of a longitudinal study for the Belgian government (www.steunpuntwvg.be). A small sample of children were randomly selected to participate in a more in depth study on mathematics learning (see also Ceulemans et al., 2015, 2017). As such, parents of 15 boys and 16 girls consented to participate with their child at the age of 24 (T1) and 48 months (T2). The mean intelligence of the children measured with Wechsler Preschool and Primary Scale of Intelligence – Third edition (WPPSI-III-NL; Wechsler, 2002; Dutch translation) was 101.33 ($SD = 12.53$). Half of the families of the children had a middle income and the other half had a high income when the research project started. The category "middle income" comprised a considerable part of the study population being a "modal" family with two working parents as manual worker or employee. Families in the "middle income" category earned between 1,501 and 3,000 euros, whereas those in the "high income" category earned more than 3,000 euros per month.

2.2 Procedure and analyses

Linear regression analyses were conducted to explore the research questions. Graphical inspection of the data revealed that error terms were normally distributed. Since not only the quantity of opportunities might be important, linear and quadratic relationships were explored. Only in case of a significant quadratic relationship was this mentioned additional to the results of the linear relationship between certain variables. Moreover, significant relationships between the opportunities were tested by taking into account the control variable (parental involvement and sensitivity).

2.3 Instruments

2.3.1 Parental talk (O-factor)

Parental talk was tested through observations and via a questionnaire. The structured play situation was used to **observe** "parental talk" (as O-factor) at 24 and 48 months. Mother and child sat on a carpet and were instructed to build a house with a set of *Duplo* blocks according to a model. After the instruction was given, parent and child were left alone in the room. The structured play was recorded for five minutes on video and all parental talk, and/or language of the children were coded manually afterward. All actions were given a score according to their frequency during the observation. The sum of all scores (which could be reduced to 18 items) resulted in the total parental talk score. Internal consistency of the data output was .63 ($M = 55.51$, $SD = 29.71$). At 24 months two experimenters achieved an averaged inter-rater reliability of .88 percentage of scores in agreement. At 48 months two experimenters achieved an averaged inter-rater reliability of .84 percentage of agreement.

All parents also completed **questionnaires**. The questionnaire on parental talk included activities related to mathematics learning at home, with 13 items related to direct activities (such as learn the right sequence of number words) and 10 items related to indirect activities (such as sing a song with numbers). Parents were asked to score all items according to their occurrence during the past month. They could choose between the options never (1), sometimes (2), or many times (3). The option "not applicable" (0) could be indicated when parents thought that their child could not perform this behavior because he/she was not yet able to do it. The original options to indicate frequency were transformed into scores ranging from zero to two per item. Cronbach's alpha was .87 (with .85 for the direct activities and .72 for the indirect activities). For the mean (M) and standard deviation (SD) of all items, see Tab. 1.

Tab. 1: Diversity in (self-reported) parental talk of todlers.

	Response options (%)					
	0	**1**	**2**	**3**	***M***	***(SD)***
Direct activities						
Use words about quantity and size	5.56	–	33.33	61,11	2.50	*(0.79)*
Use words to compare	11.11	5.56	27.78	55.56	2.28	*(1.02)*
Use number words: one, two, three	–	–	33.33	66.67	2.67	*(0.49)*
Use number words: four, . . ., ten	11.11	27.78	27.78	33.33	1.83	*(1.04)*
Say the sequence of numbers from one to ten	5.56	27.78	33.33	33.33	1.94	*(0.94)*
Learn the right sequence of number words	11.11	38.89	22.22	27.78	1.67	*(1.03)*
Learn counting or say number words using fingers	11.11	55.56	27.78	5.56	1.28	*(0.75)*

Tab. 1 (continued)

	Response options (%)					
	0	1	2	3	*M*	*(SD)*
Encourage counting	5.56	44.44	33.33	16.67	1.61	*(0.85)*
Asking “how many”	5.56	55.56	27.78	11.11	1.44	*(0.78)*
Practice counting objects	5.56	33.33	50.00	11.11	1.67	*(0.77)*
Encourage use of matching	5.56	5.56	66.67	22.22	2.06	*(0.73)*
Recognizing and naming numbers	22.22	77.78	–	–	0.78	*(0.43)*
Counting down	11.11	11.11	44.44	33.33	2.00	*(0.97)*
Indirect activities						
Name shapes	11.11	44.44	44.44	–	1.33	*(0.69)*
Sort objects on color	11.11	27.78	55.56	5.56	1.56	*(0.78)*
Sort objects on shape	11.11	50.00	33.33	5.56	1.33	*(0.77)*
Sort objects on size	11.11	72.22	16.67	–	1.06	*(0.54)*
Sing a song on numbers	11.11	33.33	27.78	27.78	1.72	*(1.02)*
Give compliments to child on using numbers	11.11	27.78	27.78	33.33	1.83	*(1.04)*
Play with magnetic numbers or number stamps	11.11	72.22	11.11	5.56	1.11	*(0.68)*
Read books with focus on numbers	16.67	33.33	44.44	5.56	1.39	*(0.85)*
Play with a dice	22.22	55.56	22.22	–	1.00	*(0.69)*
Measure ingredients	22.22	38.89	27.78	11.11	1.28	*(0.96)*

2.3.2 Parental involvement and parental sensitivity

All parents completed a questionnaire with 10 items on ***parental involvement*** from the scale “Parental Involvement in Developmental Advance (PIDA)” of the StimQ-Toddler interview (Dreyer et al., 1996) were included. These items described possible actions or activities with the child initiated by the parent in the home environment. Cronbach’s α (M = 10, *SD* = 1.41) in the present study was .71 at 24 months and .63 at 48 months. A pilot study showed that the questionnaire was easy to complete.

During observation of the structured play **parental sensitivity** was measured as well. In line with other research (e.g., Feldman & Masalha, 2010), the

Coding Interactive Behavior (CIB) system (Feldman, 1998) was used to assess the parental sensitivity during the structured play. This is a global rating system of parent–child interaction that included 42 codes rated on a scale of 1 (*low*) to 5 (*high*) that leads to eight theoretically derived parent, child, and dyadic composites on diverse aspects of parent–child interaction. For each code, the observer assigned a single score after viewing the entire interaction, and several viewings were required to complete the coding. At 24 months the coder achieved an averaged percentage of agreement of .84 with an officially trained coder by the laboratory of Feldman. At 48 months the inter-rater reliability between two observers averaged .91 with the same trained coder. The composite set of parental sensitivity indicators was used in this study. This composite set included the codes "parent acknowledgment of child signals," "maintenance of visual contact," "expression of positive affect," "appropriate vocal quality," "resourcefulness in handling child's distress or expanding the interaction," "consistency of style," and "display of an affective range that matches the infant's readiness to interact." Reliability as measured with Cronbach's α ($M = 3.96$, $SD = 0.55$) for this composite set of indicators in the present study was .87 at 24 months and .80 at 48 months.

2.3.3 Mathematics learning

Mathematics learning was assessed as an outcome variable (at 24 and 48 months) in the O-P model.

Mathematics learning at 24 months was tested with a **number discrimination task** using a manual search paradigm as described by Feigenson and Carey (2005). A wooden box (25 cm × 12.5 cm × 31.5 cm) had a slit at the front oriented toward the toddlers and an opening at the backside oriented toward the experimenter who was facing the child at an – except for the box – empty table. Parents were told that some balls would be hidden to explore how children responded to a task and that no wrong reaction existed. The task entailed three kinds of trials: a first *box empty* trial, a *more remaining* trial, and a second variant of the box empty trial, which always followed after a *more remaining* trial (see Fig. 2).

Each of the trial types was presented twice and the order of the trials was counterbalanced. Children could search through the slit for 10 seconds after each type of trial commenced. It was expected that children would search longer after the *more remaining* than after the *box empty* trials which would indicate successful discrimination. Cumulative searching time was coded. Subtracting searching time after *box empty* trials from searching time after *more remaining* trials resulted in difference scores. Reliability of the difference scores, as measured with Cronbach's α, was .79 for this task ($M = 2.42$, $SD = 1.48$).

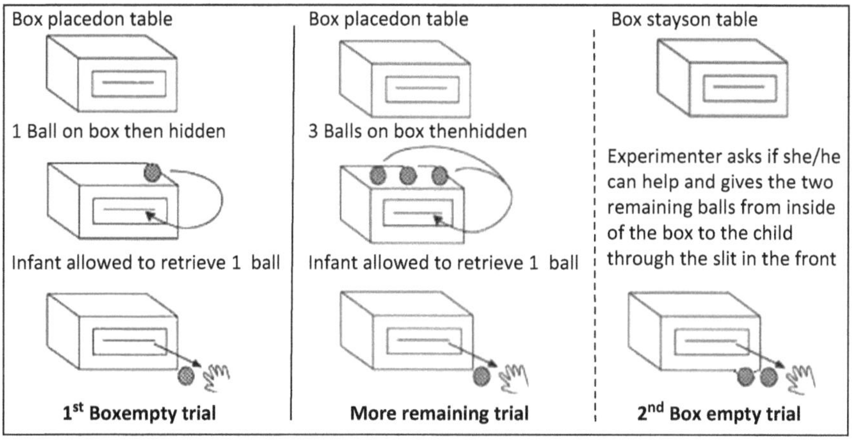

Fig. 2: Different trial types of the manual search task. Adopted from "On the limits of infants' quantification of small object arrays," by Feigenson and Carey (2005), *Cognition, 97*, p. 301. Copyright 2004 by Elsevier B.V.

Mathematics learning at 48 months was tested with three subtests of the TEDI-MATH (Grégoire et al., 2004). **Procedural counting** was tested with all eight items of TEDI-MATH where children had to count starting from one (up till 31), counting up to an upper bound (e.g., "count to 9") and/or from a lower bound (e.g., "count from 3"). Cronbach's α of the current study was .62 ($M = 1.45$, $SD = 1.61$). **Conceptual counting** was tested with all 13 items of TEDI-MATH where children had to judge the counting of linear and non-linear patterns of objects, and were asked questions about the counted amount of objects (e.g., "How many objects are there in total?"). Furthermore, they had to construct two numerical equivalent amounts of objects and use counting as a problem-solving strategy in a riddle. Cronbach's α of the current study was .76 ($M = 4.39$, $SD = 2.70$). **Calculation** was tested with all six items of TEDI-MATH where children had to solve visually supported additions and subtractions. Reliability for the current study was Cronbach's $\alpha = .73$ ($M = 1.97$, $SD = 1.80$).

3 Results

3.1 Diversity in "parental talk"

The amount of "parental talk" (assessed with a questionnaire) was positively associated with the amount of observed "parental talk" (during the manual search task) with r = .54, ($p < .001$). Table 1 gives an overview of the frequencies of the parental talk assessed via a **questionnaire** at 24 months. “Using words to tell something about the quantity or the size of objects,” “using words to express a comparison between objects,” and “using the small number words from one to three” were the most frequent activities by parents of toddlers. Direct “opportunities” to focus on mathematics learning were reported as occurring (i.e., either “sometimes” or “many times”) in about 60% of the cases. Indirect “opportunities” to focus on mathematics learning were reported as not occurring in about 60% of the cases.

During the **observation** of parental talk (the structured play) interactions using Duplo or Lego blocks occurred on average with a frequency of about 31.21 times (*SD* = 14.34, range = 0.00–58.00) during the observation which lasted 5 min, giving a rate of about 6 interactions per minute. Only once was no parental talk between mother and child observed during the structured play situation, again pointing to diversity of opportunities offered by children to enhance mathematics learning.

3.2 Diversity in mathematics learning

There was no significant association (see Tab. 2) between mathematics learning in toddlers (assessed with a number discrimination task) and in kindergarten (assessed with a counting and calculation task). Counting and calculation skills were positively and significantly associated (at 48 months).

Tab. 2: Correlations between mathematics learning measures.

Mathematics learning	1 24 month (m)	2 48 m	3 48 m	4 48 m	*M*	*(SD)*
1. Number discrimination	–				1.45	*(1.43)*
2. Procedural counting	–.03	–			1.45	*(1.61)*
3. Conceptual counting	–.03	.38*	–		4.39	*(2.70)*
4. Calculation	–.07	.41*	.57**	–	1.97	*(1.80)*

* $p \leq .05$.
** $p \leq .001$

3.3 Relation between parental talk (as “opportunity”) and mathematics learning

Table 3 provides the explorative correlations between the parental talk and all mathematics learning measures included in the current study.

Tab. 3: Correlations between opportunities and mathematics learning.

		Opportunities Parental talk		Parental involvement		Parental sensitivity	
		24 months	48 months	24 months	48 months	24 months	48 months
(Observed) parental talk	48 months	.44**	–				
Parental involvement	24 months	.22	.32	–	–	–	
	48 months	.03	.37**	.17			
Parental sensitivity	24 months	.37**	.00	–.08	–.17	–	–
	48 months	.08	.20	.46**	.24	.17	
Number discrimination	24 months	.02 .1489 .	–.14	–.16	–.05	–.10	–.11
Procedural counting	48 months	.15 .343 .434	.29	.39*	.29	.08 .231	.15
Conceptual counting	48 months	.34* .343 .434	–.00 .280	–.05	.12	.23 .231	.07
Calculation	48 months	.28 .343 .434	.40**	.29	.33*	.24 .231	.20

Note. * $p < .10$, ** $p < .05$, *** $p < .01$

At 48 months, there was a significant and positive linear relationship, $F(1,29) = 5.56$, $p = .025$, $R^2 = .161$ between parental talk and calculation which remained marginally significant in addition to the parental involvement, F_{change} $(1,28) = 3.11$, $p = .089$, $R^2_{\text{change}} = .09$.

The linear regression analysis with parental talk (observed at 24 months) as the independent variable revealed a relationship and trend of prediction for **conceptual counting** at 48 months, $F(1,27) = 3.60$, $p = .068$. In addition, although not linear $F(1,27) = 2.29$, $p = .142$, a significant quadratic (positive) relationship could be found, $F(2,26) = 3.68$, $p = .039$, $R^2 = .221$ between the parental talk at 24 months and the **calculation** skills at 48 months.

4 Discussion

Parental talk occurred on average about six times per minute during the Duplo or Lego building activity. In line with Susperreguy and Davis-Kean (2016), large differences (varying from 23 to 0 times parental talk) in the amount of math input that children received, were observed. These results indicate a substantial diversity in the amount of parental talk children experience as young children.

In contrast with the finding of Yildiz et al. (2018), in this study there was a significant relationship between parental talk operationalized via questionnaires and via observational methods in toddlers (24 months).

In line with Ramani et al. (2015) and Casey et al. (2018) parental talk was mainly involved in counting and labeling quantities. We observed that parents often focused on small number words with toddlers, whereas Gunderson and Levine (2011) revealed that the indicator "talking about large sets of objects" was the strongest predictor of mathematics learning.

Parental talk was associated with mathematics learning even controlling for parental involvement. In addition, in line with LeFevre et al. (2009), but in contrast with Yildiz et al. (2018), a significant linear relationship was found between more (observed) parental talk and better calculation in kindergarten. As such, the parental numerical language might be perceived as on opportunity factor that stimulates the child's mathematics learning in a positive way. However, it is also possible that parents who talk more about numbers or pick up more opportunities to engage with their children, do so because their children are (initially more) interested in mathematics. Children might, accordingly, provoke numerical parental talk themselves. No causal relationship could be drawn. Nonetheless, the value of parental talk could be demonstrated even when taking into account parental involvement as a plausible explaining factor.

In addition while the concurrent relationship between the constructs at kindergarten age (48 months) was linear, the relationship at toddler age (24 months) was quadratic in nature. In kindergarten this implied that more parental talk was associated with higher mathematics learning. In toddlerhood, however, it seemed that more parental talk only predicted higher mathematics learning in kindergarten to some extent, needing the appropriate engagement at the right time. At higher rates later mathematics learning declined again. This finding suggests that a child's mathematics learning might not only depend on the parental talk, and empowerment of opportunities should be within children's zone of proximal development. Future research needs to clarify this finding and the clinical relevance more in detail.

There are some limitations to this study. The first limitation is the sample size. A small sample size may lead to higher variability, leading to bias. In addition, only families with a middle or high family income were included. It would therefore be interesting for future research to also take into account low-income families to accurately investigate the influence of SES on both numerical interaction and performance.

Despite the mentioned limitations, the current study might imply that an additional focus on parental talk by agencies in support of parenting could be worthwhile. Making parents aware of the importance of numerical parental talk might empower them and stimulate mathematics learning in young children. However, it may be that education in respect of what is appropriate mathematical talk may be needed.

References

Anders, Yvonne, Rossbach, Hans-Günther, Weinert, Sabine, Ebert, Susanne, Kuger, Susanne, Lehrl, Simone & von Maurice, Jutta (2012): Home and preschool learning environments and their relations to the development of early numeracy skills. *Early Chilhood Research Quarterly* 27, 231–244.

Baten, Elk & Desoete, Annemie (2018): Mathematical (Dis)abilities within the Opportunity- Propensity Model: The Choice of Math Test Matters. *Frontiers in Psychology* 9, *Art. 667 Developmental Psychology*. doi:http://dx.doi.org:10.3389/fpsyg.2018.00667.

Benavides-Varela, Silvia, Butterworth, Brian, Burgio, Francesca, Arcara, Giorgio, Luc Angeli, Daniella & Semenza, Carlo (2016): Numerical activities and information learned at home link to the exact numercay skills in 5–6 years-old Children. Frontiers in Psychology 8, 94. doi:http://dx.doi.org/10.3389/fpsyg.2016.00094.

Blevins-Knabe, Belinda & Austin, Ann M Berghout (2016): *Early Childhood Mathematics Skill Development in the Home Environment*. Cham, Switzerland: Springer International Publishing. doi:https://doi.org/10.1007/978-3-319-43974-7.

Byrnes, James & Wasik, Barabara A. (2009): Factors predictive of mathematics achievement in kindergarten, first and third grades: An opportunity – propensity analysis. *Contemporary Educational Psychology* 34, 167–183.
Byrnes, James P. & Miller, Dana (2007): The relative importance of predictors of math and science achievement: An opportunity – propensity analysis. *Contemporary Educational Psychology* 32, 599–629.
Byrnes, James P. & Miller, Dana (2016): The growth of mathematics and reading skills in segregated and diverse schools: An opportunity-propensity analysis of a national database. *Contemporary Educational Psychology* 46, 34–51.
Byrnes, J. P. (2020): The potential utility of an opportunity-propensity framework for understanding individual and group differences in developmental outcomes: A retrospective progress report. *Developmental Review* 56. doi:http://dx.doi.org/10.1016/j.dr.2020.100911.
Casey Beth, M., Lombardi Caitlin, M, Dana, Thomson, Nguyen, Hoa Nha, Paz, M, Theriault Cote, A & Eric Dearing, E (2018): Maternal support of children's early numerical concept learning predicts preschool and first-grade math achievement. *Child Development* 89 (1), 156-173.
Ceulemans, Annelies, Baten, Elke, Loeys, Tom, Hoppenbrouwers, Karel, Titeca, Daisy, Rousseau, Sofie & Desoete, Annemie (2017): The relative importance of parental numerical opportunities, prerequisite knowledge and parent involvement as predictors for early math achievement in young children. *Interdisciplinary Education and Psychology, 17–123/* 1 (1), 6, 12 pp. http://riverapublications.com/assets/files/pdf_files/the-relative-importance-of-parental-numerical-opportunities-prerequisite-knowledge-and-parent-involv.pdf.
Ceulemans, Annelies, Titeca, Daisy, Loeys, Tom, Hoppenbrouwers, Karel, Rousseau, Sofie & Desoete, Annemie (2015): The sense of small number discrimination: The predictive value in infancy and toddlerhood for numerical competencies in kindergarten. *Learning and individual differences* 39, 150–157.
Claessens, Amy & Engel, Mimi (2013): How important is where you start? Early mathematics knowledge and later school success. *Teachers College Record* 115, 1–29.
Clements, Douglas & Sarama, Julie (2014): The importance of early years. In Slavin, R.E. (ed.): *Scinece, Technology & Mathematics (STEM)*. Thousand Oaks, CA: Corwin, 5–9.
Deary, Ian J., Whalley, Lawrence J., Lemmon, Helen, Crawford, J. R. & Starr, John M. (2000): The stability of individual differences in mental ability from childhood to old age: Follow-up of the 1932 Scottish mental survey. *Intelligence* 28 (1), 49–55.
Desoete, Annemie, Stock, Pieter, Schepens, Annemie, Baeyens, Dieter & Roeyers, Herbert (2009): Classification, Seriation, and Counting in Grades 1, 2, and 3 as Two-Year Longitudinal Predictors for Low Achieving in Numerical Facility and Arithmetical Achievement? *Journal of Psychoeducational Assessment* 27, 252–264.
Dieterich, Susan E., Assel, Michael, Swank, Paul R., Smith, Karen E. & Landry, Susan H (2006): The impact of early maternal verbal scaffolding and child language abilities on later decoding and reading comprehension skills. *Journal of School Psychology* 43, 481–494.
Dowker, Ann (2019): *Individual Differences in Arithmetic: Implications for Psychology, Neuroscience and Education, 2nd edition*. Hove: Psychology Press.
Dowker, Ann, De Smedt, Bert & Desoete, Annemie (2019): Editorial: Individual differences in arithmetical development. *Frontiers in Psychology* 10, 2–13. doi:https://doi.org/10.3389/fpsyg.2019.02672.
Dreyer, Bernard P., Medelsohn, Alan.L & Tamis-LeMonda, Catherine S (1996): Assessing the child's cognitive home environment through parental report: Reliability and validity. *Early Development & Parenting* 5, 271–287.

Duncan, Greg J & Katherine Magnuson, K. (2011): The nature and impact of early achievement skills, attention skills, and behavior problems. In Duncan, Greg J., Murnane, R. J. (eds.): *Whither Opportunity: Rising Inequality, Schools, and Children’s Life Chances*. Russel Sage, The Russell Sage Foundation New York, 47–69.

Feigenson, Lisa & Carey, Susan (2005): On the limits of infants’ quantification of small object. *Cognition* 97, 295–313.

Feldman, Robert S. (1998). Coding interactive behavior manual. (unpublished manual). Bar-Ilan University, Israel.

Feldman, Robert S & Shafiq Masalha, S. (2010): Parent–child and triadic antecedents of children’s social competence: Cultural specificity, shared process. *Developmental Psychology* 46, 455–467. doi:10.1037/a0017415.

Geary, David (2011): Consequences, characteristics, and causes of mathematical learning disabilities and persistent low achievement in mathematics. *Journal of Developmental & Behavioral Pediatrics* 32 (3), 250–263.

Grégoire, Jacques, Noël, Marie Pascale, Catherine & liVan Nieuwenhoven, C. (2004): *TEDI-MATH*. Antwerpen: Harcourt.

Gunderson, Elizabeth A & Levine, Susan C. (2011): Some types of parent number talk count more than others: Relations between parents’ input and children’s cardinal-number knowledge. *Developmental Science* 14, 1021–1032.

Hong, Sehee, Yoo, Sung-Kyung, You, Sukkyung & Chih-Chun, Wu (2010): The reciprocal relationship between parental involvement and mathematics achievement: Autoregressive cross-lagged modeling. *The Journal of Experimental Education* 78, 419–439.

Karrass, Jan & Braungart-Rieker, Julia M. (2005): Effects of shared parent infant book reading on early language acquisition. *Journal of Applied Developmental Psychology* 26, 133–148.

Kleemans, Tijs, Peeters, Marieke, Segers, Eliane & Verhoeven, Ludo (2012): Child and home predictors of early numaneracy skills in kindergarten. *Early Childhood Research Quarterly* 27, 471–477.

LeFevre, Jo-Anne, Skwarchuk, Sheri-Lynn, Smith-Chant, Brenda L., Fast, Lisa A., Kamawar, Deepthi & Bisanz, Jeffrey (2009): Home numeracy experiences and children's math performance in the early school years. *Canadian Journal of Behavioural Science* 41, 55–66.

Levine, Susan C, Suriyakham, Linda Whealton, Rowe, Meredith L, Huttenlocher, Janellen & Gunderson, Elizabeth A (2011): What counts in the development of young children's number knowledge? *Developmental Psychology* 47, 302–302.

Melhuish, Edward & Phan, Mai B. (2008): Effects of the home learning environment and preschool center experience upon literacy and numeracy development in early primary school. *Journal of Social Issues* 64, 95–114.

Missall, Kirsten, Hojnoski, Robin, Caskie, Grace I.L. & Repasky, Patrick (2015): Home numeracy environments of preschoolers: Examining relations among mathematical activities, parent mathematical beliefs, and early mathematical skills. *Early Education and Development* 26, 356–376.

Nelson, Gena & Powell, Sarah R (2018): A systematic review of longitudinal studies of mathematics difficulty. *Journal of Learning Disabilities* 51, 523–539. doi:https://doi.org/10.1177/0022219417714773.

Niklas, Frank, Cohrssen, Caroline & Tayler, Colette (2016): Improving preschooler's numerical abilities by enhancing the home numeracy environment. *Early Educaton and Development* 27, 372–383.

Ramani, Geetha B, Rowe, Meredith L, Eason, Sarah H & Leech, Kathryn A (2015): Math talk during informal learning activities in Head Start families. *Cognitive Development* 35, 15–33.
Ritchie, Stuart J & Bates, Timothy C (2013): Enduring links from childhood mathematics and reading achievement to adult socioeconomic status. *Psychological Science* 24, 5–7. doi: https://doi.org/10.1177/0956797612466268.
Segers, Eliane, Kleemans, Tijs & Verhoeven, Ludo (2015): Role of parent literacy and numeracy expectations and activities in predicting early numeracy skills. *Mathematical Thinking and Learning* 17, 219–236.
Skwarchuk, Sheri-Lynn, Sowinksi, Carla & LeFevre, Jo-Anne (2014): Formal and informal home learning activities in relation to children's early numeracy and literacy skills: The development of a home numercay model. *Journal of Experimental Child Psychology* 121, 63–84.
Stock, Pieter, Desoete, Annemie & Roeyers, Herbert (2009): Mastery of the counting principles in toddlers: A crucial step in the development of budding arithmetic abilities? *Learning and Individual Differences* 19, 419–422.
Susperreguy, Maria Ines & Davis-Kean, Pamela (2016): Maternal math talk in the home and math skills in preschool children. *Early Education and Development* 9289, 1–17.
Sy, Susan R, Gottfried, Allen W. & Gottfried, Adele Eskeles (2013): A transactional model of parental involvement and children's achievement from early childhood through adolescence. *Parenting Science and Practice* 13, 133–152.
Wang, Aubrey H., Shen, Feng & Byrnes, James P (2013): Does the opportunity–propensity framework predict the early mathematics skills of low-income pre-kindergarten children? *Contemporary Educational Psychology* 38, 259–270.
Wechsler, David (2002): *Wechsler Preschool and Primary Scale of Intelligence – Third Edition. Dutch Translation by Hendriksen,J., & Hurks, P.* Amsterdam, Nederland: Pearson.
Wilder, Sandra (2014): Effects of parental involvement on academic achievement: A meta-synthesis. Educational Review 66, 377–397.
Wilson, Anna J, Andrewes, Stuaert G., Struthers, Helena, Rowe, Victoria M., Bogdanovic, Rajna & Waldie, Karen E. (2015): Dyscalculia and dyslexia in adults: Cognitive bases of comorbidity. *Learning and Individual Differences* 37, 118–132.
Xu, Fei & Arriaga, Rosa I. (2007): Number discrimination in 10-month-old infants. *British Journal of Developmental Psychology* 25, 103–108.
Yildiz, Belde Mutaf, Delphine, Sasanguie, Bert, De Smedt & Bert Reynvoet, B. (2018): Investigating the relationship between two home numeracy measures: A questionnaire and observations during lego building and book reading. *British Journal of Developmental Psychology* 36, 354–370.
Zhou, Xin, Huang, Jim, Wang, Zhengke, Wang, Bin, Zhao, Zhenguo, Yang, Lei & Yang, Zhengzheng (2006): Parent–child interaction and children's number learning. *Early Child Development and Care* 176, 763–775.

Pixner Silvia and Dresen Verena

Number words, quantifiers, and arithmetic development with particular respect of zero

Mathematics and language. Apparently two independent fields. Why apparently? Looking at adults with mathematical knowledge, you can observe great differences in both directions in the sense of a double dissociation with regard to these two competences, that is, language and arithmetic competences. This means that there are people who have very good language competences but weak mathematical competences and vice versa. This dissociation would indicate a certain independence from language and arithmetic competences. In this chapter we would therefore like to shed light on both associations and certain independencies between language and arithmetic competences. Above all, we would like to show that both views are correct, since it always depends on the respective moment of observation in development and especially on the skills of interest. Arithmetic competences include the understanding of numbers and their relations as well as the mastery of basic arithmetic operations. Therefore, we will first look at some general associations between language and numerical competences, based on the Triple Code Model by Dehaene (1992). Then we will go into more detail about the development and the interdependence between language and numerical competences as they are in a permanent interaction. Finally, we want to give some specific insights into the associations between quantifiers and the development of cardinality before explaining the role of zero in this context. We believe that a specific knowledge of these associations is essential not only for research, but also for practical work with children with dyscalculia. This applies not only to the work with children with a migration background or a specific language development impairment and how these difficulties can be countered. It is also important to keep in mind that language can also be a resource for children with dyscalculia to better grasp numerical content.

Looking at the current leading number processing model, the Triple Code Model by Dehaene (1992), it quickly becomes clear that arithmetic competences do not represent a single homogeneous competence, but actually consist of several sub-competences that are predominantly independent of each other. This means that some sub-competences in numeracy may be more dependent on language, while others may be less dependent on language. The three representations of Dehaene's model are the verbal code, the visual Arabic code, and the

https://doi.org/10.1515/9783110661941-008

semantic code. The verbal code includes not only the spoken and written number words, but also the verbal counting and especially the verbal memorized arithmetic facts (e.g., $4 + 3 = 7$ or $3 \cdot 4 = 12$). Therefore the verbal code represents most language-dependent numerical skills. The link between language competences and the verbal code appears to be strongest at this point. Opposite to this the semantic code includes semantic knowledge about size and quantities for number comparison or estimation as well as the numerical quantities represented on the mental number line. The third representation is called the visual Arabic code and represents the numbers in Arabic format. This is mainly used for multi-digit calculation and also for parity judgments. Apart from that it must be considered that, depending on the mathematical task, the different codes are activated to different degrees or, in addition, can also be different in terms of their quality within a person. This means that someone can be very good at verbal fact retrieval, but has difficulty with procedural understanding, or vice versa. Nevertheless, language does not represent a homogeneous competence either, because it also consists of many sub-competences. It is therefore important to look in detail at the associations between language competences and arithmetic competences, as is impressively described in this book. This chapter is therefore devoted to specific or concrete number words (e.g., one, two, three), quantifiers or unspecific number words (e.g., many, some, a few), as well as the development of arithmetic competences and the influence or contribution of language.

What is the relationship between the development of arithmetic competences and language competences? Studies show a certain independence with regard to individual sub-competences, as also described by Dehaene (1992) in the Triple Code Model. According to several studies (cf: Wynn, 1992), the understanding or a certain sensitization in dealing with quantities (in sum a kind of innate number sense) is innate. Even infants without any understanding of language can discriminate between two quantities under certain conditions long before the development of language begins (Xu et al., 2005). Thus, at the beginning, the processing of numbers and quantities seems to be language-independent, which is also impressively demonstrated by behavioral studies with animals that are able to handle quantity differentiation very well (Ward & Smuts, 2007). However, if we look further along the time axis, it becomes clear quite quickly that one of the most effective ways to acquire mathematical or arithmetical knowledge is through language. Language competences not only represent one of the most important predictors of school success in general (Wagner et al., 2013), they also serve specifically as a predictor of mathematical skills (Preat et al., 2013). Additionally, it is possible to observe clear difficulties in this phase of the development of arithmetic competences in disadvantaged groups, such as children with a migration background or children with a language development impairment (Donlan et al., 2007). Therefore, this is a clear indication that language must fulfill a specific function in early numerical development. These contradictory results are not contradictory when viewed in a longitudinal section. Besides the innate number sense, knowledge of numbers and quantities is naturally acquired through learning as well as through social interaction. Children start to talk and use these new language skills to refine their numerical competences at the same time. Thus, at the beginning, children can only distinguish very large differences between two quantities with the help of their innate intuition for quantities. In the second year of life, they learn, parallel to this but independently of it, the number words. At this time, the word sequences are still isolated from quantities (Krajewski & Schneider, 2009a) – these two sources of information come together only later in the development. To be more precise, the first most spoken words up to the age of one and a half years already include a number word (Szagun, 2013). Looking at the main principles of acquiring vocabulary (mostly nouns) at this stage of development, this is not self-evident. The first principle that children follow in vocabulary acquisition is the "mutual exclusivity assumption," which means that a word is always assigned to an object (i.e., if a word has already been learned for that object, that word cannot mean another object; Markman, 1989). To use *two* for the number of bears and *two* for the number of cars would therefore contradict this principle. In addition, a second principle is used by the toddlers in this phase, the "whole object assumption." This means that the word "heard" covers the whole object and not any feature or part of the object. The word "shoe" is thus associated with the shoe as a whole and not just with the sole, the shoelace, or even the number of shoes (Markman, 1989). These principles make it a little difficult for children to grasp the concept of a number word at this early stage because the number word is an attribute of the object, not the object itself. As Gelman and Gallistel (1978) have already described in their Abstraction Principle, children recognize that things are countable and this in turn helps them to process the number words as specific words on another abstract level. This abstraction allows them to process these words differently and to refine them further. At first, it is just an empty collection of specific words without a stable order. Only with time children learn that number words follow a sequence, as Gelman and Gallistel (1978) also state in their counting principles. In parallel, one-to-one correspondence is discovered and applied. In this phase, counting is equivalent to reciting a poem. This means that there is not yet a cardinality understanding. One could also say that numbers are processed purely linguistically, without the quantity being processed as a semantic unit behind it. So the child can count four objects correctly, but cannot answer the question how many there actually are. This semantic knowledge about the number words is built up only very laboriously and slowly. In the last step, therefore, the linguistic information, that is, the number word, must be linked to the content or semantics (i.e., the quantity information). In concrete terms, this means that the child not only has to identify *one* as a number word and place it correctly in the number word series, but also must understand that exactly one object can be assigned to *one*. The so-called one-knowers (Sarnecka & Carey, 2008; Wynn, 1992) are thus children who have already understood the concept of *one* and can therefore semantically distinguish exactly between *one* and more. For example, if a child at the one-knower level is asked to give one of the presented objects (e.g., cars), exactly one is also given. If a one-knower is asked to give two objects, more objects are given, sometimes two, but sometimes more than two. It can be observed that language plays an important role in this step. The understanding of plural markings in nouns occurs at the same time as the children change to the one-knower-level (Barner et al., 2007). This is also supported by observations of Japanese children (Sarnecka et al., 2007), who do not acquire a plural marking at this time because this plural marking does not exist in Japanese. As a result, Japanese children achieve the transition to the one-knower-level much later. In this first phase of the acquisition of cardinality, the path via the linguistic concept of the plural seems to be beneficial. On the other hand, the path via the plural does not seem to be the only one, as the findings from Japan show. So there must also be at least one other way to acquire cardinality. With time, children gradually learn to differentiate *two*, *three*, *four*, and *five* and to assign them correctly. Only from *six* onward this understanding of cardinality generalizes and the children can transfer the knowledge to the next quantity. At this stage, children are called cardinality-knowers (Sarnecka et al., 2007). This acquisition of mapping the number word to the correct quantity is quite hierarchical (Sarnecka et al., 2007) and language seems to have a supportive function throughout numerical development. Studies such as Negen and Sarnecka (2012) show a clear correlation between the size of the vocabulary and the knowledge of number words. Pixner et al. (2018) also found a clear difference in vocabulary between subset-knowers (i.e., children who did not yet show any generalization in terms of cardinality) and cardinality-knowers. Phonological awareness proved to be another important language-related predictor for the prediction of basic numerical competences in a group of preschool children, as Pixner et al. (2017) were able to show. It is assumed that phonological awareness plays an important role in this phase, since the first number words are also very similar in sound, at least in German (e.g., *zwei* and *drei*). Krajewski and Schneider (2009b) state that a phonological processing deficit should affect mathematical domains that are verbally coded, while other domains, especially higher numerical processing, should not. In the following, they explain that the influence of a phonological awareness deficit on a mathematical impairment in early mathematical development can be mitigated or compensated (e.g., by training and manipulating the precise number words). Based on the number word sequence, children then learn to connect this number word sequences with the corresponding quantities. That means that good linguistic differentiation is necessary. All these studies show that language or linguistic competences, such as the size of the vocabulary but also phonological awareness, are helpful in this phase of the development of number words and counting as well as for cardinality knowledge. Knowledge of number words, counting, and cardinality are in turn an important building block for the successful development of arithmetic competences.

1 Quantifiers: Their development and their influence on cardinality

One aspect of the association between language and arithmetic skills is mathematical language. This includes terms that are strongly domain-specific and related to the mathematical context (Purpura & Reid, 2016). Understanding mathematical language is essential in school. Even "simple" quantifiers like "more" or "less" can cause massive problems in solving word problems (Dresen et al., submitted). Children usually associate "more" with an addition and "less" with a subtraction. If these terms are used inconsistently to the "anticipated" arithmetic operation, the probability of solving such problems decreases significantly. This phenomenon can be observed not only in children but also in adults. Therefore the development of an understanding of mathematical language, especially of quantifiers, is of essential importance.

Quantifiers are unspecific number words, which play a special role in the development of number words as well as cardinality. In line with Gleitman's bootstrapping theory (1990), Carey (2004) postulated that children derive the meaning of number words from their understanding of quantifiers. Resnick (1989) also assumes that children first have an imprecise association between quantifiers and the corresponding quantities, which becomes more specific and precise with the length of their experience. Quantifiers represent unspecific number words (e.g., more, many, a few) and – as specific number words – represent a quantity (Sullivan & Barner, 2011). Additionally, exact quantifiers (e.g., both) or non-exact quantifiers (e.g., some) can be distinguished. However, to understand the meaning of quantifiers, it is important to grasp semantic and pragmatic restrictions (i.e., language knowledge) as well as the quantificational meaning of each quantifier (i.e., domain-specific numerical knowledge; Dolscheid & Penke, 2018). In this sense,

Sullivan and Barner (2011) argued that to get the meaning of quantifier successfully, children need to understand that quantifiers are arranged on a shared scale. This scaling allows children to draw pragmatic conclusions about the individual significance of the quantifiers. Accordingly, Hurewitz et al. (2006) investigated linguistic similarities and differences of both specific and unspecific number words. They argued that specific number words (as, for example, *two*) are always independent from the context (e.g., *two* is always *two*). In contrast – regarding quantifiers as unspecific number words – in a set of three objects many could already be *two*, whereas in a set of thousand objects many is clearly more than *two*. At the same time, similarities regarding the embedment in the syntax are described as well. For instance, both specific and unspecific number words can be sequenced in an order (e.g., all is always more than most and 7 is always more than 4). Furthermore, findings of Dolscheid et al. (2015) demonstrated an association between unspecific quantifier knowledge and numerical skills such as specific cardinality knowledge and counting skills at the age of 4.6 years – without, however, being able to answer the extent to which these skills are interdependent.

A recent study of Dresen et al. (2020) evaluated potential associations between the acquisition of cardinality knowledge and quantifier knowledge (i.e., unspecific number words) in children. The study followed a total of 76 (34 boys and 42 girls) monolingual German-speaking children aged between 3.6 and 4.6 years at the first measurement time for two more measurement times 6 months and one year later. Children were tested with two relevant tasks: a give-N task assessing specific cardinality knowledge of numbers from 1 to 10 and a give-N task measuring unspecific quantifier knowledge (more than, less than, all, a lot of, etc.). Results clearly indicated that children's cardinality knowledge correlated over all three measuring times. Children who already had better cardinality knowledge at an earlier measurement time also performed better at later measurement times and vice versa. Therefore, it makes sense to monitor children's cardinality knowledge already at early stages of their numerical development because it is considered as one important basic numerical competence for the development of further numerical competences (e.g., Brannon & Van de Walle, 2001). A little different was the picture for the case of quantifiers. No correlation between quantifier knowledge at the first and second measurement time was found. This may indicate that the development of quantifier knowledge is not linear but may rather be influenced by other factors such as language development, for instance. It was only between the second and third measurement time that a significant correlation for quantifier knowledge was observed.

More interesting, however, were the observed associations of specific cardinality knowledge and unspecific quantifier knowledge. Quantifier knowledge assessed at the first measurement time was not associated with actual and/or future cardinality knowledge. This seems to imply that there may be no further disadvantage for children's development of the cardinality knowledge when they do not yet master the quantifiers at the ages of 3.6 to 4.6. Interestingly, however, there was a significant association between cardinality knowledge and quantifier knowledge at the second measuring time. Furthermore, there were significant bidirectional associations of cardinality knowledge and quantifier knowledge from the second to the third measurement times (when children were on average 4.4 and 4.10 of age, respectively). Importantly, mediation analyses specified that the association between cardinality knowledge at measurement times two and three was fully mediated by children's quantifier knowledge at measurement time two. Interestingly, on the other hand, there was no mediation of the development of quantifier knowledge between measurement times two and three by children's cardinality knowledge at measurement time two. Based on these results, it can be assumed that – at this particular age – quantifier knowledge seems to facilitate further development of children's cardinality knowledge. In turn, this may mean that children should have acquired an understanding of quantifiers up to this age because this helps them to further develop their cardinality knowledge.

In summary, these results show how differentiated one has to consider associations between the early development of specific cardinality knowledge and unspecific quantifier knowledge reflecting less precise numerical magnitude information. In particular, quantifier knowledge seems to facilitate cardinality knowledge at a specific age, which might indicate the specification of less precise and thus approximate magnitude representations as reflected by quantifiers.

2 Concept and characteristics of zero and negative quantifiers

As described above, the development of arithmetic competences involves many sub-competences and a wide range of influences. If we concentrate on the specific number words on the one hand and on the unspecific number words, quantifiers, on the other hand, we can discover additional specific characteristics. Both *zero* and negative quantifiers (e.g., nothing) differ from the concepts presented, although by definition they belong to these two groups. Let us first take a closer look at *zero*. As discussed at the beginning, children between the ages of 22 and 24 months learn to distinguish between *one* and more, at the same time as they mark nouns with the correct plural. This means that the understanding of quantities greater than *one* is supported by the language, thus facilitating differentiation. But what does this look like with *zero*? Also with *zero*, the following noun is marked with plural (e.g., I have *zero* apples). This phenomenon can be found in many languages. The supporting principle of quantity differentiation by plural marking is thus overridden or violated at *zero*. Concretely, this means that a quantity which is less than *one* is nevertheless marked with the plural at nouns. This may therefore be one of the explanations why *zero* is associated with a lot of difficulties in children, but also later in adults (Brysbaert, 1995; Wellman & Miller, 1986). The first, frequently described difficulty with *zero* is the implementation of arithmetic rules when calculating with zero ($2*0 = 0$ but $2 + 0 = 2$). The second difficulty has to do with the placeholder function of *zero* in multi-digit numbers. In evolutionary terms, *zero* is a relatively young number (Butterworth, 1999). Most primitive number word systems and also the symbol-x-value systems such as the Roman system did not need *zero*, since only one number was needed to represent a quantity. All of the above-mentioned number word systems begin with the number *one*, and since they do not need a placeholder due to their structure (e.g., in Roman 5 is represented as V and 10 as X), *zero* is not necessary either. In the current Arabic notation system, however, *zero* has this very important placeholder function. In concrete terms, this means that if in a multi-digit number a position is not existing (e.g., no tens), a *zero* must be entered in the missing place; otherwise, the quantity is no longer correct (e.g., there is no ten in 302, but the ten's place cannot simply be omitted, since otherwise the 302 would result in a 32, i.e., a completely wrong quantity).

Looking at these difficulties with *zero*, the question arises how the understanding of negative quantifiers, for example, no/none/nothing, looks like in child development. The concept of nothing may pose difficulties at a more general level – not only referring to the numerical value of zero. This means the understanding of the concept of no objects differs from the understanding of one or more objects as an experiment of Wynn and Chiang (1998) could show. In this experiment, 8-month-old infants were irritated when an object disappeared in a location in which this item had been shown before. This was not the fact when an object appeared where no object was before. Although young infants already have a distinct knowledge of material objects, they are not able to understand *no objects*. Only later on, children acquire the words "nothing"/"none"/"no" (objects) and their semantical meaning, without considering it as a numerical value and combining it with the symbol of *zero* (Pixner et al., 2018).

In a study by Pixner et al. (2018) 65 kindergarten children aged 4 to 5 years ($M = 4$ years and 4 month; $SD = 3$ months) were examined with regard to their understanding of small numbers and *zero* as well as their visual-spatial skills (measured with the subtest visual perception of the Visual Motor Integration (VMI), Beery, 2004), general language (measured with a standardized active vocabulary test, AWST-R, Kiese-Himmel, 2005), counting skills, knowledge of Arabic numbers, and finger knowledge. To identify children´s finger knowledge, the children were asked to present a different configuration of fingers. All quantities between 0 and 10 were asked in random order. Participating children were recruited from local public kindergartens and all of them were monolingual native German speakers. Thirty-one boys and 34 girls were included in this study. Most of the children were right-handed (81.5%). No child in this study showed an intellectual or language impairment. Significant correlations were observed between vocabulary, numeracy, finger knowledge, and counting skills, both with understanding of the cardinality of small numbers and with knowledge of *zero*. Subsequent regression analyses, however, only showed the importance of counting skills on knowledge of *zero*. General vocabulary, spatial skills, as well as the cardinality understanding of small numbers showed no independent predictive value in this regression. It is interesting to note that *zero* and the negative quantifiers, like "nothing"/"no"/"none," had a negative correlation at this age. An explanation can be derived from the general principles in the vocabulary development, specifically from the mutual exclusivity assumption, which means that if you already have a term for nothing/zero at this age, you do not need another term for this state.

3 Conclusion

In summary, it can therefore be said that language or linguistic competences, as cross-domain competences, fulfil strongly supportive functions at many stages of the development of arithmetic competences. This should be kept in mind especially for children at risk with limited linguistic development, for whom there is an additional risk that arithmetic competences will not develop in line with age. However, despite the linguistic influence, the results of some studies also show that domain-specific numerical precursor skills appear to be more important for the acquisition of cardinality understanding and zero than more cross-domain skills. An intervention should therefore cover both aspects equally.

If we look in particular at specific mathematical vocabulary, such as the quantifiers in this chapter, results indicate that understanding quantifiers contribute significantly to the acquisition of children's understanding of cardinality. However, mathematical vocabulary is also important in later life for the acquisition of various mathematical skills and the successful solution of mathematical tasks, such as mathematical word problems, and should, therefore, also be a focus in the school context.

An important and often neglected aspect is the extraordinary role of *zero*, both in acquisition and later in implementation in the mathematical sense. In order to minimize the numerous difficulties with *zero* in children as well as in adults, the corresponding arithmetic rules and also the placeholder function should not be neglected and should be repeated continuously.

References

Barner, David, Thalwitz, Dora, Wood, Justin, Yang, Shu – Ju & Carey, Susan (2007): On the relation between the acquisition of singular – plural morpho – syntax and the conceptual distinction between one and more than one. *Developmental Science* 10 (3), 365–373, doi: https://doi.org/10.1111/j.1467-7687.2007.00591.x.

Beery, Keith E., Buktenica, Norman A. & Beery, Natasha A. (2004): *The Beery –Buktenica Developmental Test of Visual-Motor Integration.* Minneapolis: Pearson.

Brannon, Elizabeth M. & Van de Walle, Gretchen A. (2001): The development of ordinal numerical competence in young children. *Cognitive Psychology* 43 (1), 53–81, doi:10.1006/cogp.2001.0756.

Brysbaert, Marc (1995): Arabic number reading: On the nature of the numerical scale and the origin of phonological recording. *Journal of Experimental Psychology: General* 124 (4), 434–452, doi:https://doi.org/10.1037/0096-3445.124.4.434.

Butterworth, Brian (1999): *The Mathematical Brain.* London: Macmillan.

Carey, Susan E. (2004): Bootstrapping & the origin of concepts. *Daedalus* 133 (1), 59–68, doi:10.1162/001152604772746701.

Dehaene, Stanislas (1992): Varieties of numerical abilities. *Cognition* 44 (1–2), 1–42, doi: https://doi.org/10.1016/0010-0277(92)90049-N.

Dolscheid, Sarah & Penke, Martina (2018): Quantifier comprehension is linked to linguistic rather than to numerical skills. Evidence from children with Down syndrome and Williams syndrome. *PLoS ONE* 13 (6), e0199743, doi:https://doi.org/10.1371/journal.pone.0199743.

Dolscheid, Sarah, Winter, Christina & Penke, Martina (2015): Counting on quantifiers: Specific links between linguistic quantifiers and number acquisition. In Airenti, Gabriella, Bruno, Bara, Sandini, Giulio (eds.): *EuroAsianPacific Jopint Conference on Cognitive Science (EAPCogSci) 2015.* Turin, 762–767.

Donlan, Chris, Cowan, Richard, Newton, Elizabeth J. & Lloyd, Delyth (2007): The role of language in mathematical development: Evidence from children with specific language impairments. *Cognition* 103, 23–33. doi:10.1016/j.cognition.2006.02.007.

Dresen, Verena, Moeller, Korbinian, Danay, Erik & Pixner, Silvia (submitted): *How Lack of Consistency and Requested Outcome Affect Primary School Children Solving Arithmetic Word Problems.* Under review.

Dresen, Verena, Moeller, Korbinian & Pixner, Silvia (2020): *Association Between Language and Early Numerical Development in Kindergarten – The Case of Quantifiers.* Under review.

Gelman, Rochel & Gallistel, C. R. (1978): *The child´s Understanding of Number*. Cambridge: MA Harvard University Press.
Gleitman, Lila R. (1990): The structural sources of word learning. *Language Acquisition* 1 (1), 3–55.
Hurewitz, Felicia, Papafragou, Anna, Gleitman, Lila & Gelman, Rochel (2006): Asymmetries in the acquisition of numbers and quantifiers. *Language Learning and Development* 2 (2), 77–96, doi:https://doi.org/10.1207/s15473341lld0202_1.
Kiese-Himmel, Christiane (2005): *AWST-R. Aktiver Wortschatz für 3-5 Jährige Kinder – Revision*. Göttingen: Hogrefe Verlag.
Krajewski, Kristin & Schneider, Wolfgang (2009a): Early development of quantity to number-word linkage as a precursor of mathematical school achievement and mathematical difficulties: Findings from a four-year longitudinal study. *Learning and Instruction* 19, 513–526.
Krajewski, Kristin & Schneider, Wolfgang (2009b): Exploring the impact of phonological awareness, visual-spatial working memory, and preschool quantity-number competencies on mathematics achievement in elementary school: Findings from a 3-year longitudinal study. *Journal of Experimental Child Psychology* 103, 516–531.
Markman, Ellen M. (1989): *Categorization and Naming in Children*. Cambridge: The MIT Press.
Negen, James & Sarnecka, Barbara W. (2012): Number-concept acquisition and general vocabulary development. *Child Development* 83 (6), 2019–2027, doi:https://doi.org/10.1111/j.1467-8624.2012.01815.x.
Pixner, Silvia, Dresen, Verena & Moeller, Korbinian (2018): Differential development of children's understanding of the cardinality of small numbers and zero. *Frontiers in Psychology* 9, doi:10.3389/fpsyg.2018.01636.
Pixner, Silvia, Kraut, Christina & Dresen, Verena (2017): Early predictors for basic numerical and magnitude competencies in preschool children – are they the same or different regarding specific subgroups?. *Psychology* 8, 271–286. doi:10.4236/psych.2017.82016.
Preat, Magda, Titeca, Daisy, Ceulemans, Annelies & Desoete, Annemie (2013): Language in the prediction of arithmetics in kindergarten and grade 1. *Learning and Individual Differences* 27, 90–96. doi:https://doi.org/10.1016/j.lindif.2013.07.003.
Purpura, David J. & Reid, Erin E. (2016): Mathematics and language: Individual and group differences in mathematical language skills in young children. *Early Childhood Research Quarterly* 36, 259–268.
Resnick, Lauren B. (1989): Developing mathematical knowledge. *American Psychologist* 44, 162–169.
Sarnecka, Barbara W. & Carey, Susan (2008): How counting represents number what children must learn and when they learn it. *Cognition* 108, 662–674. doi:10.1016/j.cognition.2008.05.007.
Sarnecka, Barbara W., Kamenskaya, Valentina G., Yamana, Yuko, Ogura, Tamiko & Yudovina, Yulia B. (2007): From grammatical number to exact numbers, early meaning of "one", "two", and "three" in English. *Cognitive Psychology* 55 (2), 136–168.
Sullivan, Jessica L. & Barner, David (2011): Number words, quantifiers, and principles of word learning. *Wiley Interdisciplinary Reviews: Cognitive Science* 2 (6), 639–645, doi:10.1002/wcs.140.
Szagun, Gisela (2013): *Sprachentwicklung beim Kind*. Weinheim: Beltz Verlag.

Wagner, H., Ehm, J.-H., Schöler, H., Schneider, W. & Hasselhorn, M. (2013): *Zusatzförderung von Kindern mit Entwicklungsrisiken. Eine Handreichung für Pädagogische Fachkräfte im Übergang vom Elementar – zum Primarbereich.* Göttingen: Hogrefe Verlag.

Ward, Camille & Smuts, Barbara B. (2007): Quantity – based judgments in the domestic dog (Canis lupus familiaris). *Animal Cognition* 10, 71–80. doi:10.1007/s10071-006-0042en7.

Wellman, Henry M. & Miller, Kevin F. (1986): Thinking about nothing: Preservice elementary school teachers´concept of zero. *Journal for Research in Mathematics Education* 14 (3), 147–155.

Wynn, Karen (1992): Children's acquisition of the number words and the counting system. *Cognitive Psychology* 24 (2), 220–251, doi:https://doi.org/10.1016/0010-0285(92)90008-P.

Wynn, Karen & Chiang, Wen-Chi (1998): Limits to infants'. *Psychological Science* 9 (6), 448–455, doi:https://doi.org/10.1111/1467-9280.00084.

Xu, Fei, Spelke, Elizabeth S. & Goddard, Sydney (2005): Number sense in human infants. *Developmental Science* 8 (1), 88–101, doi:10.1111/j.1467-7687.2005.00395.x.

III Multilingualism and mathematical learning

Sarit Ashkenazi and Nitza Mark-Zigdon

Directionality of number space associations in Hebrew-speaking children: Evidence from number line estimation

It is widely believed that quantity manipulation and understanding is an innate human ability, which is preverbal and based upon the mapping between space and quantities (number-space associations) (Dehaene, 1992, 2009; Dehaene et al., 2003, 1999; de Hevia & Spelke, 2010). Infants are born with preverbal approximate number sense (ANS). Later, with maturation and schooling, an exact symbolic quantity system develops, supported by the preverbal ANS as well as verbal abilities (Halberda et al., 2008; Libertus et al., 2011).

The preverbal representation of quantity (ANS) is believed to be needed across the life span mostly for estimation tasks. During numerical estimation, a quantity (symbolic or non-symbolic) is translated into an abstract code of that quantity in the form of the mental number line (Dehaene, 1992, 2009; Dehaene et al., 2003). The mental number line has a few documented characteristics, including its left to right directionality (Dehaene et al., 1993), as well as its arrangement; it is initially logarithmic, and becomes linear with maturation and schooling (Siegler & Booth, 2004; Siegler & Opfer, 2003). Logarithmic representation of the number line includes overestimation of small numbers and underestimation of larger numbers toward the end of the scale. By contrast, linear representation includes equal distances between quantities, regardless of their size.

1 Number line estimation tasks

One popular estimation task, aimed at examining the averbal numerical representations, is the number line estimation task (or the number to position task) (Siegler & Opfer, 2003). In the number line estimation task, a number is presented above a number line, with 0 at one end and 100 or 1,000 at its other end. The participant is instructed to place a number spatially on the line. Corresponding to

---

Note: Dr. Yossef Arzuan, of blessed memory, was part of the research team and helped design the experiment and performed analysis of the data.

https://doi.org/10.1515/9783110661941-009

the development of the mental number line (see the previous section), and with maturation, children's estimation shifts from logarithmic to linear (Siegler & Booth, 2004; Siegler & Opfer, 2003). The shift between logarithmic and linear representations is related to familiarity with the range (i.e., a child can present a linear representation for a small and familiar range, and a logarithmic representation for a larger range) and also to age. Specifically, most second graders will show logarithmic representation in an unfamiliar range (up to 1,000), with linear representation in the familiar range (until 100). However, most sixth graders, like adults, will use linear representation regardless of range (Siegler & Booth, 2004; Siegler & Opfer, 2003). However, logarithmic representation can be found in adults due to cultural invention. For example, the Mundurucu, an Amazonian indigenous tribe with no formal education and with a minimal number word system, who at all ages map symbolic and non-symbolic numbers onto a logarithmic scale (Dehaene et al., 2008). Furthermore, a linear tendency is also related to individual differences in numerical abilities. Participants with better numerical abilities will show preference for linear representation over logarithmic representation (Booth & Siegler, 2006). Hence, even if number line estimations should be based upon averbal spatial representations, they are shaped by educational level, culture, and individual differences.

Most of the studies that used the number line estimation task tested school-age children or adults (Siegler & Opfer, 2003). Only a few studies have tested preschoolers (Berteletti et al., 2010; Siegler & Booth, 2004). For example, Siegler and Booth (2004) examined 5- to 6-year-old children, in the range between 0 and 100, and found a logarithmic representation of that range. Berteletti et al. (2010) tested even younger preschoolers, beginning from the age of 3.5 years, in the ranges of 1–10, 1–20, and 0–100. They discovered that with development, children's estimates shifted from logarithmic to linear in the smaller number range. Estimation accuracy was correlated with knowledge of Arabic numerals and numerical order.

2 Do number line tasks reflect pure averbal numerical estimations?

Lately, there has been debate about the nature of the number to position task. Barth and colleagues (Barth & Paladino, 2011; Slusser & Barth, 2017; Slusser et al., 2013) suggest that the classical number to position task is based upon proportion–judgment strategies, rather than non-symbolic representations. In the number to position task participants are asked to place a number on a number line, and to do it using a base ten understanding of the number line and the distance between the number and reference point on the line (e.g., beginning, mid-point, and end of the line). Proportion estimation strategies involve focusing on a part of the line, thereby producing an estimation bias that is reflected in more precise estimations closer to the reference point, as compared to farther away from it. Accordingly, the shift between logarithmic and linear representation can be explained alternatively by the shift between judgment according to one reference point (the beginning of the line), to judgment according to two reference points (beginning and end of the line), and later on judgment according to three reference points (beginning, mid-point, and end point) (Slusser et al., 2013). Specifically, the first graders used the beginning of the line as a single reference point, while the second graders used both endpoints, and typically by the third grade children's strategy was similar to adults, as they used multiple reference points, including both endpoints and the midpoint. One major support for that view is that across age groups, 1 cycle function and 2 cycle function functions explain the data better (in relation to percentage of explained variance), compared to logarithmic and linear functions (Barth & Paladino, 2011).

3 Number space associations: Are they culturally driven or innate?

A variety of evidence, aside from the number line task, supports number space associations (de Hevia & Spelke, 2010; Hubbard et al., 2005; Wood et al., 2008). For example, in the parity judgment task, participants make their responses more quickly on the right for larger numbers and on the left for smaller numbers (Dehaene et al., 1993). This effect is called Spatial Numerical Association of Response Codes (SNARC). The SNARC effect is the main support for the assumption that the mental number line orients from left (small numbers) to right (large numbers) (Hubbard et al., 2005; Wood et al., 2008).

While the SNARC effect can be found in various tasks and situations (Wood et al., 2008), there is cultural variation in its direction and even its appearance (for review, see Göbel et al., 2011). For example, Dehaene et al. (1993) described an absence or even reversed SNARC effect in Iranian subjects, who read and write Farsi from right to left. Shaki et al. (2009) tested three groups of participants that differed in number and reading directions. (1) Hebrew-speaking participants, who read numbers from left to right, and Hebrew words from right to left; (2) Canadians (words and numbers are oriented from left to right); and (3) Palestinians (words and numbers are oriented from right to left). As expected, Canadians show associations between left-side space with small numbers, and right-side space with large numbers. By contrast, Palestinians showed reversed associations. Importantly, the Hebrew-speaking participants showed no associations between space and numbers. These results suggest that reading habits, for both words and numbers, contribute to the spatial representation of numbers (Shaki et al., 2009). Like the cultural variation of the SNARC effect, a cultural variation was found in finger counting direction (Lindemann et al., 2011). While most Western individuals started counting with the left hand (from left to right), most Middle Eastern (Iranian) respondents preferred to start counting with the right hand (from right to left) (Lindemann et al., 2011).

Lastly, right to left number space associations were found in English-speaking preschoolers prior to formal reading education. For example, young children showed left to right number space mapping during a non-symbolic numerosity comparison in the form of spatial–numerical congruity (quicker reaction times to smaller sets presented on the left side of the screen, and to larger ones presented on the right side) (Patro & Haman, 2012). Moreover, preschoolers in an implicit, color discrimination task showed the SNARC effect (represent right to left mapping of number to space; see previous paragraphs) (Hoffmann et al., 2013; Patro & Haman, 2012).

Notably, cultural variation in the direction of number space associations can start very early, before the formal acquisition of reading habits (Göbel et al., 2018; McCrink et al., 2018; McCrink & Opfer, 2014; Nuerk et al., 2015). In a recent study, American and Israeli toddler–caregiver dyads were tested in a situation of natural interactions (the infants were 2 years old). In the case of ordering spatial structures English-speaking American caregivers were more likely to use left to right spatial structuring, while Hebrew-speaking Israeli parents were more likely to use right to left spatial structuring. The authors concluded that spatial structure biases exhibited by caregivers are a potential route for the development of spatial biases in early childhood, before experiencing any formal education (McCrink et al., 2018).

4 The current study

The current study tested number space associations using the number line estimation task. We tested a group of young children (4–6 years old), before formal experience with reading. We used a familiar range for this age group, 1–9. In the first step, we verified our assumption that 1–9 was a familiar range for young children. Formal number knowledge was tested. Only children with formal number knowledge for the range of 1–9 were included in our sample.

All the children in our sample are native Hebrew speakers; Hebrew is read from right to left. This was found to modulate the direction of number space associations, in a task that implicitly connects numbers to space (e.g., the parity task), in adults (Göbel et al., 2018; McCrink et al., 2018; Nuerk et al., 2015; Shaki et al., 2009). Hence, the main goal of the present study was to test directionality of number space associations in Hebrew-speaking children. Estimations were to be given on number lines oriented from left to right (as expected in Western society) or right to left (similar to the directionality of number space associations for Hebrew-speaking preschoolers) (Shaki et al., 2012). We expected to find better estimations for right to left orientation of the number line than for left to right orientation. This effect should be modulated by children's age. Hence, younger children will show greater preference for right to left orientation than older children.

An additional goal of the present study was to test the effect of presentation on estimations. Magnitudes were presented as symbolic (Arabic numeral) or non-symbolic (dots). We expected that estimations would be modified by presentation. This effect should be modulated by children's age. It was suggested that younger children possess two separate representations for symbolic and non-symbolic quantities, while older children possess a united representation (Kolkman et al., 2013). Hence, we expected that the effect of presentation on estimation should be greater in younger children than in older children.

5 Method

5.1 Participants

Fifty-three subjects from three kindergartens in the central district of Israel participated in the study. The children's mean age was 51.43 months, SD 5.78 months, ranging between 41 and 60 months. All subjects were native Hebrew speakers and did not know how to read (an alphabet); 21 of the participants were female and 32 were male.

5.2 Procedure

Prior to conducting the study, the required approvals were obtained from the Ministry of Education and from the parents of the children for their participation in the study.

Each of the subjects was examined individually in a separate room. The tests were conducted in two sessions, each session lasting about 20 min. All subjects started the test with the pre-knowledge tasks. The order of tasks for each subject was randomly selected. As for the number line tasks, subjects were divided randomly into two groups. One group started with a symbolic number line, and the other group with a non-symbolic number line. Each group was divided into subgroups.

5.3 Task

The study consisted of three parts: (1) pre-knowledge tasks; (2) formal knowledge tasks; and (3) number line tasks. The aim of the pre-knowledge tasks was to test the required knowledge as a pre-condition for participating in the number line tasks.

5.3.1 Pre-knowledge tasks

The coding of tasks was done on a trial-by-trial basis. The score of 1 was given if the child performed with no errors, and a 0 was given for error trial. We averaged each of the repetitions in every task. Each of the tasks was tested four times.

5.3.1.1 Counting

The children were instructed to count freely up to ten.

5.3.1.2 Matching

The children were instructed to match between cards with quantities, and cards with numbers from 1 to 10. The numbers were selected randomly for each child.

5.3.1.3 Ordering

The children were instructed to arrange cards with numbers from 1 to 10.

5.3.1.4 Size ratio

The children were presented with two cards with consecutive numbers. They were asked to indicate which was the larger or the smaller number.

5.3.1.5 Concepts

5.3.1.5.1 Size
Big – small, bigger – smaller.

5.3.1.5.2 Quantity
Much – few, more – less.

5.3.1.5.3 Space
Close – far. For the absolute concepts, two pictures were presented to the children, and for the relational concepts three pictures were presented. The children were asked to point out the picture that fits the concept about which they were asked. The cards contained images of fruits (grapes or bananas) animals (dogs and horses) and objects (houses).

5.3.2 Formal knowledge tasks

5.3.2.1 Cardinality

5.3.2.1.1 The "How many" task
A brown rabbit doll was presented on a table. The experimenter said: "The rabbit likes cucumbers, so I'll give him some.

Please tell me how many cucumbers I should give him?" The amounts were varied randomly from 1 to 5.

5.3.2.1.2 "Give –N" task
A gray rabbit doll was presented on a table. The experimenter said: "The rabbit likes carrots very much. Please give him X carrots." The amounts were varied randomly from 1 to 5.

5.3.2.2 Adding one and subtracting one
The purpose of the next two tasks was to test whether the child knows the principle of a constant difference of 1 between consecutive numbers. The experimenter placed cards with pictures of carrots next to the rabbit and asks the child to count how many carrots the rabbit has. The experimenter repeated the

question several times until he was sure the child remembered how many carrots the rabbit has. Then the experimenter covered the cards and asked the child: "Do you remember how many carrots the rabbit has?" If the child didn't remember, the experimenter allows him to count again. Then he said: "now I give/take one carrot from the rabbit." Then, he added or subtracted one carrot. Then he asked the child without allowing him to count: "How many carrots will be left for the rabbit now?"

5.3.2.3 Order relations

Two rabbits are on the table, one brown and one gray. The experimenter gave one of the rabbits a quantity of cards with pictures of cucumbers (from 1 to 5). The other rabbit got a smaller (−1) or larger (+1) amount of cucumbers than the first rabbit. The experimenter asked the child to count and say how many carrots each rabbit has. Then he asked the child: "What can you do to make the gray rabbit have the same number of cucumbers as the brown rabbit, and vice versa." The child could see and count the quantities of the cucumbers for the two rabbits.

5.3.3 Number line estimation task

Included four subtasks: (1) symbolic orientation from left to right; (2) symbolic orientation from right to left; (3) non-symbolic orientation from left to right; and (4) non-symbolic orientation from right to left. The order of subtasks was selected randomly. Each trial was given on separate sheets of paper with a 13.5 cm. line on it, with one end labeled as 1 or () and the other end labeled as 9 or (). The children got four cards with numbers 3, 4, 6, and 7, one at a time, for symbolic and non-symbolic presentation and for congruent and incongruent directions. They were requested to locate the number by a vertical hatch mark on each sheet of paper. Before performing the task, the children performed practice trials with the numbers 1 and 8. The experimenter practiced the task with the children until he was sure the child understood what to do.

5.4 Stimuli

5.4.1 Cubes

In the matching, cardinality, adding one, subtracting one, and order relation tasks, we used cubes. In each task different color cubes were selected, to avoid the possibility of children attaching numerical values to certain colors. All the cubes were about 1 × 1 cm in size.

5.4.2 Tabs

In the Matching and Ordering tasks, numeral tabs from 1 to 10 were used. In the size, quantity, and space tasks, the images on each tab were different and contained vegetables or fruits (lettuce, onions, carrots, potatoes, grapes, and bananas). All tabs were about 10 × 8 cm. in size.

5.4.3 Dolls

Brown and Gray Rabbit dolls about 10 × 20 cm. in size.

6 Results

6.1 Pre-knowledge

All the children in the sample were able to count to ten, count objects until 7 (as a minimum), and understood numerical concepts of size, quantity, and space. All of the sample could match quantity to Arabic numerals. Most of our sample could understand ratio relations (98%). Most of the children could order groups by size (94%).

6.2 Number line understanding

The entire sample understood the cardinality principle. However, formal understanding of the adding one principle was found in some of the children (73%). An even smaller proportion of the children were able to understand the subtracting one principle (34%). Similarly, only some of the children understood the role of numbers in representing ordinal relations (82%). Please note that none of the variables were related to age (see Tab. 1).

Tab. 1: Descriptive statistics for: (A) Pre-knowledge and (B) Formal number line understanding. The correlations between scores in these tests and age failed to reach significance (see the last column for the correlations).

Variable	Minimum	Maximum	Mean	Std. Dev	Age correlations
A. Matching	1	1	1	0	
Counting	1	1	1	0	
Ratio	0	1	.98	.14	
Ordering	.40	1	.94	.14	.05
Mathematical concepts	1	1	1	0	
B. Cardinality	1	1	1	0	
Adding one	.04	1	.73	.28	−0.6
Subtracting one	0	1	.35	.28	−.01
Order relations	.40	1	.82	.26	.04

6.3 Number line

6.3.1 Functions fit

The fit statistics of linear, logarithmic, 1 cycle (0.8), and 2 cycle (0.8) functions were computed to analyze the pattern of estimates. We computed these fit statistics for each individual child. We first aimed to exclude children whose fit statistics were negative or less than 0.55 (see the profile analysis).

6.3.1.1 Profile analysis

We examined the profiles of children related to their function fit statistics. To characterize profiles we used the linear fit, examining whether linear fit for each of the study conditions was greater than 0.55 (symbolic congruent (left to right), symbolic incongruent (right to left), non-symbolic congruent (left to right), non-symbolic incongruent (right to left)). Children in profile 1 (N = 5) had only 1 condition with fit greater than 0.55. Children in profile 2 (N = 7) had 2 conditions with fit greater than 0.55. Children in profile 3 (N = 12) had 3 conditions with fit greater than 0.55. Children in profile 4 (N = 30) had 4 conditions with fit greater than 0.55. We excluded children from profiles 1 and 2 from the other analysis.

6.3.1.2 ANCOVA for individual level fits

A three-way analysis of covariance (ANCOVA) with a mean r value of fit to power analysis was computed, with function (1 cycle / 2 cycle/ linear / logarithmic), presentation type (symbolic/ non-symbolic), direction (congruent from left to right/ incongruent from right to left) and the age of the children serving as a covariant. The effect of presentation reached significance [$F(1, 39) = 9.22$, partial $\eta^2 = .19$, $p < .01$]. The fit was better for symbolic (mean = 0.91, SD = 0.18) than for non-symbolic (mean = 0.78, SD = 0.39). The effect of directionality was also significant [$F(1, 39) = 5.27$, partial $\eta^2 = .12$, $p < .05$]. The fit was better for incongruent direction (right to left) (mean = 0.86, SD = 0.26) than for congruent (left to right) direction (mean = 0.81, SD = 0.32). The interaction between presentation and direction was also significant [$F(1, 39) = 4.64$, partial $\eta^2 = .11$, $p < .05$]. The differences between congruent and incongruent fit were not significant in the symbolic presentation (mean = 0.90, SD = 0.14, mean = 0.9, SD = 0.13 for congruent and incongruent respectively, $t(41) = -.18$, $p = .86$). However, in the non-symbolic presentation, fit for incongruent presentation (mean = 0.83, SD = 0.23) was better than for congruent presentation (mean = 0.71, SD = 0.39), with $t(41) = -2.4$, $p < .05$.

The effect of age interacted with presentation [$F(1, 39) = 7.6$, partial $\eta^2 = .16$, $p < .01$] and direction [$F(1, 39) = 4.43$, partial $\eta^2 = .10$, $p < .05$]. The interaction between age and congruency was also significant [$F(1, 50) = 6.10$, partial $\eta^2 = .1$, $p < .05$]. The correlation between age and congruent (left to right) direction was significant $r(40) = .40$, $p = .009$. Fit was better as age increased. However, there were no significant correlations for the incongruent (right to left) direction $r(40) = .10$, $p = .52$. Similarly, the correlation between age and non-symbolic direction was significant, $r(40) = .41$, $p = .007$. Fit was better as age increased. However, there were no significant correlations for the symbolic representation r (40) = $-.09$, $p = .56$. Please note that the effects of function and interaction with function were not significant (see Fig. 1).

6.3.1.3 Percentage of absolute error

Children's estimation accuracy was computed as percentage of absolute error (PE). This was calculated with the following equation (Siegler & Booth, 2004):

$$PE = \frac{estimate - target\ number}{scale\ of\ estimates} \times 100$$

A two-way analysis of covariance (ANCOVA) with mean PE was computed, with presentation type (symbolic/non-symbolic), direction (congruent/incongruent), and the age of the children serving as a covariant. The effect of direction reached significance [$F(41) = 7.39$, partial $\eta^2 = .15$, $p < .05$]. PE were larger

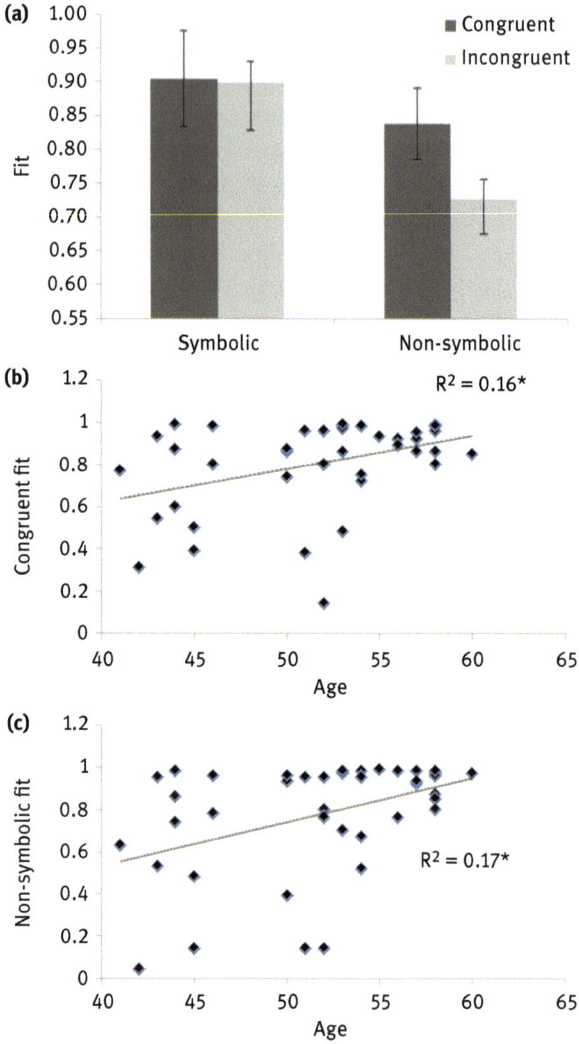

Fig. 1: (a) Function fit as a function of presentation (symbolic or non-symbolic) and direction (congruent from left to right or incongruent from right to left). In the symbolic representation the fit was very large regardless of direction. However, in the non-symbolic representation, fit was larger for incongruent representation than for congruent representation. Correlations between age and function fit in the congruent condition left panel (b) and non-symbolic representation right panel (c). Both are positively correlated, as age increases the fit is better both in the congruent condition and in the non-symbolic presentation. The correlations between incongruent direction and symbolic representation and age were small and non-significant.

in the congruent direction (left to right) (mean = 0.56, SD = 2.04) than the incongruent direction (right to left) (mean = −0.34, SD 2.43). This effect was modulated by age [$F(1, 41) = 6.27$, partial $\eta^2 = .13$, $p < .05$]. Age correlated with incongruent direction (right to left) errors, r (41) = .37, p = .015, but not with congruent direction (left to right), r (41) = −.063, p = .69.

The interaction between symbolic presentation and age was marginally significant [$F(1, 41) = 3.47$, partial $\eta^2 = .08$, $p = .07$] (see Fig. 2).

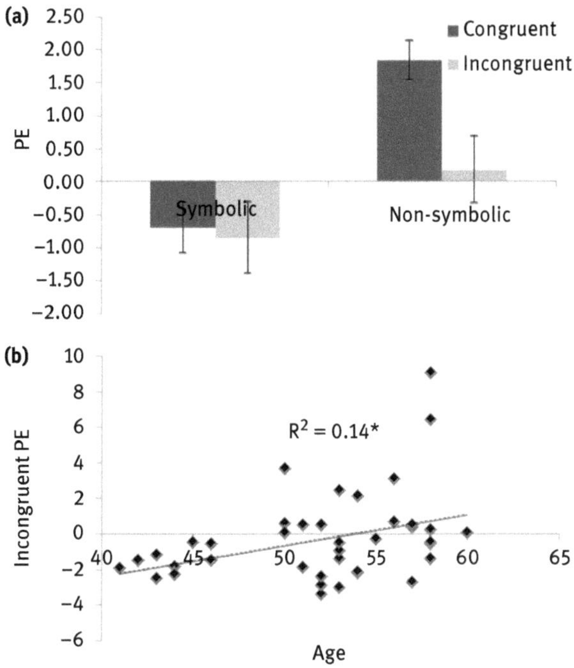

Fig. 2: Children's estimation accuracy was computed as percentage of absolute error (PE). (a) PE as a function of presentation (symbolic or non-symbolic) and direction (congruent from left to right or incongruent from right to left). PE were larger in the congruent direction than the incongruent direction. (b) Age correlated with incongruent direction errors, but did not correlate with congruent direction errors.

7 Discussion

The present study examined a number line estimation task, in a young group of Hebrew-speaking children (starting from 4 years old). Due to the young age of the children, we used a familiar range for that age: 1–9. Two conditions were manipulated: (1) presentation: symbolic or non-symbolic (Arabic numerals or dots), and (2) directionality: congruent to the mental number line (from left to right), or opposite to the classical number line (from right to left).

We first examined general numerical abilities, as a prerequisite for participation of children in our sample. We found that the entire group had basic understanding of quantities: they were able to count freely, and all of them could count objects and understood mathematical concepts. This finding confirmed our assumption that for all the children in our group, 1–9 was quite a familiar range. However, in relation to the number line estimation task, 23% of the sample had no number line representation (linear, logarithmic, 1 cycle, or 2 cycle) in that familiar range, reflecting the dissociation of formal number knowledge and number space associations in young children.

We calculated functions fit (linear, logarithmic, 1 cycle or 2 cycle) on the individual level. We did not discover any preference (in terms of explained variability) for a specific function or interaction with specific function and the other variables. This finding may possibly be due to the small number of data points to estimate function fits in the present study. Hence, we will not refer to differences between functions fits from now on.

Two variables explained estimations. First, directionality: in the condition of number line orientation from right to left, we found better fits and lower estimation error rates than for orientation of the number line from left to right. Fit of left to right orientation improved with age. There was no such correlation with right to left orientation. In what follows, we suggest that Hebrew-speaking children possess a cultural bias for number space associations. Second, presentation affected estimation. Symbolic presentation had better fits than non-symbolic presentation. Estimations for non-symbolic presentation improved with age. There was no such correlation with symbolic presentation. In what follows, we suggest that young children possess two separate systems for processing of exact symbolic information and approximate non-symbolic information.

7.1 Dissociation of formal number knowledge and number space associations, in young children

Most studies on number line estimation tested children beginning from 5 years old (Siegler & Booth, 2004). To the best of our knowledge, only one other study tested young children (Berteletti et al., 2010), as in the present study. Similar to the present findings, 27% of their sample had no representation of small range (1–10). The present study has shown, for the first time, that children with formal knowledge of numbers could lack number space associations. However, formal knowledge of (1) order relations and (2) adding and subtracting one rules was not fully developed in our sample, hinting at possible relations between these formal number line abilities and number space associations.

7.2 Directionality of number space associations in Hebrew-speaking children

In the present study, Hebrew-speaking children had better and more accurate fit when the number line orientation was from right to left, compared to the left to right orientation. It has been previously found that Hebrew-speaking adult participants differ from English-speaking participants in their number space associations (Shaki et al., 2009). While English-speaking participants associate small numbers with the left-side space and large number with the right-side space, Hebrew-speaking participants showed reversed associations or no associations (but see Feldman et al., 2019; Zohar-Shai et al., 2017).

This cultural directionality was associated with reading habits (Hebrew words are oriented from right to left). Specifically, the associations between number and space originate from directional counting preference (Göbel et al., 2018). Initially, there is a strong relation between reading direction and directional counting preference (i.e., most of the Hebrew-speaking participants start counting from the right, while most of the English-speaking participants start counting from the left), and later, directional counting preference determines the direction of number space associations (Göbel et al., 2011).

Please note, however, that we tested young children (as young as 4 years old) before formal reading education. How can cultural orientations emerge in young children? (Nuerk et al., 2015). According to Shaki et al. (2012), counting biases exist before reading acquisition in preschoolers, and were found to be modified by early reading experience (Shaki et al., 2012). About 72.9% of the Hebrew-speaking preschoolers started counting from right to left. Importantly, the rates dropped significantly to 55.8% of the Hebrew-speaking children attending school (Shaki et al., 2012). A common explanation for acquiring cultural counting bias, prior to school, is children's observation of adult's actions. Children watch adults reading stories to them, and observe them pointing at words and turning pages (Göbel et al., 2018). Moreover, it has been found that adults present spatial structures from left to right or from right to left according to their culture, demonstrating it to infants as young as 2 years old (McCrink et al., 2018).

Cultural bias for number space associations was found in several implicit number space association tasks, such as parity judgment (Dehaene et al., 1993). This task does not deliberately require participants to present numbers spatially.

However, there are no indications, so far, of cultural bias for number space associations, in a task that requires participants to represent numbers spatially, such as the number line estimation task. Here, for the first time, we revealed a preference for orientation of the number line from right to left in Hebrew-speaking children.

Please note that similar to the present finding, which demonstrates cultural bias in a number line estimation task in young children, the number line estimation task was found to be sensitive to cultural bias due to language structure of multi-digit number words. In German, most multi-digit numbers are inverted (e.g., 48 → "eight-and-forty"). However, there is no inversion in Italian number words. Accordingly, in the first grade, it was found that Italian children were more accurate in number line estimations than German-speaking children (Helmreich et al., 2011).

7.3 Dissociations between symbolic and non-symbolic processing in young children

The current results demonstrated that estimations of spatial locations of number are more precise when numbers are represented as Arabic numerals rather than as dots. This indicates that in young children, two numerical systems exist next to each other: (1) exact symbolic, (2) approximate non-symbolic. Kolkman et al. (2013) have already tested the effect of symbolic and non-symbolic presentation on estimations in a number line task in young children. They tested it, along with other tasks, longitudinally in 4- to 6-year-old children. Specifically, the authors tested whether the data can be explained best by a three-factor model that includes symbolic, non-symbolic, and mapping factors, or one united model. For 4- to 5-year-olds, the three-actor model best explained the data, while at the age of 6 the united model best explained the data. Accordingly, the authors concluded that the developmental courses of non-symbolic and symbolic skills are all separate at a younger age and integrated by the first grade.

The finding of more precise estimations when numbers are represented as Arabic numerals instead of dots is not in line with the view that number line estimation reflects activation of the mental number line (Dehaene, 1992; Dehaene et al., 2003, 1999; Halberda & Feigenson, 2008; Halberda et al., 2008). According to the triple code model (Dehaene, 1992), the mental number line is a-modal, and representation of quantity on the mental number line is based upon translation of a specific number representation to an abstract quantity. Hence, contrary to the present results, estimations should not be modulated by presentation notation (symbolic or non-symbolic). Please note, however, that contrary to the triple code model that suggests only an abstract representation of quantity (Dehaene, 1992) Cohen Kadosh et al. (2007) suggested that the right parietal lobe possesses a notation-specific representation of symbolic and non-symbolic presentations. This view can shed light on the differences between the precise representation of Arabic numerals and the approximate representation of dots that was discovered in the present study.

Lastly, the triple code model (Dehaene, 1992) as well as the findings of Cohen Kadosh et al. (2007) refers to processing by a developed adult brain, rather than a child's developing brain. Hence, it might be only a developmental phase of two numerical systems for processing of exact symbolic and approximate non-symbolic quantities, that will later be united into one system (Kolkman et al., 2013).

8 Conclusions and limitations

The present study tested number line estimation in young Hebrew-speaking children. Innovatively, we found more precise estimations when the number line was orienting from right to left. This finding is in line with the implicit number space association task that found cultural bias of preferences for right to left directions of the number line in Hebrew-speaking adults. This result in Hebrew-speaking adults is associated with reading habits; accordingly, we assumed that the explanation is parallel with the present study. However, we did not directly compare differences between number space associations between Hebrew-speaking children (Hebrew words are oriented from right to left) and English-speaking children (words are oriented from left to right). Hence, it will be important, in future studies, to examine directly number space associations in the number line estimation task of children from different cultures.

We found more precise estimations when numbers are represented as Arabic numerals rather than dots. Accordingly, we suggested that in young children, two numerical systems exist, one next to the other: (1) exact symbolic, (2) approximate non-symbolic. It would be important to see whether preferences for symbolic representation in number line estimation change developmentally, and whether these two systems integrate into one united system in later development.

References

Barth, Hilary C. & Paladino, Annie M. (2011): The development of numerical estimation: Evidence against a representational shift. *Developmental Science* 14 (1), 125–135. doi:10.1111/j.1467-7687.2010.00962.x.

Berteletti, Illaria, Lucangeli, Daniela, Piazza, Manuela, Dehaene, Stanislas & Zorzi, Marco (2010): Numerical estimation in preschoolers. *Developmental Psychology* 46 (2), 545–551. doi:10.1037/a0017887.

Booth, J. L. & Siegler, R. S. (2006): Developmental and individual differences in pure numerical estimation. *Developmental Psychology* 42 (1), 189.

Cohen Kadosh, Roi, Cohen Kadosh, Kathrin, Kaas, Amanda, Henik, Avishai & Goebel, Rainer (2007): Notation-dependent and-independent representations of numbers in the parietal lobes. *Neuron* 53 (2), 307–314. doi:10.1016/j.neuron.2006.12.025.

de Hevia, Maria D. & Spelke, Elizabeth S. (2010): Number-space mapping in human infants. *Psychological Science* 21 (5), 653–660. doi:10.1177/0956797610366091.

Dehaene, Stanislas (1992): Varieties of numerical abilities. *Cognition* 44 (1–2), 1–42.

Dehaene, Stanislas (2009): Origins of mathematical intuitions: The case of arithmetic. *Annals of the New York Academy of Sciences* 1156, doi:10.1111/j.1749-6632.2009.04469.x.

Dehaene, Stanislas, Bossini, Serge & Giraux, Pascal (1993): The mental representation of parity and number magnitude. *Journal of Experimental Psychology: General* 122 (3), 371–396. doi:10.1037/0096-3445.122.3.371.

Dehaene, Stanislas, Izard, Veronique, Spelke, Elizabeth & Pica, Pierre (2008): Log or linear? Distinct intuitions of the number scale in Western and Amazonian indigene cultures. *Science* 320 (5880), 1217–1220. doi:10.1126/science.1156540.

Dehaene, Stanislas, Piazza, Manuela, Pinel, Philippe & Cohen, Laurent (2003): Three parietal circuits for number processing. *Cog Neuropsychol* 20, doi:10.1080/02643290244000239.

Dehaene, Stanislas, Spelke, Elizabeth, Pinel, Philippe, Stanescu, Ruxandra & Tsivkin, Sanna (1999): Sources of mathematical thinking: behavioral and brain-imaging evidence. *Science* 284 (5416), 970–974.

Feldman, Anat, Oscar-Strom, Yafit, Tzelgov, Joseph & Berger, Andrea (2019): Spatial–numerical association of response code effect as a window to mental representation of magnitude in long-term memory among Hebrew-speaking children. *Journal of Experimental Child Psychology* 181, 102–109.

Göbel, Silke, McCrink, Koleen, Fischer, Martin & Shaki, Samuel (2018): Observation of directional storybook reading influences young children's counting direction. *Journal of Experimental Child Psychology* 166, 49–66. doi:https://doi.org/10.1016/j.jecp.2017.08.001.

Göbel, Silke, Shaki, Samuel & Fischer, Martin (2011): The cultural number line: A review of cultural and linguistic influences on the development of number processing. *Journal of Cross-Cultural Psychology* 42 (4), 543–565.

Halberda, Justin & Feigenson, Lisa (2008): Developmental change in the acuity of the" Number Sense": The Approximate Number System in 3-, 4-, 5-, and 6-year-olds and adults. *Developmental Psychology* 44 (5), 1457.

Halberda, Justin, Mazzocco, Michele & Feigenson, Lisa (2008): Individual differences in non-verbal number acuity correlate with maths achievement. *Nature* 455 (7213), 665.

Helmreich, Iris, Zuber, Julia, Pixner, Silvia, Kaufmann, Liane, Nuerk, Hans-Christoph & Moeller, Korbinian (2011): Language effects on children's nonverbal number line estimations. *Journal of Cross-Cultural Psychology* 42 (4), 598–613.
Hoffmann, Danielle, Hornung, Caroline, Martin, Romain & Schiltz, Christine (2013): Developing number–space associations: SNARC effects using a color discrimination task in 5-year-olds. *Journal of Experimental Child Psychology* 116 (4), 775–791.
Hubbard, Edward, Piazza, Manuela, Pinel, Philippe & Dehaene, Stanislas (2005): Interactions between number and space in parietal cortex. *Nature Reviews. Neuroscience* 6 (6), 435–448. doi:10.1038/nrn1684.
Kolkman, Meijke, Kroesbergen, Evelyn & Leseman, Paul (2013): Early numerical development and the role of non-symbolic and symbolic skills. *Learning and Instruction* 25, 95–103.
Libertus, Melissa, Feigenson, Lisa & Halberda, Justin (2011): Preschool acuity of the approximate number system correlates with school math ability. *Developmental Science* 14 (6), 1292–1300.
Lindemann, Oliver, Alipour, Ahmad & Fischer, Martin (2011): Finger counting habits in middle eastern and western individuals: an online survey. *Journal of Cross-Cultural Psychology* 42 (4), 566–578.
McCrink, Koleen, Caldera, Christina & Shaki, Samuel (2018): The early construction of spatial attention: Culture, space, and gesture in parent–child interactions. *Child Development* 89 (4), 1141–1156.
McCrink, Koleen & Opfer, John E. (2014): Development of spatial-numerical associations. *Current Directions in Psychological Science* 23 (6), 439–445.
Nuerk, Hans Christoph, Patro, Katarzyna, Cress, Ulrike, Schild, Ulrike, Friedrich, Claudia & Göbel, Silke (2015): How space-number associations may be created in preliterate children: Six distinct mechanisms. *Frontiers in Psychology* 6, 215.
Patro, Katarzyna & Haman, Maciej (2012): The spatial–numerical congruity effect in preschoolers. *Journal of Experimental Child Psychology* 111 (3), 534–542.
Shaki, Samuel, Fischer, Martin & Petrusic, William (2009): Reading habits for both words and numbers contribute to the SNARC effect. *Psychonomic Bulletin & Review* 16 (2), 328–331.
Shaki, Samuel, Fischer, Martin H. & Göbel, Silke (2012): Direction counts: A comparative study of spatially directional counting biases in cultures with different reading directions. *Journal of Experimental Child Psychology* 112 (2), 275–281. doi:https://doi.org/10.1016/j.jecp.2011.12.005.
Siegler, Robert S. & Booth, Julie L. (2004): Development of numerical estimation in young children. *Child Development* 75 (2), 428–444. doi:10.1111/j.1467-8624.2004.00684.x.
Siegler, Robert S. & Opfer, John E. (2003): The development of numerical estimation: Evidence for multiple representations of numerical quantity. *Psychological Science* 14 (3), 237–243.
Slusser, Emily & Barth, Hilary (2017): Intuitive proportion judgment in number-line estimation: Converging evidence from multiple tasks. *Journal of Experimental child Psychology* 162, 181–198. doi:10.1016/j.jecp.2017.04.010.
Slusser, Emily B., Santiago, Rachel T. & Barth, Hilary C. (2013): Developmental change in numerical estimation. *Journal of Experimental Psychology. General* 142 (1), 193–208. doi:10.1037/a0028560.

Wood, Guilherme, Willmes, Klaus, Nuerk, Hans Christoph & Fischer, Martin (2008): On the cognitive link between space and number: A meta-analysis of the SNARC effect. *Psychology Science*.
Zohar-Shai, Bar, Tzelgov, Joseph, Karni, Avi & Rubinsten, Orly (2017): It does exist! A left-to-right spatial–numerical association of response codes (SNARC) effect among native Hebrew speakers. *Journal of Experimental Psychology: Human Perception and Performance* 43 (4), 719.

Helga Klein

Exact number representations in first and second language

One of the major questions in the field of cognitive psychology is the extent to which our thought is dependent on, or formed by, the language we speak. In the mid-1900s, proponents of the linguistic relativity principle claimed that different languages with distinct grammatical properties and lexicons would have a major impact on the way the native speakers of that language perceived reality. This idea was based on the work of the anthropologists Sapir (1949), and Whorf (1956), and named the "Sapir-Whorf-Hypothesis" by Hoijer (1971). The opposite view is expressed by the theory of cultural universality (Au, 1983), meaning that basic concepts innate to human beings can be found in every culture irrespective of linguistic differences.

The concept of number seems to be a good example for a theory of cultural universality at first sight, as all known cultures have developed at least some number words, and even pre-verbal infants and animals are able to single out the larger of two sets based on the respective number of items. The term "numerosity" was used by Dehaene (1997) for the awareness of quantity. Yet, it is still not clear whether nature has provided us with the concept of exact number or if this is a cultural acquirement based on the acquisition of verbal counting procedures.

This chapter will review evidence supporting the language relativity hypothesis for the instance of exact number representations in a small number range (up to 10); other chapters in this book focus on the linguistic specificities of multi-digit number word systems and other aspects of mathematics Bahnmüller, this volume; Dowker, this volume). Presenting studies from different fields, this chapter will propose that the concept of exact numerosity is based on natural language, and furthermore that linguistic specificities even put constraints on the form of exact numerosity representations. The first focus is on the finding that grammatical properties shape the development of the concepts for one versus two, three, and more. Second, studies that describe a representational change in adults who learn a new number word system (including symbols for numerosities higher than four or five) will be presented. Third, differences in arithmetic fact retrieval in both first and second language will be reviewed. These findings will be discussed in the light of the "access-deficit-hypothesis" regarding developmental dyscalculia, suggesting that children with mathematical difficulties may have a problem in accessing number magnitude from symbols (e.g., presenting with longer response times

https://doi.org/10.1515/9783110661941-010

or less matured reaction time patterns) rather than in processing quantities per se (Rousselle & Noël 2007). The chapter concludes with the proposal that using canonical finger configurations provide a language-independent means of symbolizing exact quantities for children presenting with problems in this area.

1 The role of language for the ontogenetic development of exact numerosity concepts in early childhood

The mapping of number words to their corresponding numerosities develops gradually: First, children understand the cardinality meaning of the number word "one," then they differentiate "two" from other number words, next comes "three," and at some point the relationship between counting and numerosity is understood and the cardinality principle generalizes to all number words in the counting sequence. Wynn (1990, 1992) named these levels 0-knowers, 1-knowers, 2-knowers, 3-knowers, and cardinality-principle-knowers. According to Wynn (1990), children acquire the last stage at an approximate age of 3.5 years and use counting to refer the numerosities of sets only from this stage onward. She argued that mapping number words and their respective numerosities is a kind of associative learning process which is possible because children already know that number words correspond to specific, unique, and inherently ordered numerosities, even before they map each word to its numerosity (Wynn, 1992). The reason why children learn to map smaller number words to their numerosity first would be the higher frequency of smaller numbers in the children's environment (Wynn, 1990). However, there may also be language-related aspects at play. The scientific controversy regarding the role of language for the first developmental steps toward an exact number representation is reviewed in the next sections.

1.1 The language-irrelevant hypothesis

According to the language-irrelevant hypothesis, the dominant role in acquiring the concept of cardinality should lie in recognizing the equivalence of numerosities of sets via one-to-one correspondence. As first prominent proponents of this view, Gelman and Gallistel (1978) stated that preschool children already understand the principles underlying counting due to their correspondence to this pre-verbal numerosity coding mechanism, and young children's failure to execute the counting process successfully would occur only because of high performance demands (Greeno et al., 1984; but see Le Corre et al., 2006 for opposite results). According to this view, mapping number words and their respective numerosities is just a kind of associative learning. The reason why children learn to map smaller number words first should be the higher frequency of smaller numbers in the children's environment. Butterworth (2010) also postulated that pre-counting children and animals possess an ordered sequence of numerosity concepts (numerosity coding). Furthermore, in this pre-verbal numerosity coding it is assumed that the individual is able to establish numerical equivalence of two sets through one-to-one correspondence. He emphasized this view with the following statement: "The concept of fiveness pre-exists acquisition of the knowledge that the word five refers to the numerosity fiveness" (p. 538).

1.2 The strong language hypothesis

The perspective opposing the language-irrelevant hypothesis is held by the strong language hypothesis (see Brannon & Van de Walle, 2001 for an overview of this controversy). This strong language view is characterized by the assumption that the concepts of exact number and even abstract numerosity can develop only as a consequence of cardinality understanding, which in turn has been shown to result from acquiring the verbal counting process (Wynn, 1990, 1992). For example, Brannon and Van de Walle (2001) showed that 2-year-old children could not make successful ordinality judgments (determining which of two sets contains more items after being asked which one was "the winner") as long as they had not acquired any number word knowledge.

Most researchers in favor of the strong language hypothesis argue that for success in their tasks, children have to explicitly draw their attention to the numerosity of the stimuli. Furthermore, they postulate that number becomes a salient dimension of the environment to which children will consciously attend only after they have developed at least some cardinality understanding (Brannon & van der Walle, 2001; Rousselle et al., 2004).

1.3 The weak language hypothesis

An intermediate view within this debate is represented by the weak language hypothesis (Brannon & Van de Walle, 2001). The most important representative of the weak language hypothesis is the bootstrapping theory (Carey, 2001; Spelke & Tsivkin, 2001; see next paragraph). Similarly to the strong language hypothesis, this theory states that children are not able to represent large exact numerosities before they have acquired the cardinality meaning of numbers (Sarnecka & Carey, 2008).

"Bootstrapping" describes the processes by which children learn the mapping of number words to their corresponding numerosities (Wynn, 1990, 1992) building on the following cognitive prerequisites: (i) the ability to attentively track up to three objects in parallel via representing each item with a separate *symbol* (object file tracking system or OTS), (ii) the ability to learn meaningless ordered word lists (like hickory-dickory-dock or one-two-three-four), and (iii) the ability to understand quantity markers in language such as the singular/plural distinction. The role of quantity markers will specifically be reviewed in Section 1.4.

The most important difference between the strong and the weak language hypotheses is that according to the bootstrapping view, children can represent small numerosities up to three exactly before they have learned how to count.

Evidence for the weak and strong language hypotheses comes from studies showing that in some tasks toddlers don't seem to know that number words refer to specific numerosities (Condry & Spelke, 2008; Sarnecka & Gelman, 2004), which would not be expected by the language-irrelevant hypothesis. More specifically, Sarnecka and Gelman (2004) showed that children without full cardinality understanding did not know that equal sets must have the same number word, although they judged that the application of unmapped number words changes when numerosity changes (by adding or taking away an object of a set). Furthermore, Condry and Spelke (2008) found that young children did not know that a large unmapped number word (like "eight") continues to apply to a set whose members are rearranged or that a specific number word ceases to apply if the set is increased by one, doubled, or halved.

1.4 The role of grammar for the development of cardinality understanding

The bootstrapping theory explicitly states that the ability to understand singular/plural markers is a prerequisite for the mapping of number words to their corresponding numerosities (see above).

Koudier et al. (2006) investigated the developmental sequence of singular/plural distinctions in 20- to 36-month-old infants using a preferential-looking paradigm. In their experiment, children were presented with sets containing either one item or more than one item concurrent to hearing different sentences asking them to look: These sentences marked number either only with noun morphology ("Look at the blicketS"/"Look at the blicket") or redundantly with noun morphology, lexical quantifiers, and verb morphology ("Look, there ARE SOME blicketS"/"Look, there IS A blicket"; Koudier et al., 2006: (1). What they found was that 20-month-old infants were not able to look at the set corresponding to the sentence, 24-month-old children looked at the corresponding set only in the redundant sentence condition, and the 36-month-olds succeeded even in the noun morphology only condition. The authors concluded that "infants first come to understand the semantic force of the singular/plural distinction in the months just before their 2nd birthday" (Koudier et al., 2006: 2).

Similarly, Barner et al. (2007) could show that children up to the age of 20 months failed not only to comprehend singular-plural morpho-syntactic cues in a manual search task, but did not even distinguish sets of 1 from 4, although they were able to do so within the limit of object-based attention (e.g., 1 vs. 3). By the age of 22 months, some children were able to distinguish 1 from 4 in both verbal and nonverbal trials. Most interestingly, success in the manual search task in the age group of 22–24 months was due to the children who were beginning to produce plural nouns according to parental reports, indicating that the linguistic and conceptual abilities became available at around the same time.

These studies confirm that a specific grammatical aspect, namely, the number of different plural markers used within a sentence, and the individual ability to understand and use plural markers, play a role in the development of exact number representations. Interestingly, this effect can be found among children of different ages within one language community, and also in cross-language comparison studies.

Languages differ in the number of plural markers, with some using hardly any – like Japanese. Sarnecka et al. (2007) compared children learning English, Russian (which uses plural markers), and Japanese within the age range of 2-1/2 to 3-1/2 years to match the one from Wynn's (1990) study. Children were asked to complete counting tasks and a task where children are required to give a specific amount when asked, also known as a Give-N task. Their main finding was that more English and Russian learners than Japanese learners were 1-knowers, 2-knowers, 3-knowers, and cardinality-principle-knowers. Their interpretation was that the learning of the exact number concepts of one, two, and three is rather supported by the conceptual framework of grammatical number than implicit understanding of integers.

These developmental studies (Koudier et al., 2006; Barner et al., 2007; Sarnecka et al., 2007) strongly support the notion of linguistic constraints on the development of numerical concepts. On the other hand, the object file tracking system, which allows us and even pre- and nonverbal groups to individuate up to three in most children or four in most adults objects in parallel, seems to have posed constraints on the development of grammatical aspects regarding number morphology. Franzon et al. (2019) compared the number morphology of 218 different languages. The possibilities of morphological expressions for number values included few (paucal), two (dual), three (trial), and possibly sometimes even four (quadral) options. They concluded that nonverbal numerical cognition, specifically the object tracking system, constitutes a core part of language when it comes to numerical expressions. These two notions may seem contradictory at first sight, posing a "chicken-and-egg-problem": How can numerical concepts, which seemingly rely on language properties for their individual development, have constrained the cultural development of the same language properties? However, one should take into consideration that the mechanisms underlying ontogenetic and phylogenetic (including cultural) development need not be identical.

The next section will focus on observations obtained in adults whose first language contains no words for numerosities larger than four or five, investigations of their restricted numerical concepts, and the effect of learning a more elaborate number word system in a second language. These studies can be seen as providing crucial evidence for the language dependency of number concepts, even if the results of the developmental studies presented so far may not be conclusive.

2 The impact of number word systems on exact number representations

Strong evidence for language determining thought at least in one instance comes from observations in two independent Amazonian cultures with very restricted counting systems, namely, the Pirahã (Gordon, 2004) and the Mundurukú (Dehaene et al., 2008; Pica et al., 2004), and from deaf individuals who do not have access to a usable model for spoken or signed language, but live in a numerate culture (Spaepen et al., 2010).

Members of the Pirahã tribe use only number words for one, two, and many (Gordon, 2004). In informal observations they show no recursive use of their restricted count system, which means that they do not combine these words to depict larger quantities. They use fingers to supplement oral enumeration, but inaccurately even for numbers smaller than five. Furthermore, the word for one is sometimes also used to denote small quantities such as two or three, appearing to mean "roughly one." More formal examinations revealed that their restricted number word system limits their ability to exactly enumerate sets exceeding two or three items. For tasks requiring additional cognitive processing, their performance deteriorated even in this small number range. Gordon's (2004) interpretation of these results was that the Pirahã used magnitude estimation to solve the tasks, meaning that they did not use counting even in the smallest number range. He concluded that humans who are not exposed to a number word system cannot represent exact quantities even for medium-sized sets of four or five.

The Mundurukú lack number words for quantities beyond five, but are able to compare and add large quantities far larger than their naming range (Pica et al., 2004). However, they fail in exact arithmetic tasks using numbers larger than four or five. The authors come to the same conclusion as Gordon (2004), namely, that language plays a special role in the emergence of exact number representations during development. They also state that the availability of number names may not be sufficient to promote a mental representation of exact number, but that the crucial factor would be the existence of a counting routine – which the Mundurukú do not have.

Even stronger evidence for the role of counting for the development of exact numerical representations comes from observations of deaf adults living in a numerate society, leading to the same conclusions as the two studies described above. The deaf individuals investigated by Spaepen and colleagues (2010) live in Nicaragua. Despite having no language model at hand, they have access to other aspects of culture which may foster the development of number concepts, for example, observing other people's hand gestures. These deaf adults have developed their own gestures for communication, called homesigns, which they also use to communicate about number. The four observed adult men showed no congenital cognitive deficits, performed as well as hearing siblings and friends on mental rotation tasks, held jobs, made money, and interacted socially with hearing friends and relatives. However, the homesigners did not consistently extend the correct number of finger for set sizes larger than three. Furthermore, they did not always correctly match the number of items in one set to a target set containing more than three items. Overall, the homesigners failed to appreciate the one-to-one correspondence guaranteeing numerical equivalence. The authors concluded that individuals who lack input from a conventional language structure, do not spontaneously develop representations of exact numerosities above the subitizing range (three or four items, which can be individualized in parallel due to the object tracking system). They speculated that although the homesigners at least partially mastered the monetary system (identifying currency and rating relative values), their numerical homesigns are not embedded in accounting routine. Thus, the homesigners lacked summary symbols for each integer representing the cardinal value and the principle of the successor function (that each natural number n has a successor that is exactly $n + 1$).

However, the question remains what happens if adult humans without a counting routine, without number words for numerosities beyond five and thus without concepts for exact quantities larger than that are exposed to a more elaborate number word system. If language really is the constituting factor for exact numerical representations, experience with a counting system should be sufficient for that even in adults.

Dehaene and colleagues (2008) tested 33 Mundurukú children and adults of varying exposure to formal education and thus with varied exposure to the Portuguese number word system. Using number line tasks in different modalities (sets of dots, sequences of tones, Mundurukú number word [composites], and Portuguese number words for bilinguals), they could show that mere exposure to the Portuguese counting system was not enough to evoke an exact (linear) magnitude representation on a number line task. Only the individuals with the highest number of years of formal education – who had experience with addition and subtraction procedures – were able to provide with exact responses for Portuguese number words. The authors concluded that experience with arithmetic and measurement yields the intuition that all consecutive numbers are separated by the same interval + 1. However, even the most educated Mundurukú participants of this study did not extend this principle to the Mundurukú number words, speculatively because this cultural device "does not emphasize measurement or invariance by addition and subtraction as defining features of number, contrary to Western numerical systems" (Deheane et al., 2008: 1219).

So far, it is safe to conclude that there is a very strong argument that the development of exact numerical representations is dependent on language. For the smallest number range, developmental studies imply a crucial supporting role of morphological number markers (most importantly, plural markers). For numbers beyond the subitizing range, a counting routine seems to be necessary to develop a concept of the successor function of numbers, which in turn allows for exact number representations. However, the presented evidence from bilingual Mundurukú speakers implies that this may not be the whole story and that experience with adding and subtracting could play an important role as well. Hence, the next section will explore the relation between language and arithmetic.

3 Arithmetic fact retrieval in first and second language

Since the 1990s more and more scientific evidence points to the language-sensitivity – if not language dependency – of arithmetic fact retrieval (Bernardo, 2001). Interestingly, one of the first experimental studies showed that a first-language advantage in bilinguals is usually only found for arithmetic (response times and accuracy), but not for the manipulation of number words in general (Frenck-Mestre & Vaid, 1993). Arithmetic facts were usually highly overlearned and automatized only in the first language (L1) and showed weaker associations between problems and answers in the second language (L2).

3.1 Language-sensitivity of arithmetic fact retrieval depends on language of instruction

However, it is important to note that this language-sensitivity is not constituted by an advantage of L1 in general, but that the crucial factor is the language of first arithmetic instruction and/or the language of training (Bernardo, 2001; see also Saalbach et al., 2013). Bernardo (2001) noted that this effect is not consistent with abstract, format-independent number fact representations. However, it fits into the encoding complex model by Campbell and Clark (1988). Saalbach and colleagues (2013) pointed out that it is also in line with the triple-code model by Dehaene and Cohen (1997), which assumes an abstract, format-independent representation of numerical quantity, but not of arithmetic operations.

3.2 Language effects occur in trained exact, but not in trained approximate, arithmetic operations

Spelke and Tsivkin (2001) observed the same training-language effects for arithmetic operation in adult bilinguals, but added an important new specification. Notably, they found language switching costs (slower responses for problems presented in the untrained language) for trained exact arithmetic operations (like repeatedly adding 54 or 63 to a given number, addition in base 6 or base 8), but not for trained approximate arithmetic operations (like approximation of cube roots, approximation of logarithmic bases). They proposed that natural language contributes to the representation of large, exact numbers but not to our biologically inherited approximate number representations and offered a possible explanation for this special role of language within the domain of numerical representations. The authors argued that language was the first evolutionary developed representational system allowing for the combination of input from different modalities (e.g. spatial layouts and smell). They further elaborated that the two known biological (and thus language-independent) core systems for exact number representations, namely, an approximate system for large numerosities and a small number system (see above), both fail to represent all the complementary aspects of number. The "large approximate system fails to represent each member of a set as a persisting individual," whereas the "small number system fails to represent a group of individuals explicitly as a set" (Spelke & Tsivkin, 2001: 82). They suggested that the natural counting system may allow humans to combine these two distinct types of representation into a representation of sets of individuals whose cardinality increases as new individuals are added to the set. As language was necessary to link the two limited systems, the new hybrid system – which captures the benefits and overcomes the limits of each language-independent system – depends on language.

In summary, Spelke and Tsivkin (2001) suggest that the symbolic nature of the verbal counting system brings together the distinct features of small exact numerosity processing and large number approximation. This theory provides an explanation for all of the phenomena described so far in this chapter. However, at least for the context of bilingual or multilingual instruction, the question remains what happens if individuals have been overlearning arithmetic facts in a second language for years.

3.3 Developmental change of language effects for arithmetic fact retrieval

Van Rinsveld et al. (2015) investigated this phenomenon in Luxembourg, where the German-speaking individuals usually attend the first six years of primary school with instruction in their mother tongue, but then switch to French-language secondary education. Implementing a cross-sectional study design using five different age groups from the seventh grade to adulthood, they could show that simple addition fact retrieval was almost equally accurate after extended practice in French, but still faster in German. For addition including a carry procedure, even the young adults were faster and more accurate in German, although the performance differences between the two languages diminished somewhat.

So in the case of highly efficient bilinguals who can solve simple arithmetic facts almost equally well in two languages, the question remains whether they will build new associations within the numerical fact networks in the second language, or become more and more efficient in translating from one language into the other (Lin et al., 2012; Wang et al., 2007), or will use a different strategy altogether?

3.4 Which strategy is used for arithmetic fact retrieval in L2?

The possible relations between arithmetic fact representations in different languages and strategies used for fact retrieval in L2 have been investigated using functional magnetic resonance imaging (fMRI) studies (Van Rinsveld et al., 2017) and EEG studies using event-related potentials (ERP: e.g., Salillas & Wicha, 2012). fMRI studies measure differences in oxygenated and de-oxygenated blood flow in the brain and thus allow for three-dimensional localizations of active regions in the functioning human brain with high spatial, but low temporal resolution (Windhorst & Johannson, 2013). Event-related potentials use EEG to measure electrical brain activity following specific "events" and have a very high temporal, but low spatial, resolution (Windhorst & Johannson, 2013).

These studies foster the conclusion that arithmetic in L2 is mediated by the Arabic digit code and not the verbal code of L1 (as expected if participants would translate the problems into L1). Specifically, van Rinsveld and colleagues (2017: 27) asked participants to solve visually and auditively presented arithmetic problems in L1 and L2 in the scanner and observed additional brain activation in the auditive task in the L2 condition compared to the other conditions in occipito-temporal areas (known to underlie the processing of Arabic digits) and in the precuneus area (important for visual imagery). They concluded that arithmetic in L2 may be performed "with the help of a mental visual support, such as via imagining the heard numbers in their visual symbolic form."

Summing up, both the development of large, exact number concepts and the mental manipulation of exact quantities (arithmetic) seem to be language-dependent. Furthermore, diminishing the costs of mental arithmetic in L2 (or more specifically in a language different from the one used in first mathematical instruction at school) appears to rely on switching to another representational format altogether, namely, to the visual symbolic code of Arabic digits. As the next section will show, this might generate even more problems for bilingual learners who already have mathematical difficulties.

3.5 The role of symbolic number processing for arithmetic

In 2007, Rouselle, and Noël presented a theory claiming that a core deficit in children with developmental dyscalculia may not lie in their approximate number system and thus in processing numerosity per se, but in their ability to access number magnitude from symbols. This access deficit hypothesis has triggered extensive research tapping the differential roles of symbolic and non-symbolic numerical magnitude processing skills. A review of these studies concluded that there is indeed consistent and robust evidence across studies and different age groups for a correlation between weak performance in symbolic numerical tasks (using Arabic digits) and low math achievement, whereas conflicting findings have been reported for respective non-symbolic formats (De Smedt et al., 2013). This stronger relation between math achievement and symbolic vs. non-symbolic number processing means that children with mathematical difficulties will most likely be less efficient in accessing the meaning of Arabic digits – namely their exact numerical value – compared to their peers. An even more pronounced problem will hence arise if they have to handle arithmetic tasks in multiple languages. This situation may be relevant for bilingual children whose caregivers cannot provide them with help in mathematics in the instructional language, or for children whose instructional language in mathematics changes during their educational years. As pointed out above, the most efficient strategy in such situations seems to be mentally switching to the Arabic code rather than translating problems into L1 – which, however, may not be possible to the same extent for children with difficulties in mathematics.

The question is then: Could children with problems in accessing the exact quantity of Arabic digits be provided with another set of symbols to bypass language-related difficulties in dealing with numbers?

4 Finger representations as an alternative set of symbols

During the last decade, neurocognitive research produced growing evidence that finger counting results in finger-based numerical representations in the sense of embodied cognition (Fischer & Brugger, 2011) with a sub-base 5 (Domahs et al., 2008) which can still be found in healthy adults (Domahs et al., 2010). Several studies also showed that finger gnosis, the ability to differentiate between the fingers (Gerstmann, 1940), predicts calculation abilities even if general cognitive and motor abilities are controlled for (see Berteletti & Booth, 2015 for a review).

The reasons for these connections between finger representations and number representations are still unclear. Some authors suggest that there are anatomically neighboring brain areas for finger- and number processing; others also suggest a functional connection (see Berteletti & Booth, 2015 for an overview of this discussion). Finger-based numerical representations can be seen as a prototypical example of embodied cognition, hence the question arises which aspect of finger counting may be crucial for this: the one-to-one correspondence of fingers and cardinal value, the additional somatosensory route of perception, or the automatic processing of canonical finger patterns (Brissiaud, 1992).

Priming studies in adults showed that canonical finger patterns for showing numbers (e.g., all 5 fingers of one hand and 2 fingers of the other hand to show 7) were processed as exact magnitudes like Arabic numbers, whereas non-canonical finger patterns (e.g., 3 fingers of one hand and 4 fingers of the other hand to show 7) were processed as approximate magnitudes like dot patterns (Di Luca, Lefevre & Pesenti, 2010; Di Luca & Pesenti, 2008). This finding means that canonical finger patterns may have the same symbolic character as Arabic digits, providing a possible alternative route to exact numerical quantity representations. This interpretation was confirmed by Krinzinger et al. (2011) in a developmental fMRI study. Berteletti and Booth (2015: 8) also concluded that educational practices should encourage the use of fingers "as a functional link between numerical quantities and their symbolic representation." Another advantage of this may well be that fingers provide an external support for arithmetic problems, decreasing the working memory load and therefore increasing the efficiency of mental calculation (Berteletti & Booth, 2015).

5 Summary and discussion

In summary, several different effects of language on the processing of exact number can hardly be explained by a theory of cultural universality (Au, 1983).

First, the number of grammatical plural markers used in a language influences the speed with which children acquire the exact number concepts of one, two, and three (see Section 1.4). Second, adult humans who speak a language lacking words for exact numbers higher than three or four do not present with exact number concepts even in nonverbal tasks (see Section 2). Both lines of evidence defy the view of the language-irrelevant hypothesis, but rather support the claim that the acquisition of concepts of large, exact numerosities (or cardinality) is language-driven and relies on learning respective linguistic representations (or symbols).

Third, the mental manipulation of exact numbers (or arithmetic) is also language-dependent, as arithmetic fact retrieval is usually faster and more accurate in the language of instruction (see Section 3). Interestingly, more efficient arithmetic fact retrieval in a secondary language seems to correspond to a stronger reliance on the visual imagery of Arabic digits and thus to a change in the modality of mental representations rather than on faster translational processes or building a new mnemonic network of arithmetic facts altogether. As the core deficit of children with low math achievement seems to lie in symbolic, exact number processing rather than in non-symbolic, approximate number processing (see Section 3.5), it is safe to assume that doing mental arithmetic in a secondary language should pose even larger difficulties for these children compared to their typically developing peers.

A solution to this problem may lie in the explicit use of canonical finger configurations to depict exact number in formal and informal educational settings, as they seem to possess the same symbolic characteristics as Arabic digits and may thus provide children with an additional developmental route to exact number representations (see Section 4). In conclusion, the automatized processing of canonical finger patterns might enhance symbolic (and therefore exact) processing of numerical magnitudes especially in children presenting with language-related difficulties in mathematics or exhibiting less efficient processing of Arabic digits.

References

Au, Terry Kit-Fong (1983): Chinese and English counterfactuals: The Sapir–Whorf hypothesis revisited. *Cognition* 15, 155–187.

Barner, David, Thalwitz, Dora, Wood, Justin, Yang, Shu-Ju & Carey, Susan (2007): On the relation between the acquisition of singular-plural morpho-syntax and the conceptual distinction between one and more than one. *Developmental Science* 10, 365–373.

Bernardo, Ana B. (2001): Asymmetric activation of number codes in bilinguals: Further evidence for the encoding complex model of number processing. *Memory & Cognition* 29 (7), 968–976.

Berteletti, Ilaria & Booth, James R. (2015): Perceiving fingers in single-digit arithmetic problems. *Frontiers in Psychology* 6, 226. doi:10.3389/fpsyg.2015.00226.

Brannon, Elizabeth M. & Van de Walle, Gretchen A. (2001): The development of ordinal numerical competence in young children. *Cognitive Psychology* 43, 53–81.

Brissiaud, Remi (1992): A toll for number construction: Finger symbol sets. In Bideaud, J., Meljac, C., Fischer, J. P. (Eds.): *Pathways to Number: Children's Developing Numerical Abilities*. Hillsdale, NJ: Lawrence Erlbaum 41–65.

Butterworth, Brian (2010): Foundational numerical capacities and the origins of dyscalculia. *Trends in Cognitive Sciences* 14, 534–541.
Campbell, Jamie I.D. & Clark, James M. (1988): An encoding-complex view of cognitive number processing: Comment on McCloskey, Sokol, andGoodman (1986). *Journal of Experimental Psychology: General* 117, 204–214.
Carey, Susan (2001): Cognitive foundations of arithmetic: Evolution and ontogenesis. *Mind & Language* 16, 37–55.
Condry, Kirsten F. & Spelke, Elizabeth S. (2008): The development of language and abstract concepts: The case of natural numbers. *Journal of Experimental Psychology: General* 137, 22–38.
De Smedt, Bert, Noël, Marie-Pascale, Gilmore, Camilla & Ansari, Daniel (2013): How do symbolic and non-symbolic numerical magnitude processing skills relate to individual differences in children's mathematical skills? A review of evidence from brain and behaviour. *Trends in Neuroscience and Education* 2 (2), 48–55.
Dehaene, Stanislas (1997): *The Number Sense: How the Mind Creates Mathematics*. New York, NY: Oxford University Press.
Dehaene, Stanislas & Cohen, Laurent (1997): Cerebral pathways for calculation: Double dissociation between rote verbal and quantitative knowledge of arithmetic. *Cortex* 33 (2), 219–250. doi:doi.org/10.1016/S0010-9452(08)70002-9.
Dehaene, Stanislas, Izard, Veronique, Spelke, Elizabeth & Pica, Pierre (2008): Log or linear? Distinct intuitions of the number scale in western and amazonian indigene cultures. *Science* 320, 1217. doi:10.1126/science.1156540.
Di Luca, S., Lefèvre, N. & Pesenti, M. (2010): Place and summation coding for canonical and non-canonical finger numeral representations. *Cognition* 117, 95–100.
Di Luca, Samuel & Pesenti, Mauro (2008): Masked priming effect with canonical finger numeral configurations. *Experimental Brain Research* 185, 27–39.
Domahs, Frank, Krinzinger, Helga & Willmes, Klaus (2008): Mind the gap between both hands: Evidence for internal finger-based number representations in children's mental calculation. *Cortex* 44, 359–367.
Domahs, Frank, Moeller, Korbinian, Huber, Stefan, Willmes, Klaus & Nuerk, Hans-Christoph (2010): Embodied numerosity: Implicit hand-based representations influence symbolic number processing across cultures. *Cognition* 116, 251–266.
Fischer, Martin H. & Brugger, Peter (2011): When digits help digits: Spatial-numerical associations point to finger counting as prime example of embodied cognition. *Frontiers In Psychology* 2 (260). doi:10.3389/fpsyg.2011.00260.
Franzon, Francesca, Zanini, Chiara & Rugani, Rosa (2019): Do non-verbal number systems shape grammar? Numerical cognition and number morphology compared. *Mind & Language* 34, 37–58.
Frenck-Mestre, Cheryl & Vaid, Jyotsna (1993): Activation of number facts in bilinguals. *Memory & Cognition* 21 (6), 809–818.
Gelman, Rochel & Gallistel, Charles R. (1978): *The Child's Understanding of Number*. Cambridge, Mass.: Harvard University Press.
Gerstmann, Josef (1940): Syndrome of finger agnosia, disorientation for right and left, agraphia and acalculia. *Archives of Neurology and Psychiatry* 44, 398–408.
Gordon, Peter (2004): Numerical cognition without words: Evidence from Amazonia. *Science* 306, 496–499.

Greeno, James G., Riley, Mary S. & Gelman, Rochel (1984): Conceptual competence and children's counting. *Cognitive Psychology* 16, 94–143.
Hoijer, Harry (1971): The Sapir-Whorf-hypothesis. In Hoyer, Harry (ed.): *Language in Culture. Conference on the Interrelations of Language and Other Aspects of Culture. Seventh Impression.* Chicago: Chicago University Press, 1971, 92–105.
Koudier, Sid, Halberda, Justin, Wood, Justin & Carey, Susan (2006): Acquisition of English number marking: The singular-plural distinction. *Language Learning and Development* 2, 1–25.
Krinzinger, Helga, Koten, Jan Willem, Horoufchin, Houpand, Kohn, Nils, Arndt6, Dominique, Sahr, Katleen, Konrad1, Kerstin & Willmes, Klaus (2011): The role of finger representations and saccades for number processing: An fMRI study in children. *Frontiers in Psychology.* doi:10.3389/fpsyg.2011.00373.
Le Corre, Mathieu, Van de Walle, Gretchen, Brannon, Elizabeth M. & Carey, Susan (2006): Re-visiting the competence/performance debate in the acquisition of the counting principles. *Cognitive Psychology* 52, 130–169.
Lin, J.-F.L., Imada, T. & Kuhl, P.K. (2012): Mental addition in bilinguals: An fMRI study of task-related and performance-related activation. *Cerebral Cortex* 22 (8), 1851–1861.
Pica, Pierre, Lemer, Cathy, Izard, Véronique & Dehaene, Stanislas (2004): Exact and approximate arithmetic in an amazonian indigene group. *Science* 306, 499–503.
Rousselle, Laurence & Noël, Marie-Pascale (2007): Basic numerical skills in children with mathematics learning disabilities: A comparison of symbolic vs non-symbolic number magnitude processing. *Cognition* 102 (3), 361–395.
Rousselle, Laurence, Palmers, Emmanuelle & Noël, Marie-Pascale (2004): Magnitude comparison in preschoolers: What counts? Influence of perceptual variables. *Journal of Experimental Child Psychology* 87, 57–84.
Saalbach, Henrik, Eckstein, Doris, Andri, Nicoletta, Hobi, Reto & Grabner, Roland H. (2013): When language of instruction and language of application differ: Cognitive costs of bilingual mathematics learning. *Learning and Instruction* 26, 36–44.
Salillas, Elena, & Wicha, Nicole Y. (2012): Early learning shapes the memory networks for arithmetic: Evidence from brain potentials in bilinguals. Psychological Science 23(7), 745–55.
Sapir, Edward (1949): In Mandelbaum, D. (Ed.): *Selected Writings of Edward Sapir*. Berkeley, CA: University of California Press. 1–617
Sarnecka, Barbara W. & Carey, Susan (2008): How counting represents number: What children must learn and when they learn it. *Cognition* 108, 662–674.
Sarnecka, Barbara W. & Gelman, Susan A. (2004): Six does not just mean a lot: Preschoolers see number words as specific. *Cognition* 92, 329–352.
Sarnecka, Barbara W., Kamenskaya, Valentina G., Yamana, Yuko, Ogura, Tamiko & Yudovina, Yulia B. (2007): From grammatical number to exact numbers: Early meanings of 'one', 'two', and 'three' in English, Russian, and Japanese. *Cognitive Psychology* 55, 136–168.
Spaepen, Elizabet, Coppola, Marie, Spelke, Elizabeth S., Carey, Susan E. & Goldin-Meadow, Susan (2010). Number without a language model. *Proceedings of the National Academy of Sciences of the United States of America* 108, 3163–3168.
Spelke, Elizabeth.S. & Tsivkin, Sanna (2001): Language and number: A bilingual training study. *Cognition* 78, 45–88.
Van Rinsveld, Amandine, Brunner, Martin, Landerl, Karin, Schiltz, Christine & Ugen, Sonja (2015): The relation between language and arithmetic in bilinguals: Insights from

different stages of language acquisition. *Frontiers in Psychology* 6 (265). doi:10.3389/fpsyg.2015.00265.

Van Rinsveld, Amandine, Dricot, Laurence, Guillaume, Mathieu, Rossion, Bruno & Schiltz, Christine (2017): Mental arithmetic in the bilingual brain: Language matters. *Neuropsychologia* 101, 17–29.

Wang, Yue, Lin, Lotus, Kuhl, Patricia & Hirsch, Joy (2007): Mathematical and linguistic processing differs between native and second languages: An fMRI study. *Brain Imaging Behaviour* 1 (3–4), 68–82.

Whorf, Benjamin Lee (1956): In Carroll, John B. (Ed.): *Language, Thought, and Reality: Selected Writings of Benjamin Lee Whorf*. Cambridge, MA: MIT Press. 1–294

Windhorst, Uwe & Johannson, Hakan (2013): *Modern Techniques in Neuroscience Research*. Berlin, Heidelberg: Springer.

Wynn, Karen (1990): Children's understanding of counting. *Cognition* 36, 155–193.

Wynn, Karen (1992): Children's acquisition of the number words and the counting system. *Cognitive Psychology* 24, 220–251.

Sophie Martini, Christine Schiltz, Antoine Fischbach, and Sonja Ugen

Identifying math and reading difficulties of multilingual children: Effects of different cut-offs and reference groups

1 Introduction

An increasing number of students speak a language at home that differs from the language of instruction at school (L2 students). It is estimated that roughly half of the children in the world learn to read in a language other than their home language and are taught in an L2 (McBride-Chang, 2004). In this chapter, we use L2 to refer to all other languages children learn or speak that are different from the main language of instruction. Some children, regardless of which language they speak at home, encounter severe problems with reading or math and may have a specific learning disorder (SLD). Identifying students with SLD can be challenging, as it often co-occurs with other disorders (Fletcher et al., 2018). Identifying L2 students with SLD can be even more challenging, because these students' L2 proficiency often develops differently from students who do speak the language of instruction at home (L1 students), and weak L2 language proficiency has to be ruled out as a cause of low achievement on diagnostic tests administered in an L2 (American Psychiatric Association, 2013). Although, diagnostic criteria and normed diagnostic tests for SLD exist, their norms are often based on an overrepresentation of L1 students. Therefore, these norms may not be accurate for L2 students due to their different language development from L1 students. In this chapter we will explain what SLD is and why it is especially challenging to identify L2 students with SLD.

Early diagnosis and identification of SLD are paramount and screeners may help determine which students to refer for further diagnostic testing. In this chapter, we do not aim to identify and label students with a clinical diagnosis, such as SLD as specified in the *DSM-5* (American Psychiatric Association, 2013). Rather, we aim to identify students who are in the lowest achievement groups of large-scale tests and may need to be referred for further diagnostic SLD testing. We will use the terms *learning difficulties* (LDs), *reading difficulties* (RD),

Funding: The present research was supported by a PRIDE grant (n°10921377) of the Luxembourg National Research Fund (FNR).

https://doi.org/10.1515/9783110661941-011

and *math difficulties* (MD) to refer to these groups. Concretely, we explore what would happen if a large-scale math and reading test would be used as a screener for children at risk for MD and RD in a multilingual education setting. We investigate how different cut-off settings, that is, cut-offs at different percentiles and with different language reference groups, impact the profile of students characterized as having LD. Lastly, we discuss the implications of our findings in relation to (diagnostic) testing in general.

1.1 What is SLD?

SLD is listed in the *DSM-5* as a neurodevelopmental disorder (American Psychiatric Association, 2013). There are different subtypes of SLD, namely, with impairment in reading, written expression, or in math. Each subtype can manifest itself in different ways. For example, SLD with impairment in reading may correspond to a diagnosis of impairments in word reading accuracy, reading rate or fluency, and/or reading comprehension. For impairments in math, students may have problems with number sense, memorization of arithmetic facts, accurate or fluent calculation, and/or correct math reasoning. To start a diagnostic process, students' achievement should be substantially lower than expected for their age and this lower achievement should persist for at least six months. An SLD diagnosis can be given only after a child has been tested individually, using appropriate standardized tests. Furthermore, an SLD should not be explained by a low proficiency in the language of instruction (American Psychiatric Association, 2013: 66–67). Impairments in math and in reading are assumed to have equal prevalence rates (e.g., Geary, 1993) of less than 10% (Desoete et al., 2004; Gross-Tsur et al., 1996).

1.2 Identification of SLD

Identifying students with SLD may be complicated in general, but it is even more so for those who do not speak the language of instruction at home while being tested in the language of instruction, as low proficiency in the language of instruction has to be excluded as a potential cause. Students are usually tested in their language of instruction, as that is the language in which they learned how to read/write and calculate. Due to L2 students' different language development compared to L1 students, they may have lower language of instruction proficiency than their L1 peers. Diagnostic tests are rarely normed with L2 students as a separate reference group, which can lead to both over-identification (e.g. Cummins, 1984, cited in Limbos & Geva, 2001) and under-identification (e.g. Limbos & Geva, 2001) of L2 students. On the one hand, L2 students may thus have lower proficiency in the language of instruction than L1 students, and this lower proficiency may be classified as an SLD according to the test norms. On the other hand, if it is assumed that students' difficulties are caused by a low L2 proficiency, difficulties caused by a possible SLD may be missed. The interaction between language of instruction proficiency and a possible SLD is especially important to consider when diagnosing reading disorders. However, math learning is also highly influenced by language, for example, counting and transcoding (Kempert et al., 2019). Hence, language proficiency also impacts the diagnosis of learning disorder in mathematics.

It is assumed that the prevalence of SLD is the same for L2 students as for L1 students (Letts, 2011), yet on large-scale achievement tests L2 students often lag behind their L1 peers. Furthermore, L2 students are often referred to as a homogeneous group, though that may not be accurate and may mask differential characteristics between L2 students in terms of home language types and SES (Jang et al., 2013). Considering differences in home languages and SES is important, because they underlie reading comprehension in an L2 (Geva & Wiener, 2014). Socioeconomic status (SES) is a proxy for the resources students have at their disposal and can comprise, for example, parental education, income, or possessions at home (Lenkeit et al., 2018). SES may be related to math and reading development, as SES affects students' language development (Hammer et al., 2014; Hoff, 2006, 2013), their numerical abilities (Mejias & Schiltz, 2013), and academic achievement (e.g., Pace et al., 2017; Paetsch et al., 2015).

1.3 Cut-offs to screen for SLD

A wide variety of cut-offs are used to screen for SLD. Screening tests are often group-based and identify students with low performance. It is quite common to label students in the lowest achievement group of a large-scale test as (potentially) being at risk for developing SLD. A certain percentile (cut-off) on the frequency distribution of the test scores is defined and students at or below this percentile are flagged as at risk for developing SLD. For instance a cut-off at the 10th percentile means that 10% of the test takers will be labeled at risk for developing SLD or as having learning difficulties. The test score corresponding to the 10th percentile is the cut-off score. However, there is no consensus on which cut-off should be used to classify this lowest-achieving group with difficulties, especially in multilingual populations. In reading studies, the 25th percentile is often used, whereas the 10th percentile is very common in math research. It is generally accepted that cut-offs above the 25th percentile are undesirable, as too many students with average ability would be flagged (Fletcher et al., 2018). Therefore, when examining MD or RD, these two cut-offs are the most obvious ones. Although, the 10th percentile is probably most suitable to screen children for SLD as it appears to be more stable across time (Braeuning et al., 2020). Additionally, less than 10% of the population is estimated to have SLD (Desoete et al., 2004; Gross-Tsur et al., 1996). However, the 25th percentile may still be very useful to find the group of children who need extra support, but who may not have SLD.

Besides which cut-offs are most suitable, there is also no consensus on which *reference group* should be used for norming diagnostic tests. Combined, the cut-off and the reference group used are called "cut-off setting" in this chapter. Theoretically, norms should be based on a reference group that is representative of and comparable to the characteristics of the tested student, at least for those characteristics that may influence students' performance (e.g., age, gender, grade, proficiency in the test language) (American Educational Research Association et al., 2014). In the following section, we will present three different reference groups and describe how their use affects the identification of students with SLD.

(1) Sometimes the whole sample that is available is used as a reference group, without differentiating for students' characteristics such as SES or home language (such as for the *Tempo-Test-Rekenen* (TTR); an arithmetic test) (de Vos, 1992). If this sample is representative of the population, this may lead to L1 students with SLD performing above the cut-offs and L2 students without SLD below the cut-off, due to the difference in language of instruction development.
(2) In countries or regions with one main language of instruction, which is the L1 for most students, tests may be normed based on (an overrepresentation of) students for whom the language of instruction is their L1 (e.g., Leysen et al., 2018). This could lead to norms that identify the expected proportion of L1 students with difficulties, but a larger number of L2 students as their test scores are often lower.
(3) Lastly, it has been suggested that students who speak a language other than the language of instruction should be compared to each other, instead of to the whole sample or their L1 peers (Bedore & Pena, 2008). This is done to try to compare students to other students with similar backgrounds, and thus similar development of the language of instruction development, to each other. Ideally, this would lead to the expected and similar proportions of L1 and L2 students below the cut-offs, as the same proportion of students is expected to have SLD, regardless of language background. This way of norming is, for example, implemented for the *Diagnostischer Rechtschreibtest*-tests (e.g. Müller, 2004a, 2004b) that have norms for German home language students, and for students who speak another language than German at home. This norming

does not take into account the specific L2s of the students, though different students' L2s may affect language of instruction development differently (e.g., Geva & Wiener, 2014).

1.4 The present study

The present chapter focuses on the effect of setting criteria for the screening of learning difficulties by using a complete population dataset. We do not focus on clinical SLD, as the tests used in this chapter were not primarily designed to identify SLD; however, they can be considered as potential screeners to identify children who have math or reading difficulties and may be at risk of developing SLD. These group-based tests can only be considered as global achievement indicators of reading and math. They cannot give precise information on the cognitive processes underlying reading or calculation which would be necessary to diagnose SLD. Most screening tests are group-based (e.g., Mejias et al., 2019; Nosworthy et al., 2013) followed by a more precise, diagnostic, individual follow-up.

Most cut-offs on standardized large-scale tests used to identify potential math and reading difficulties are based on samples that consist of students who speak the language of instruction at home. The present study aims to investigate the impact of different cut-offs and reference groups on the number of children being considered as having reading difficulties (RD) and math difficulties (MD) in relation to third graders' linguistic backgrounds. In this chapter we look at RD and MD separately. Due to space limitations, the analyses for combined RD and MD could not be included in this chapter.

More precisely, we investigated the effect of using different cut-offs on L2 students (of different home language groups) and bilingual students (L1 and another language at home) compared to L1 students (who speak the language of instruction at home) in the multilingual educational setting of Luxembourg. In Luxembourg, there are currently no standardized diagnostic tests that are normed on the country's multilingual population, and that take the trilingual educational system into account. In the Luxembourgish school system, kindergarten is taught in Luxembourgish. From grade 1 onward, students acquire formal literacy skills in German, while learning this language; German functions as the main language of instruction at the same time. The Luxembourgish student population is also very multilingual, as the majority of students speak a language other than Luxembourgish or German at home. The main other languages spoken at home are Portuguese, French, and South Slavic languages (Ministry of National Education, Children and Youth, Department of Statistics and Analysis, 2018). Though this level of multilingualism at a national level may be quite exceptional, other countries (for example Canada) (Martel et al., 2011) are facing an increasingly diverse school population, especially in urban areas (see, for example, the sample from Geva & Yaghoub Zadeh (2006)). We compared different ways of setting cut-offs to identify students with MD and RD. For that reason, we used a dataset that comprises the population and chose the two most common cut-offs (10th and 25th percentiles) and three different reference groups (whole sample, native (L1), and within each home language group).

We aimed to answer the following research question: What is the effect of different cut-off settings on cut-off scores and consequently on the amount and the characteristics of students classified as having MD and RD?

2 Method

2.1 Participants

The data were collected as part of the Luxembourgish National School Monitoring Programme in 2016. The tests in this program are administered to all students enrolled in Luxembourgish state-funded schools that follow the national curriculum. The overall sample included 5367 third-grade students. This is the vast majority of third graders in Luxembourg, as there are few private schools (Ministry of National Education, Children and Youth, Department of Statistics and Analysis, 2018). In this analysis, we excluded students whose math, German reading comprehension (RC), and/or German listening comprehension (not used in this study) scores were missing ($N = 169$) and/or whose sex indication was missing ($N = 87$). The final sample contained 5111 students (49.7% girls).

Students were divided into six home language groups (HLGs), based on which language(s) the student reported to speak with their two primary caretakers, for example, their mother and father (see Tab. 1). These HLGs comprise all students in the Luxembourgish public school system that speak these languages at home and are therefore representative. For four HLGs, both parents mainly speak the same language to their child: Luxembourgish, French, Portuguese, or South Slavic. Students in the South Slavic group speak South Slavic languages of former Yugoslavia with their parents, that is, Bosnian, Croatian, Macedonian, Montenegrin, Serbian, and Serbo-Croatian. For two HLGs, parents mainly speak two different languages to their child: Luxembourgish and French, and Luxembourgish and Portuguese. Luxembourgish and German were grouped together for the establishment of the home language groups, as these languages are linguistically very close (Serva & Petroni, 2008) and are considered L1 students in

Tab. 1: Overview of the cut-off settings, the reference groups, and the number of students per cut-off setting. For the six home language groups (HLGs) the mean HISEI scores and number of students whose HISEI score is missing are listed.

Cut-off settings	Reference groups	N	%	Cut-off percentiles		HISEI			
						Mean	Std. dev.	No. of missings	%Missings in HLG
Population	Whole sample	5111	100	10	25				
HLG	Luxembourgish	1568	31	10	25	53	15	167	11
	French	415	8	10	25	54	15	57	14
	Portuguese	1195	23	10	25	34	12	256	21
	South Slavic	219	4	10	25	37	13	55	25
	Luxembourgish-French	259	5	10	25	54	14	26	10
	Luxembourgish-Portuguese	208	4	10	25	43	14	36	17
Native	Luxembourgish HLG	1568		10	25				

this study. Additionally, German-speaking children (with both parents) constitute only 1.3% (N = 67) of the sample. The six HLGs constitute 75.6% of the sample; the remaining 24.4% of students speak various other languages at home and were excluded from the analyses. These students' language backgrounds are too diverse to be grouped together, but too small to be able to calculate reliable cut-offs for each home language background separately.

Students' SES was based on the professions caretakers reported on the background questionnaire. The profession of the caretaker that ranks highest on the international socioeconomic index of occupational status (HISEI) scale (Ganzeboom et al., 1992) was taken as the SES-indicator. An overview of the mean HISEI scores per HLG can be found in Tab. 1.

2.2 Instruments

Competency Measures. The school monitoring program in grade 3 consisted of three paper-pencil-based competency tests: mathematics (divided over two sessions), German reading comprehension, and German listening comprehension (not reported here). These tests were standardized within cohorts as well as between cohorts of different years. Students' parents also filled out a background questionnaire, on which students' SES scores were based. The competency tests were taken at students' schools. Each test session lasted for 50 min. The items of all competency tests were scaled with a unidimensional Rasch model. The resulting estimates were converted to standardized scores (Lorphelin et al., 2014).

German Reading Comprehension (RC). The German RC test took 50 minutes. In this test, "[closed and half-open] items mainly address two sub-competencies: (a) locating and understanding written information, and (b) interpreting written information and applying reading strategies" (Sonnleitner et al., 2014: 8). For this measure we used the scale as it is used for school monitoring purposes with a reliability of .892.

Math. The math test as used in the school monitoring program comprised "two content domains: (a) numbers and operations, and (b) space and shape. Item development further covers two contexts (applied vs. not applied)" (Sonnleitner et al., 2014: 8). A subset of the math items was used for this study: only items that test students' ability to do calculations were included. The selected math items were as language-free as possible; that is, students could solve them without reading an explanation and did not need to know specific terms in German to solve the tasks. This was done so that students' math achievement is affected by reading comprehension and German-language proficiency as little as possible. Of the 75 items in the math test, 38 items met the criteria for this scale. The final scale contained 35 items, because three items had to be removed due to poor model fit (Fischbach et al., 2014). The reliability of this math scale is .605.

2.3 Groupings

This chapter explores the impact of two different cut-offs (percentiles 10 and 25) for RC and math achievement for three different reference groups. (1) The first reference group is the whole sample; that is, the cut-off scores have been calculated based on the achievement of all students who took part in the test. (2) The second reference group was within the home language groups; that is, the cut-offs were based on the performance within each of the 6 HLGs. (3) The third reference group was the native language group. In an educational system where there is one dominant language that is used as the language of instruction and most students are raised in this language, diagnostic and large-scale tests are often standardized based on a supposedly mostly native, monolingual, majority language group. In this context, the native reference group is the Luxembourgish HLG.

Students who performed below the cut-offs (percentile 10 or 25) were classified as having potential learning difficulties. Due to rounding up or down to whole percentages, and because a number of students may have the same test score, the actual number of students below the cut-off based on the HLGs and the native reference group may be slightly higher or lower than the expected 10 and 25%. An overview of these cut-off settings can be found in Tab. 1.

2.4 Procedure

Data were collected in five testing sessions. The tests were administered by the students' teachers, who had received instruction manuals on how to administer these tests. The national school monitoring program has a proper legal basis and has been approved by the national committee for data protection. Parents were asked to fill out a background questionnaire on, for example, their educational and occupational backgrounds and country of birth. All students and their parents or legal guardians were duly informed before the data collection and had the possibility to opt out. All statistical analyses were performed with anonymized data.

3 Results

In this section we will discuss the differences in prevalence of math and reading difficulties for the six different HLGs. We will first focus on the exact cut-off scores for all six cut-off settings (Section 3.1). In a second step, we will examine the prevalence of MD and RD for the six HLGs (Section 3.2).

3.1 Cut-off scores

Before looking at the cut-offs for MD and RD, and consequently the prevalence of these difficulties for students from the six different HLGs, differences in mean scores between different HLGs are described in Tab. 2. Differences in mean scores are related to cut-off scores and number of students below the cut-offs, depending on the reference group that is chosen. On average, students in the Luxembourgish and Luxembourgish-French HLGs have the highest German RC and math scores. The other four HLGs have lower mean scores.

For the native reference group and the whole sample reference group, at the 10th and 25th percentile cut-offs for both math and German RC, the scores

Tab. 2: The mean scores and cut-off scores per cut-off setting for all reference groups.

Cut-off settings	Reference groups	Mean scores per reference group		Cut-off scores per percentile			
				10		25	
		Math	RC	Math	RC	Math	RC
1) Whole sample	Whole sample	501	501	382	325	433	397
2) HLG	Luxembourgish	522	584	395	397	457	493
	French	504	465	368	325	433	384
	Portuguese	475	415	368	291	421	341
	South Slavic	483	467	354	341	421	384
	Luxembourgish-French	514	541	395	384	445	451
	Luxembourgish- Portuguese	461	446	354	291	408	356
3) Native	Luxembourgish HLG	522	584	395	397	457	493

based on the native reference group are always higher than the whole sample reference group. These differences are bigger for German RC than for math, and for the 25th percentile than for the 10th percentile; see Tab. 2.

For the cut-off scores within the HLGs as reference groups, there is a cut-off score for each HLG. In this case, the proportion of students identified as having difficulties is the same for each HLG at each cut-off, but the cut-off score will differ depending on the distribution of math and RC scores; the cut-off scores differ within each HLG.

Table 2 shows that for the cut-off scores with the HLGs as reference group, the cut-off scores for the Luxembourgish HLG are higher than the cut-off scores for the other home language groups. The only exception is the math 10th percentile cut-off score of the Luxembourgish-French HLG (395), which is equal to the Luxembourgish HLG (395) cut-off. Overall, the differences between the MD cut-offs are smaller than the differences between the RC cut-offs of the six HLGs, and the differences between the 10th percentile cut-offs are smaller than the 25th percentile cut-offs. However, the differences between cut-off scores for the different HLGs can be large: for example 152 points between the Luxembourgish (493) and Portuguese (341) HLGs at the 25th percentile for RD.

For most HLGs, the cut-off score for math is higher than the one for RC at the same percentile. Moreover, some HLGs have the same or similar cut-off scores for one subject, but not for the other. This implies that L2 and language proficiency interact differently, depending on the HLG. For example, the 10th percentile cut-off scores for math are equal for the French and Portuguese HLGs, namely, 368, but different for RC, namely, 325 and 291. Additionally, for the 25th percentile cut-off scores, the Portuguese and South Slavic HLGs' math cut-off scores is 421, but their RC cut-off scores are 43 points apart.

In sum, the six different cut-off settings result in different cut-off scores. As a consequence of these differences in cut-off scores, the prevalence of math and reading difficulties varies per cut-off setting.

3.2 Prevalence of MD and RD

The prevalence of math and reading difficulties varies per cut-off setting. For the cut-off settings with the whole sample and within HLGs as reference groups, respectively 10 or 25% of the students (either for the whole sample or for the six HLGs) are classified as having difficulties. However, for the cut-off settings based on the native reference group, only the proportion of students in the Luxembourgish HLG is fixed at 10 or 25%, but not for the other five HLGs. Therefore, the percentage of students with MD and RD in the other five HLGs may be higher or lower than 10 and 25 percent respectively.

If the students with difficulties were distributed equally over all HLGs, 25 or 10% respectively of students in all home language groups would be classified as having difficulties for the two cut-off settings based on the whole sample and the two settings based on the Luxembourgish home language group. However, as is shown in Fig. 1 (and Appendix A), this is not the case.

For the cut-offs with the HLGs as reference groups, the percentage of students below the cut-offs is fixed for each HLG at 10 or 25 percent respectively. Therefore, the cut-offs based on the HLGs as reference group are not discussed in this section. However, the cut-off scores differ between HLGs, which is discussed in Section 3.1.

3.2.1 Whole sample reference group

25th percentile

For MD, the Luxembourgish and Luxembourgish-French HLGs have fewer students with MD than the expected 25%: 19 and 21; see Fig. 1, a2. The French home language group would have 27% of students with MD, slightly more than expected. The Portuguese, South Slavic, and Luxembourgish-Portuguese HLGs have the most students that are identified with MD: 34, 36, and 40% respectively. For RD

(Fig. 1, a1), the difference in proportion of students per HLG that are classified as having difficulties are larger than for MD. Again, the Luxembourgish and Luxembourgish-French HLGs have fewer students with RD than the expected 25%: only 10 and 13%. The French and South Slavic HLGs have 30 and 31% of students with RD, which is slightly more than the expected 25%. However, the Portuguese and Luxembourgish-Portuguese HLGs have more students with RD than expected, namely, 48 and 37%.

10th percentile

For MD, the Luxembourgish and Luxembourgish-French HLGs have fewer students with MD than the expected 10%: 7 and 6%, as shown in Fig. 1, a2. The French HLG would have 11% of students with MD, slightly more than expected. The Portuguese, South Slavic, and Luxembourgish-Portuguese HLGs have the most students that are identified with MD: 13, 13, and 14% respectively. This pattern is in line with the one found for the 25th percentile whole sample cut-off. For RD, there is a bigger difference in the proportion of students classified as having difficulties than for MD.

Again, the Luxembourgish and Luxembourgish-French HLGs have the smallest proportion of students with RD (Fig. 1, a1), namely, 4 and 5%. The French HLG has 11% students with RD, slightly more than the expected 10%; the South Slavic HLG has 9% students with RD, slightly below the expected 10%. Similar to MD, the HLGs with the highest proportion of students with difficulties are the Portuguese and Luxembourgish-Portuguese HLGs: 21 and 20%.

3.2.2 Native reference group

25th percentile

As this cut-off setting is based on the Luxembourgish HLG, this group has 26% (not 25, due to duplicate scores and rounding up to a whole percentage) of students with MD. The other HLGs have a higher proportion of students with MD. The French and Luxembourgish-French HLGs are closest to the expected 25% of students with MD, namely, 34 and 29%; see Fig. 1, b2. Similar to the 25th percentile cut-off with the whole sample reference group, the Portuguese, South Slavic, and Luxembourgish-Portuguese HLGs have the highest proportions of students with MD: 46, 45, and 52%.

For RD, all other HLGs have more students with RD than the reference group. The Luxembourgish-French HLG has the lowest proportion of students with RD after the Luxembourgish HLG: 38%. Three groups have a similar proportion of

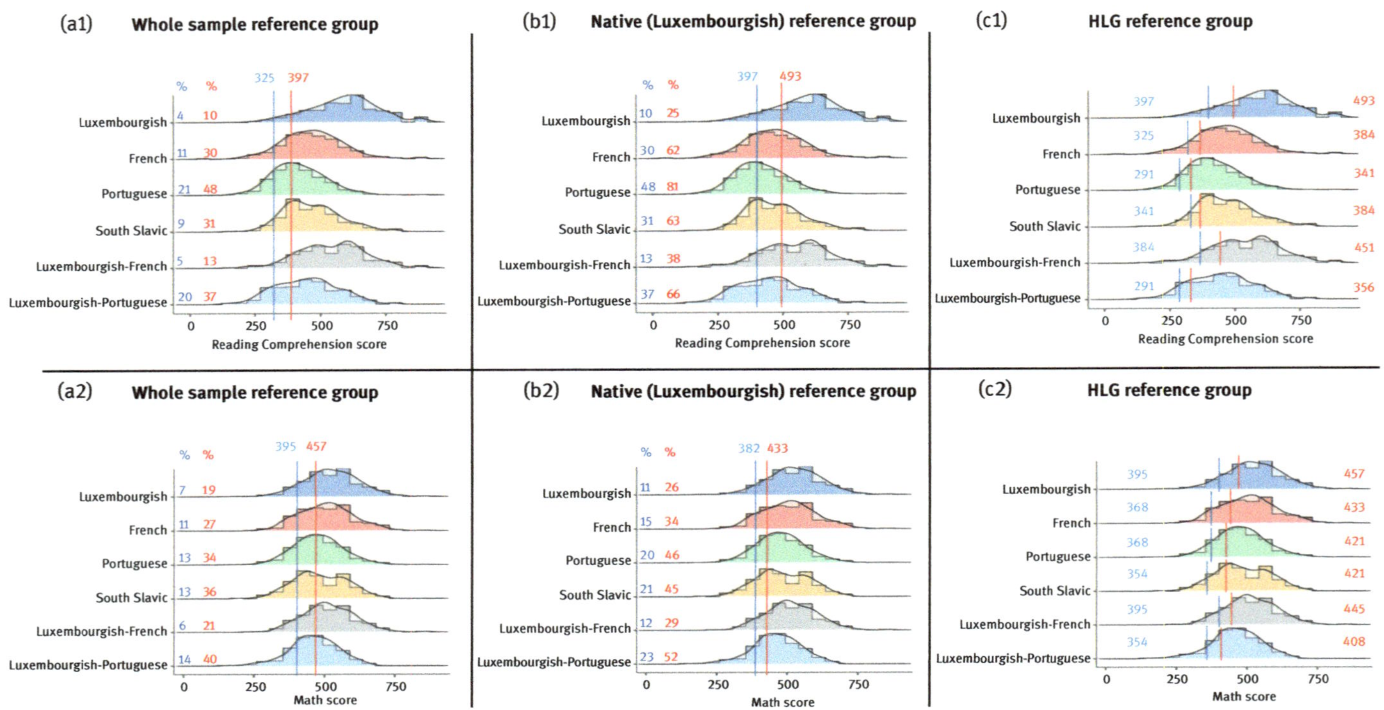

Fig. 1: Cut-off scores for reading comprehension (a1, b1, c1) and math scores (a2, b2, c2) for the whole sample (a1, a2), native (Luxembourgish) (b1, b2), and HLG reference groups (c1, c2). Per reference group, the cut-off score at the 10th percentile is shown in blue and the cut-off score at the 25th percentile in red. For the whole and native language reference groups, percentages of students below the 10th and the 25th percentiles are indicated in blue and red.

students with RD, the French, South Slavic, and Luxembourgish-Portuguese groups: 62, 63, and 66%. The group with most students with RD for this cut-off setting is the Portuguese home language group with 81%.

10th percentile

As this setting is based on the Luxembourgish HLG, 11% of the students in the Luxembourgish HLG are classified as having MD. The Luxembourgish-French HLG have a slightly higher proportion of students with MD, namely, 12%. The group with the next lowest proportion of MD are the French HLG with 15%. The other three HLGs, that is, the Portuguese, South Slavic, and the Luxembourgish-Portuguese, have the highest proportion of students with MD: 20, 21, and 23%. For RD, again the Luxembourgish HLG has the lowest proportion of students with MD, 10%, as expected for this cut-off. The Luxembourgish-French HLG has a similar proportion of students with RD, 13%. The French, South Slavic, and Luxembourgish-Portuguese HLGs have similar prevalence of RD: that is, 30, 31, and 37%. The HLG with most RD is the Portuguese HLG, namely, 48%.

In sum, for the four cut-off settings discussed above (whole sample reference group, both 10th and 25th percentiles, and native reference group, both 10th and 25th percentiles), the Luxembourgish HLG has the smallest proportion of students with both MD and RD; in all other HLGs, MD and RD are more prevalent. Sometimes, difficulties are only slightly more frequent than the Luxembourgish HLG than in other HLGs, for example, MD for the Luxembourgish-French and French HLGs, for most of the four cut-off settings discussed. However, for one of the other HLGs, the proportion of students classified with MD is twice as high as in the Luxembourgish HLG, namely, for the Portuguese HLG. Though the pattern of HLGs with more or less MD and RD is similar for these four cut-offs, the actual proportion of students classified as having difficulties varies, as the exact cut-off score differs (see Tab. 2)

3.3 Consistency of identified students across cut-off settings

It is interesting to examine how many students are classified as having difficulties when using all three reference groups, versus how many are only classified as such for one or two reference groups. This will provide further insights into how the different cut-off settings work for the different HLGs.

3.3.1 RD prevalence across the cut-offs

Across the different cut-off settings, a different pattern emerges for the six HLGs (see Tab. 3). At the 10th percentile, about 10% of students in the French, Portuguese, South Slavic, and Luxembourgish-Portuguese HLGs are consistently below the cut-off for all three reference groups, at the 25th percentile, that is, 25%. The Luxembourgish and Luxembourgish-French HLGs have a lower proportion of students who are below all three cut-offs than the other HLGs. Additionally, these two HLGs are the only ones with students who are only below the HLG and below the native reference group cut-offs. In the French, Portuguese, South Slavic, and Luxembourgish-Portuguese HLGs, there are students below the native and below the whole sample cut-offs; this is not the case for the Luxembourgish and Luxembourgish-French HLGs. Additionally, for these four HLGs many students are only below the native reference group cut-offs: 19, 27, 22, and 17% of students at the 10th percentile and 32, 33, 32, and 29% at the 25th percentile.

3.3.2 MD prevalence across the cut-offs

Similar to RD, a different pattern emerges for the six HLGs across the different cut-off settings. At the 10th and 25th percentiles, about 10 or 25% of students in the French, Portuguese, South Slavic, and Luxembourgish-Portuguese HLGs are below the cut-off for all three reference groups; see Tab. 4. The Luxembourgish and Luxembourgish-French HLGs have a lower proportion of students who are below all three cut-offs. As for RD, these two HLGs are the only groups with students below the HLG and the native reference group cut-offs, but above the whole sample cut-offs. In the French, Portuguese, South Slavic, and Luxembourgish-Portuguese HLGs, there are students below the native and whole sample cut-offs but above the within HLG reference group cut-off; this is not the case for the Luxembourgish and Luxembourgish-French HLG. Additionally, for these four HLGs many students are only below the native reference group cut-offs: 4, 7, 8, and 9% of students at the 10th percentile and 7, 12, 10, and 12% at the 25th percentile.

4 Discussion

In this study we aimed to answer the question: What is the effect of different cut-off settings on the classification of students with MD and RD in a multilingual student population? Concretely, we examined the impact of different language-group

Tab. 3: Percentage (%) of students below one, two, or three reference groups (RG) cut-offs per home language group (HLG) for reading.

		HLG						
		Luxembourgish	French	Portuguese	South Slavic	Luxembourgish-French	Luxembourgish-Portuguese	Overall
10th percentile								
Below three RG cut-offs	%	3.8	9.2	11.0	8.7	4.6	10.6	7.3
Below HLG and below native RG cut-offs	%	6.3	0.0	0.0	0.5	4.2	0.0	2.9
Below whole sample and below native RG cut-offs	%	0.0	1.7	9.9	0.0	0.0	9.1	3.7
Below native RG cut-off only	%	0.0	19.0	27.2	21.9	4.2	16.8	12.9
Not below any cut-off	%	89.9	70.1	51.9	68.9	86.9	63.5	73.2
25th percentile								
Below three RG cut-offs	%	10.1	25.5	26.4	27.9	13.1	27.4	19.0
Below HLG and below native RG cut-offs	%	14.8	0.0	0.0	0.0	12.7	0.0	6.9
Below whole sample and below native RG cut-offs	%	0.0	4.3	21.7	3.2	0.0	9.1	7.8
Below native RG cut-off only	%	0.0	32.0	32.5	32.0	12.4	29.3	17.7
Not below any cut-off	%	75.1	38.1	19.4	37.0	61.8	34.1	48.6

Tab. 4: Percentage (%) of students below one, two, or three reference groups (RG) cut-offs per home language group (HLG) for math.

		Luxembourgish	French	Portuguese	South Slavic	Luxembourgish-French	Luxembourgish-Portuguese	Overall
10th percentile								
Below three RG cut-offs	%	6.6	10.8	9.6	10.5	6.2	9.1	8.3
Below HLG and below native RG cut-offs	%	4.5	0.0	0.0	0.0	3.1	0.0	2.0
Below whole sample and below native RG cut-offs	%	0.0	0.0	2.8	2.3	0.0	4.8	1.3
Below native RG cut-off only	%	0.0	4.3	7.1	8.2	3.1	8.7	3.8
Not below any cut-off	%	88.9	84.8	80.4	79.0	87.6	77.4	84.6
25th percentile								
Below three RG cut-offs	%	19.4	23.1	23.8	23.7	20.5	23.1	21.7
Below HLG and native RG cut-offs	%	7.0	0.0	0.0	0.0	4.6	0.0	3.1
Below whole sample and below native RG cut-offs	%	0.0	3.6	10.5	11.9	0.0	16.8	5.2
Below native RG cut-off only	%	0.0	7.0	11.5	9.6	3.9	12.0	5.8
Not below any cut-off	%	73.7	66.3	54.1	54.8	71.0	48.1	64.2

related criteria of cut-off scores and consequently the number of students below the cut-offs and thus classified as having MD or RD. The associated number and characteristics of students classified as potentially having MD and RD. We used three different cut-off settings, that is, whole sample, HLG, and native reference group, and found that they resulted in different cut-off scores and hence in large variations in the number of students classified as having potential MD and RD.

Generally, the cut-off scores based on the performance of the native reference group were higher than those based on the whole sample. This means that the performance-level below which a student is considered as having MD or RD is higher when natives are used as the reference group compared to the whole sample. As a consequence, more students in general would be identified as having MD and RD with the native group as reference for the cut-off. Within the whole sample and the native reference group, there were differences between the cut-off scores for math and reading comprehension. For the whole sample, the cut-off scores for math were higher than those for reading, whereas for the native reference group the reading cut-off scores were higher than for math. This indicates that the Luxembourgish native reference group had an advantage in the more language-dependent task, that is, reading comprehension, as it was the only group who took the tests in their L1. The comparisons between the cut-off scores within each of the six HLGs showed that there were large differences between cut-off scores of HLGs. Furthermore, for most HLGs the math cut-off score was higher than the RC cut-off score, except for the Luxembourgish HLG, meaning that generally performance-levels under which students were considered as having RD were relatively lower than those for MD. This may be due to the math scale used in this study, as only items with very low language requirements, which did not require students to read and understand a word problem, were included.

As the cut-off scores differed per cut-off setting, the prevalence rate of MD and RD consequently varied between the different groups per cut-off setting too. Moreover, there were large differences in MD and RD prevalence rate between HLGs for the cut-offs based on the whole sample and native reference groups. For these reference groups' cut-offs, the difference between the Luxembourgish and Portuguese HLGs was most pronounced: MD was twice as prevalent in the Portuguese HLG, and RD five times as frequent as in the Luxembourgish HLG.

4.1 Over- and under-identification

In the literature, both over- and under-identification of L2 students with RD and MD have been reported (e.g. Cummins, 1984 cited in Limbos & Geva, 2001). In this study, when the population and native reference groups are used to set cut-offs, more students from the French, Portuguese, and South Slavic HLGs than the expected 10 and 25% are classified as having difficulties, especially for German reading. On the other hand, when the population reference group is used, it is very likely that students with an LD would remain unidentified within the (native) Luxembourgish HLG.

When the HLGs are used as reference groups the proportion of students with difficulties is the same across HLGs, but the actual cut-off scores vastly differ, up to 150 points, which equals 1.5 standard deviation. Overall, only a minority of students is consistently below the 10th or 25th percentile respectively for math or reading for all three reference groups. For the Luxembourgish HLG, the students who are not always classified as having difficulties perform above the population cut-off, but below the native/HLG reference group cut-off scores (that is the same for this HLG). In contrast, for the Portuguese HLG, for example, students who are only sometimes classified as having difficulties are above the HLG cut-off, but below the native or population reference group cut-off scores (or both). Therefore, a combination of cut-off settings while considering students' linguistic backgrounds may be most effective to identify students with LDs. Knowing a student's score in relation to the whole sample or native reference group, as well as in comparison to a group of students with a similar background, may help in deciding what follow-up testing or interventions should be taken.

Two HLGs speak two languages at home, of which one is Luxembourgish, yet their cut-off scores and prevalence of MD and RD differ substantially from each other. For the Luxembourgish-French HLG, the cut-off scores within HLG are close to the Luxembourgish HLG, as well as the proportion and ratio of MD and RD. Both these HLGs have relatively high mean SES. In contrast, the Luxembourgish-Portuguese HLG is more like the Portuguese HLG, in terms of mean SES. It also resembles the Portuguese HLG's cut-off scores based within the HLG reference group, and prevalence and ratio of MD and RD. The Luxembourgish-French HLG thus seems to benefit from speaking Luxembourgish at home, while this is not the case for the Luxembourgish-Portuguese group, who have similar cut-off scores and MD and RD prevalence to the Portuguese HLG. This might be indirectly related to SES.

This chapter did not investigate the influence of SES in combination with HLG on math and reading achievement in detail. Previous research with grade-three data from the Luxembourgish National School Monitoring Programme showed that the differences in math and German RC achievement are for a large part explained by home language and SES: students who speak Luxembourgish or German at home and who have high SES have higher mean achievement for both RC and math compared to students with average or low SES and who do not speak either of these languages at home. National School Monitoring data further have shown that the effect of home language background is bigger for RC than for math and the effect of SES is similar for math and RC (Muller et al., 2015). Furthermore, school monitoring data has shown that that students' German RC in grade 3 was predictive for their German RC in grade 9 (Sonnleitner et al., 2018). However, the predictive values differed for students with high and low SES and between different HLGs. Students with high SES (top quartile) and students who speak Luxembourgish at home were more likely to have high RC and math in later grades. This is in line with the findings of the present study and implies that the students who do not speak Luxembourgish at home and/or come from HLGs with lower mean SES are more likely to fall below cut-offs and have low math and/or reading achievement.

4.2 Screening for difficulties

Screening procedures should identify students that need further diagnostic testing (Fletcher et al., 2018, p. 63). Helping students with a deficit as early as possible usually has the best outcome (Heckman, 2008). For these tests, missing an at-risk child is a bigger problem than having a false positive (i.e., identifying a child that does not have difficulties). Generally, it is indeed less harmful to identify more children than necessary, than missing children who would need further diagnostic testing and help. On the other hand, having too many "false positives" is counterproductive, as valuable resources would be spent on students who may not need it.

Which cut-offs (or a combination of cut-offs) are useful depends on the purpose of the screener: to only select the students who are likely to have a clinical SLD, then the 10th percentile is most suitable. Moreover, at the 10th percentile for the whole sample and HLG reference groups, the prevalence of MD and RD is similar, which is what is expected for clinical SLD in math and reading (e.g. Geary, 1993). However, all students below the 25th percentile might benefit from extra instruction, even though they might not have a clinical SLD. When the whole sample or the native reference groups are used, this may lead to many "false positive" students for SLD; therefore, a comparison with the within HLG cut-off score would be useful to decide what kind of support students need.

In this chapter, we did not discuss the prevalence of MD and RD simultaneously occurring in the same students, due to space limitations. This would be important to consider, as the underlying causes of the learning difficulties may be different in isolated or combined LDs. Therefore, when choosing follow-up tests for a clinical SLD diagnosis or non-clinical LD identification and the following choice of adequate interventions, this should also be considered.

4.3 Limitations and implications

While a large-scale group-based standardized test can serve as a screener for students at risk of having SLD, these tests are neither designed for nor sufficiently fine-grained to detect clinical SLD. Nevertheless, the results of this study using population data to examine the impact of cut-off scores and reference groups to screen for SLD also have an impact on other settings in which tests are used, such as for diagnostic tests. For instance, the choice of the reference group to establish norms and/or the language of the instruction should be considered to make the diagnostic process more accurate and fair, especially for L2 children. For the SLD diagnosis, several individual follow-up testing sessions with a psychological diagnostic specialist are necessary. For RD, for instance, a potential follow-up could be to investigate reading skills at the word-level, as RD can comprise word decoding or recognition and/or spelling deficiency (Fletcher et al., 2018). Typically L2 students do not differ from L1 students on the level of decoding (Limbos & Geva, 2001), so this type of supplementary assessment could help distinguish between a predominantly German proficiency problem and a clinical RD (e.g., Lesaux & Siegel, 2003).

For the purpose of this study on LDs, we refined the math scale in this study, which artificially decreased the large achievement differences between HLGs usually found (e.g. Muller et al., 2015; Sonnleitner et al., 2018). The adjustment of the math scale was to minimize the language required to solve the items on the math test. This minimization of language explains why the differences between students from six different HLGs were smaller for math achievement than for RC, whereas in previous studies with National School Monitoring data and PISA data these differences are larger. Our findings thus imply that using a subset of language-reduced standardized, large-scale test items may be more fine-grained as a screener for MD in a multilingual setting and should be researched further.

In countries in which national standardized tests are implemented, these tests could potentially serve as a screener for students who are at risk of falling behind. As both math and RC in the language of instruction depend on language of instruction proficiency, strengthening students' language of instruction skills may be helpful for students and may help decrease performance differences and associated cut-off score differences between HLGs. Ertel et al. (2019) found that for Portuguese HLG students in Luxembourg, differences in German and Luxembourgish language skills were the only language-related factor that differed significantly between low-achieving students who passed second grade and those who were retained. Differences in Portuguese skills were statistically insignificant. In general, providing students who have low RC and/or math achievement in grade 3 with an extended training in the language of instruction could help improve their reading comprehension. Therefore, the large-scale test used in this study or in similar settings could potentially be used to identify students with difficulties, so that they can receive extra help to catch up as soon as possible, to prevent low long-term achievement.

In this study we found that two different cut-offs, that is, the 10th and 25th percentile and three different reference groups (i.e., whole sample, HLGs, native reference group), resulted in different cut-off scores and consequently different proportions of students identified with MD and RD. Practitioners should be aware of all of the caveats of using standardized tests and their corresponding norms when using them.

Appendix A: Math and Reading difficulties per home language group

	25th percentile									10th percentile							
	Whole sample reference group									Whole sample reference group							
	No MD		MD		No RD		RD			No MD		MD		No RD		RD	
	N	% within HL	N	% within HL	N	% within HL	N	% within HL		N	% within HL	N	% within HL	N	% within HL	N	% within HL
Luxembourgish	1264	81	304	19	1409	90	159	10	Luxembourgish	1465	93	103	7	1508	96	60	4
French	304	73	111	27	291	70	124	30	French	370	89	45	11	370	89	45	11
Portuguese	785	66	410	34	620	52	575	48	Portuguese	1046	88	149	13	945	79	250	21
South Slavic	141	64	78	36	151	69	68	31	South Slavic	191	87	28	13	200	91	19	9
Luxembourgish and French	206	80	53	21	225	87	34	13	Luxembourgish and French	243	94	16	6	247	95	12	5
Luxembourgish and Portuguese	125	60	83	40	132	64	76	37	Luxembourgish and Portuguese	179	86	29	14	167	80	41	20

	Native reference group									Native reference group							
	No MD		MD		No RD		RD			No MD		MD		No RD		RD	
	N	% within HL	N	% within HL	N	% within HL	N	% within HL		N	% within HL	N	% within HL	N	% within HL	N	% within HL
Luxembourgish	1155	74	413	26	1177	75	391	25	Luxembourgish	1394	89	174	11	1409	90	159	10
French	275	66	140	34	158	38	257	62	French	352	85	63	15	291	70	124	30
Portuguese	647	54	548	46	232	19	963	81	Portuguese	961	80	234	20	620	52	575	48
South Slavic	120	55	99	45	81	37	138	63	South Slavic	173	79	46	21	151	69	68	31
Luxembourgish and French	184	71	75	29	160	62	99	38	Luxembourgish and French	227	88	32	12	225	87	34	13
Luxembourgish and Portuguese	100	48	108	52	71	34	137	66	Luxembourgish and Portuguese	161	77	47	23	132	64	76	37

(continued)

(continued)

	HLG reference group									HLG reference group							
	No MD		MD		No RD		RD			No MD		MD		No RD		RD	
	N	% within HL	N	% within HL	N	% within HL	N	% within HL		N	% within HL	N	% within HL	N	% within HL	N	% within HL
Luxembourgish	1155	74	413	26	1177	75	391	25	Luxembourgish	1394	89	174	11	1409	90	159	10
French	319	77	96	23	309	75	106	26	French	370	89	45	11	377	91	38	9
Portuguese	910	76	285	24	879	74	316	26	Portuguese	1080	90	115	10	1063	89	132	11
South Slavic	167	76	52	24	158	72	61	28	South Slavic	196	90	23	11	199	91	20	9
Luxembourgish and French	194	75	65	25	192	74	67	26	Luxembourgish and French	235	91	24	9	236	91	23	9
Luxembourgish and Portuguese	160	77	48	23	151	73	57	27	Luxembourgish and Portuguese	189	91	19	9	186	89	22	11

Appendix B – List of abbreviations

HLG	Home language group
L1	First language
L2	Second language
LDs	Learning difficulties
MD	Math difficulties
RC	Reading comprehension
RD	Reading difficulties
SLD	Specific learning disorder

References

American Educational Research Association, American Psychological Association, National Council on Measurement in Education & Joint Committee on Standards for Educational and Psychological Testing (U.S) (2014): *Standards for Educational and Psychological Testing*. Washington, DC: American Educational Research Association.

American Psychiatric Association (2013): *Diagnostic and Statistical Manual of Mental Disorders (DSM-5®)*. American Psychiatric Association. Washington, DC & London, England. https://economie.gouv.cg/sites/default/files/webform/pdf-diagnostic-and-statistical-manual-of-mental-disorders-5th-editio-american-psychiatric-association-pdf-download-free-book-f6714d9.pdf

Bedore, Lisa M. & Pena, Elizabeth D. (2008): Assessment of bilingual children for identification of language impairment: Current findings and implications for practice. *International Journal of Bilingual Education and Bilingualism* 11 (1), 1–29.

Braeuning, David, Lambert, Katherina, Hirsch, Stefa, Schils, Trudie, Borghans, Lex, Nagengast, Benjamin & Moeller, Korbinian. 2020. Abstract: Diagnose von Rechenschwäche bei Grundschulkindern – Evaluation gängiger Kriterien und Vorgehensweisen. In. Potsdam.

Cummins, Jim. (1984): *Bilingualism and Special Education: Issues in Assessment and Pedagogy*. 6, Taylor & Francis Group. Vol. 6. Clevedon: Multilingual Matters, 1984.

Desoete, Annemie, Roeyers, Herbert & De Clercq, Armand (2004): Children with mathematics learning disabilities in Belgium. *Journal of Learning Disabilities* 37 (1), 50–61.

Ertel, Cintia, Alieva, Aigul, Hornung, Caroline & Schiltz, Christine (2019): *The Effect of Grade Retention on Reading Skills of immigrant children in multilingual elementary school. A longitudinal study*. Esch-sur-Alzette.

Fischbach, Antoine, Ugen, Sonja & Martin, Romain (eds.). 2014. ÉpStan technical report. University of Luxembourg.

Fletcher, Jack M., Reid Lyon, G., Fuchs, Lynn S. & Barnes, Marcia A. (2018): *Learning Disabilities, Second Edition: From Identification to Intervention*. New York, NY: Guilford Publications.

Ganzeboom, Harry BG, De Graaf, Paul M. & Treiman, Donald J. (1992): A standard international socio-economic index of occupational status. *Social Science Research* 21 (1), 1–56.

Geary, David C. (1993): Mathematical disabilities: Cognitive, neuropsychological, and genetic components. *Psychological Bulletin* 114 (2), 345–362. doi:https://doi.org/10.1037/0033-2909.114.2.345.
Geva, Esther & Wiener, Judith (2014): *Psychological Assessment of Culturally and Linguistically Diverse Children and Adolescents: A Practitioner's Guide*. New York, NY: Springer Publishing Company.
Geva, Esther & Zadeh, Zohreh Yaghoub (2006): Reading efficiency in native english-speaking and english-as-a-second-language children: The role of oral proficiency and underlying cognitive-linguistic processes. *Scientific Studies of Reading* 10 (1), 31–57. doi:https://doi.org/10.1207/s1532799xssr1001_3.
Gross-Tsur, Varda, Manor, Orly & Shalev, Ruth S. (1996): Developmental dyscalculia: Prevalence and demographic features. *Developmental Medicine & Child Neurology* 38 (1), 25–33.
Hammer, Carol Scheffner, Hoff, Erika, Uchikoshi, Yuuko, Gillanders, Cristina, Castro, Dina & Sandilos, Lia E. (2014): The language and literacy development of young dual language learners: A critical review. *Early Childhood Research Quarterly* 29 (4), 715–733. doi: https://doi.org/10.1016/j.ecresq.2014.05.008.
Heckman, James J. 2008. Schools, skills, and synapses. NBER Working Paper No. 14064. *National Bureau of Economic Research*.
Hoff, Erika (2006): How social contexts support and shape language development. *Developmental Review* 26 (1), 55–88. doi:https://doi.org/10.1016/j.dr.2005.11.002.
Hoff, Erika (2013): Interpreting the early language trajectories of children from Low SES and language minority homes: Implications for closing achievement gaps. *Developmental Psychology* 49 (1), 4–14. doi:https://doi.org/10.1037/a0027238.
Jang, Eunice Eunhee, Dunlop, Maggie, Wagner, Maryam, Kim, Youn-Hee & Zhimei, Gu. (2013): Elementary school ELLs' reading skill profiles using cognitive diagnosis modeling: Roles of length of residence and home language environment. *Language Learning* 63 (3), 400–436.
Kempert, Sebastian, Schalk, Lennart & Saalbach, Henrik (2019): Übersichtsartikel: Sprache als Werkzeug des Lernens: Ein Überblick zu den kommunikativen und kognitiven Funktionen der Sprache und deren Bedeutung für den fachlichen Wissenserwerb. *Psychologie in Erziehung und Unterricht* 66 (3), 176–195. doi:https://doi.org/10.2378/PEU2018.art19d.
Lenkeit, Jenny, Schwippert, Knut & Knigge, Michel (2018): Configurations of multiple disparities in reading performance: Longitudinal observations across France, Germany, Sweden and the United Kingdom. *Assessment in Education: Principles, Policy & Practice* 25 (1), 52–86. doi:https://doi.org/10.1080/0969594X.2017.1309352.
Lesaux, Nonie K. & Siegel, Linda S. (2003): The development of reading in children who speak English as a second language. *Developmental Psychology* 39 (6), 1005.
Letts, Carolyn (2011): Communication impairment in a multilingual context. In Ellis, Sue, McCarthy, Elspeth (eds.): *Applied Linguistics and Primary School Teaching*. Cambridge: Cambridge UP. doi:10.1017/CBO9780511921605.026.
Leysen, Heleen, Van den Broek, Wim, Keuning, Jos, Noé, Marjolein & Geudens, Astrid (2018): Vlaamse Normering van de Drie-Minuten-Toets en AVI-toetskaarten van 2009. Thomas More: Antwerpen.

Limbos, Marjolaine M. & Geva, Esther (2001): Accuracy of teacher assessments of second-language students at risk for reading disability. *Journal of Learning Disabilities* 34 (2), 136–151.

Lorphelin, Dalia, Keller, Ulrich, Fischbach, Antoine & Brunner, Martin (2014): Data processing, analyses, and reporting. In Fischbach, Antoine, Ugen, Sonja, Martin, Romain (eds.): *ÉpStan Technical Report.* Luxembourg: University of Luxembourg, LUCET, 12–29.

Martel, Laurent, Malenfant, Erin Caron, Morency, Jean-Dominique, Lebel, André, Bélanger, Alain & Bastien, Nicolas (2011): Projected trends to 2031 for the Canadian labour force. *Canadian Economic Observer* 24, 6.

McBride-Chang, Catherine (2004): *Children's Literacy Development.* London, United Kingdom: Taylor & Francis Group. http://ebookcentral.proquest.com/lib/unilu-ebooks/detail.action?docID=746361 (8 July, 2020)

Mejias, Sandrine, Muller, Claire & Schiltz, Christine (2019): Assessing mathematical school readiness. *Frontiers in Psychology* 10, 1173. doi:https://doi.org/10.3389/fpsyg.2019.01173.

Mejias, Sandrine & Schiltz, Christine. 2013. Estimation abilities of large numerosities in Kindergartners. 4. 1–12. https://doi.org/10.3389/fpsyg.2013.00518.

Ministry of National Education, Children and Youth, Department of Statistics and Analysis. 2018. The Key Figures of the National Education – Statistics and Indicators 2016/2017.

Muller, Claire, Reichert, Monique, Gamo, Sylvie, Hoffmann, Danielle, Hornung, Caroline, Sonnleitner, Philipp & Martin, Romain (2015). Kompetenzunterschiede aufgrund des Schülerhintergrundes. *Épreuves Standardisées: Bildungsmonitoring für Luxemburg. Nationaler Bericht 2011 bis 2013*, 34–56.

Müller, Rudolf. 2004a. Diagnostischer Rechtschreibtest für 3. *Klassen (DRT-3).* Göttingen: Hogrefe.

Müller, Rudolf (2004b): *Diagnostischer Rechtschreibtest für 1. Klassen: DRT 1; Manual.* Beltz Test.

Nosworthy, Nadia, Bugden, Stephanie, Archibald, Lisa, Evans, Barrie & Ansari, Daniel (2013): A two-minute paper-and-pencil test of symbolic and nonsymbolic numerical magnitude processing explains variability in primary school children's arithmetic competence. (Ed.) Kevin Paterson. *PLoS ONE* 8 (7), e67918. doi:https://doi.org/10.1371/journal.pone.0067918.

Pace, Amy, Luo, Rufan, Hirsh-Pasek, Kathy & Golinkoff, Roberta Michnick (2017): Identifying pathways between socioeconomic status and language development. *Annual Review of Linguistics* 3 (1), 285–308. doi:https://doi.org/10.1146/annurev-linguistics-011516-034226.

Paetsch, Jennifer, Felbrich, Anja & Stanat, Petra (2015): Der Zusammenhang von sprachlichen und mathematischen Kompetenzen bei Kindern mit Deutsch als Zweitsprache. *Zeitschrift für Pädagogische Psychologie* 29 (1), 19–29. doi:https://doi.org/10.1024/1010-0652/a000142.

Serva, Maurizio & Petroni, Filippo (2008): Indo-European languages tree by Levenshtein distance. *EPL (Europhysics Letters)* 81 (6), 68005. doi:https://doi.org/10.1209/0295-5075/81/68005.

Sonnleitner, Philipp, Krämer, Charlotte, Gamo, Sylvie, Reichert, Monique, Muller, Claire, Keller, Ulrich & Ugen, Sonja (2018): Étude longitudinal des compétence des élèves. Évolution en compréhension écrite en allemand et en mathematiques entre la classe de 3e et la classe de 9e. In University of Luxemourg, Luxembourg Centre for Educational Testing & SCRIPT (eds.): *Rapport national sur l'Éducation au Luxembourg 2018*. Esch-sur-Alzette: University of Luxembourg, Luxembourg Centre for Educational Testing and SCRIPT. 39–58.

Sonnleitner, Philipp, Reichert, Monique & Ugen, Sonja (2014): Item development and test compilation. In Fischbach, Antoine, Ugen, Sonja, Martin, Romain (eds.): *ÉpStan Technical Report*. Luxembourg: University of Luxembourg, Luxembourg Centre for Educational Testing and SCRIPT. 5–11.

Vos, Teije de (1992): *Tempo-Test-Rekenen: test voor het vaststellen van het rekenvaardigheidsniveau der elementaire bewerkingen (automatisering) voor het basis- en voortgezet onderwijs: Handleiding*. Berkhout. "Swets & Zeitlinger"; Lisse.

IV Vision, hearing, and speech language impairments

Kirsten Schuchardt and Claudia Mähler

Numerical competencies in preschoolers with language difficulties

School children with specific language disorders (SLI) often experience massive learning difficulties that concern not only literacy but also numeracy. Since preschool basic numerical precursor competencies have a great influence on the later development of arithmetic at school, this chapter is interested in potential early difficulties in counting skills, numerical knowledge, understanding of quantities, and early arithmetic skills. Given the close link between learning difficulties and working memory, a second question is whether these potential early difficulties can be associated with functional problems of working memory.

One of early childhood's central developmental tasks lies in the development of language. Yet, not every child achieves the milestones of language development smoothly. Specific language disorders rank among the most frequently occurring developmental dysfunctions during childhood and adolescence, with a total incidence between 5% and 8%. Boys are affected three times as often as girls (Tomblin et al., 1997). The relevant individuals typically display anomalies in language acquisition which do not result from cognitive deficits, physical illness, impaired hearing, or lack of stimuli due to unfavorable or stressful surroundings. SLI is defined by a considerable deviation from normal speech and language development, both in quantity and in quality. Language production as well as language comprehension may be affected (World Health Organization, 2011). The most severe effects manifest themselves in the acquisition of grammatical structures, but also pragmatic competence may be affected (Leonard, 2014). Frequently, articulatory deficits can be detected; however, an isolated functional impairment of articulation does not justify the diagnosis of SLI (Leonard, 2014). Speech anomalies resulting from certain illnesses (i.e., autism) will be excluded from this consideration. These cases are rather referred to as unspecific or secondary language development impairments.

Because of their speech difficulties, children suffering from SLI stand out at an early age. Language delay is a typical sign, along with a relatively small vocabulary and a late usage of phrases of two or more words (Desmarais et al., 2008). This initial deficit in language acquisition will further increase over the developmental course. While affected children show progress in language acquisition to some extent and are capable of understanding and producing simple sentences over the course of their development, they are never going to reach the level of individuals unaffected by SLI. Oftentimes, a number of accompanying

https://doi.org/10.1515/9783110661941-012

difficulties will develop as a result of the language deficit, especially emotional and social issues (Yew & O'Kearney, 2013); anomalies in the development of motoric abilities (Sanjeevan et al., 2015) and attention deficit hyperactivity disorder (Beitchman et al., 1996, 2001) are characteristic for these children.

1 School performance of SLI-affected children

With the onset of schooling, usually extensive learning difficulties arise, since language competence is a prerequisite for the understanding and application of content knowledge. SLI frequently goes hand in hand with an impaired acquisition of reading and writing (Joye et al., 2018). According to estimates, about 25–75% of children affected by SLI will also develop dyslexia (Tomblin et al., 2000; Catts et al., 2005; McArthur et al., 2000). Moreover, the affected children also experience difficulties in the subject of mathematics. In comparison to unaffected children of their age group, they possess significantly lower mathematic capabilities (Durkin et al., 2015). This discrepancy becomes greater as their time in school progresses (Durkin et al., 2013). Children suffering from SLI exhibit considerable problems in counting, both forward and backward, but also in the estimation of quantities and in the comprehension of positional notation (Cowan et al., 2005; Donlan et al., 2007; Fazio, 1996; Nys et al., 2013). Regarding numeracy skills, they are much slower and more prone to mistakes (Cowan et al., 2005). Furthermore, recalling mathematical facts from memory appears to be challenging (Cowan, 2014; Cowan et al., 2005; Fazio, 1996). On the other hand, children with SLI seem to understand mathematical rules and regularities equally well as unaffected children from the same grade (Donlan et al., 2007).

Based on research findings regarding difficulties in the subject of mathematics, this present study poses the central question whether these problems surface in school for the first time, or whether the children have already experienced difficulties with numerical basic skills in the preschool context (cf. Donlan et al., 2007).

2 Preschool-level basic numerical skills

School starters already possess considerable amounts of knowledge of quantities and numbers, which they have acquired during their preschool time (Mähler et al., 2017). This knowledge facilitates a successful performance in their early mathematics instruction. At the point of school enrollment, however, the individual knowledge levels differ substantially (Mähler et al., 2017). While some children manage calculations within the number range up to 100 with ease, other children are unable to count to ten. It can be assumed that SLI-affected children start school with an already disadvantageous learning predisposition in the fields of written language acquisition and numeracy, as a result of their extensive speech and language developmental deficits. Examples of relevant precursor competencies in the context of school numeracy are the ability to comprehend a numerical series, the ability to count, associating numbers to quantities, and the recognition of quantitative relations (Passolunghi et al., 2015). The model of number–quantity connection (Krajewski et al., 2013) describes a developmental progress on three levels: Level (1) comprises the differentiation of easily distinguishable quantities, the recital of numerical series, as well as the imitation of the counting procedure. Subsequently, level (2) establishes the connection between quantities and numbers. Two performances initiate the mental conceptualization of quantity: the internalization of the number's ordinal aspect, as well as the realization of the one-to-one assignment while audibly counting. Distinguishing merely between "little" and "a lot," this notion of quantity is yet a rather unprecise one, but it will further evolve into a precise concept of number (assignment of numbers and their corresponding quantities). Level (2) provides the children with the insight, that quantities can be changed by adding amounts or to subtracting amounts from those (= part-whole concept). If both insights are now combined on level (3), the students will have attained a sophisticated understanding of numerical relations, enabling them to express in numbers both partial quantities and differences in quantity.

3 Working memory and numerical competence

Aside from the influential role language skills have, the development of basic numerical competence is also heavily impacted by general cognitive functions. Among these, the operability of working memory is a decisive factor in the development of preschool skills (Friso-van de Bos et al., 2013; Schuchardt et al., 2014). Working memory can be described as a system of short-term storage and simultaneous processing of information, prior to permanent storage in long-term memory. Therefore, working memory is involved in every single instance of information processing and can be construed as a sort of bottleneck of cognitive capacity (Süß, 2001). According to Baddeley (1986), working memory consists of a cross-modal central executive, which comprises two subordinate, modality-specific components which are limited in capacity. The first component is the phonological loop for verbal and auditory information; the second is the visuo-spatial sketchpad, responsible for visual patterns and spatial layout (see Fig. 1). Following Baddeley's model, the central executive carries out the functions of control, monitoring, and coordination, such as the coordination between subsystems during simultaneous information processing, selection of and switching between different strategies of retrieval, management of selective attention, as well as retrieval and manipulation of long-term memory information. In a further elaboration of his model, Baddeley (2000) postulated the episodic buffer as a fourth component. This instance's purpose lies in combining differently coded information (from perception, from other subsystems of working memory, and from long-term memory) into a coherent whole, before transferring this information to long-term memory. Empirical evidence has not yet been provided for this structural addition; so far, only few studies on this subject have been published.

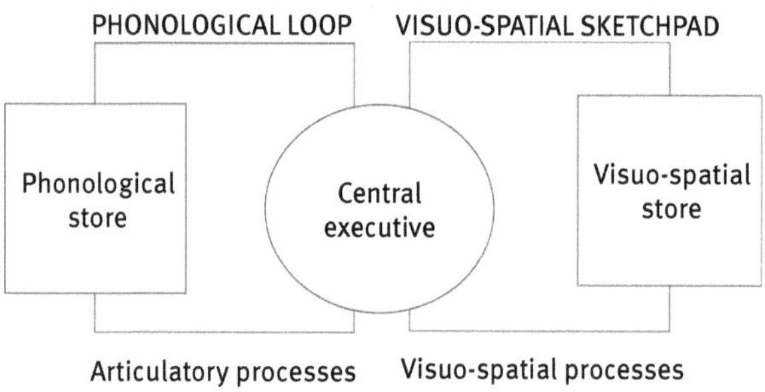

Fig. 1: Working memory model according to Baddeley (1986).

Overall, working memory is regarded as the active memory constituent, which can take in only a limited amount of information. Hence, the operability of working memory constitutes the limiting resource on which an individual's cognitive performance depends. In this context, interindividual differences exist in the size of processed units, and the processing speed of the respective working memories. Both phonological loop and visuo-spatial sketchpad are easily differentiable at an age of four already (Alloway et al., 2006). The differentiation of the central executive cannot be reliably detected before the age of five (Alloway et al., 2006; Michalczyk et al., 2013). For the development of preschool numerical competencies, particularly the visuo-spatial sketchpad has been proven to be a central influencing factor (Krajewski & Schneider, 2009;

Kyttälä et al., 2003; Schuchardt et al., 2014). As Preßler et al. (2013) pointed out in their study, preschool children with an impaired visuo-spatial sketchpad displayed poor mathematical competencies immediately before and three months subsequent to the onset of schooling. This matches the findings on children suffering from dyscalculia, where particularly dysfunctions of working memory's visuo-spatial component have been identified as a decisive causal factor for a reduced numeracy performance (Klesczewski et al., 2018; Schuchardt et al., 2008).

4 Working memory and SLI

Working memory difficulties anomalies are also diagnosed in children affected by SLI. Accordingly, a large portion of studies attests a substantial functional deficit of phonological loop and central executive, while impairments of the visuo-spatial sketchpad have been observed rather infrequently (Archibald & Gathercole, 2006; Schuchardt et al., 2013; Marton & Schwartz, 2003; Montgomery & Evans, 2009; Riccio et al., 2007). Schuchardt et al. (2013) examined elementary school children with dyslexia, as well as with a combination of deficits (dyslexia and dyscalculia), focusing on working memory functions. Moreover, half of the test subjects were affected by SLI. Whereas children with a comorbidity of dyslexia and dyscalculia exhibited a functionally impaired visuo-spatial sketchpad, children additionally affected by SLI possessed a properly operating visuo-spatial working memory. Thus, these children's numerical shortcomings appear to be rather a consequence of speech difficulties than attributable to malfunctions of the visuo-spatial sketchpad.

5 Research issue

On the basis of research findings on the subject of SLI-affected children experiencing difficulties in mathematics, the principal question (1) arises, whether anomalies regarding preschool basic numeracy competence become apparent prior to schooling? For this, the following categories of competence undergo a closer examination: counting ability, knowledge of numbers, comprehension of quantities, and basic numeracy. It is assumed that preschoolers affected by SLI will, in comparison to their unaffected peers, exhibit a less developed basic numerical competence in all categories.

Moreover, an interesting task (2) lies in exploring the causes of possible developmental deficits in the field of basic numeracy. There are two conceivable explanatory approaches: On the one hand, the study results of Schuchardt et al. (2013) suggest that the present language deficits are responsible for the children's failure to progress at an age-appropriate level in basic numerical learning (language deficit hypothesis). In order to verify this hypothesis, the children affected by SLI will be compared to a group of younger children on the same level of speech development. If the development of basic numerical competencies proceeds analogously to the language level, the performances of these two groups should attain a similar level.

On the other hand, it is also conceivable that, beyond the language deficits, an additional cause in the form of a cognitive working memory deficit exists. In the latter case, the deficit would be attributed to the visuo-spatial sketchpad (working memory deficit hypothesis). Hence, should anomalies of the visuo-spatial sketchpad appear compared to the control group of the same age, these anomalies could relate to the deficient numeracy development. However, the case of age-appropriate intact visuo-spatial working memory functions would further consolidate the language deficit hypothesis.

6 Method

6.1 Sample and research design

Within the frame of a three-group design, the group of SLI-affected children ($n = 25$) is contrasted to two control groups: (1) children of the same chronological age without signs of language deficits (CA, $n = 25$), and (2) a group of children on the same level of language development as the SLI-affected group, that is, of the same language maturity (LA, $n = 25$). All participating children are native speakers of German and have an IQ of ≥80 (CPM; Bulheller & Häcker, 2002). The children of the SLI group come from two speech therapy kindergartens, the children of both control groups from regular kindergartens. The latter constitute partial random samples from a study on differential developmental courses of cognitive competencies during preschool and elementary school age (Differentielle Entwicklungsverläufe kognitiver Kompetenzen im Vor- und Grundschulalter). The parallelization of language maturity has been conducted on the basis of raw score from the active vocabulary test (AWST-R; Kiese-Himmel, 2005), as well as based on raw score of the subtest morphological rule formation, which is part of the language development test for children (SETK 3–5; Grimm, 2001) in order to assess grammatical competence. Table 1 contains the sample parameters of all three groups. Here, it becomes apparent that group SLI shows unambiguously substandard performances in the fields of vocabulary and grammar, which would be expected for the age of four.

Tab. 1: Means (SDs) for descriptive characteristics of subgroups.

	SLI (n = 25)	CA (n = 25)	LA (n = 25)
Sex(m/f)	20/05	20/5	14/11
Age (years)	5;4	5;3	4;0
IQ	106.70 (8.01)	105.56 (8.66)	104.32 (7.63)
Vocabulary (RW)	41.09 (8.57)	55.84 (10.09)	40.40 (11.05)
Vocabulary (*T*-Score)	23.81 (4.45)	52.24 (13.79)	50.28 (9.75)
Grammar (RW)	18.82 (6.56)	26.44 (4.48)	17.04 (7.85)
Grammar (*T*-Score)	31.79 (4.15)	49.90 (14.77)	50.72 (10.61)

Note: SLI = specific language disorders; CA = chronological age; LA = language age.

6.2 Instruments

6.2.1 Numerical competencies

The implicated battery of tasks has been developed for the age group between three and six years (retest-reliability rtt = .95). Computers have been used as interface for all tasks. *Counting abilities* have been tested via two subtests. For *audible counting* (Cronbach's alpha = .95), the child is instructed to count a sequence of numbers up to 25. In a further step, the child is asked to count from 58 to 72. The latter sequence is meant to test the child's ability to start counting from any given point within the numerical series. In *counting objects* (Cronbach's alpha = .83), the child is instructed to count 11 given quantities, which are being visually presented (yellow stars on a blue background) in succession. The child is asked to point a finger at the individual objects while audible counting them. The number of stars varies between three and twenty-one. The task assesses the degree to which the child already masters basic counting principles, such as one-to-one correspondence, stable order, and cardinality. For the survey regarding *numeral knowledge*, the two following tasks have been designed. For the *denomination of Arabic numerals* (Cronbach's alpha = .93), a task with 17 items captures the ability of transferring Arabic numerals into words. The numerals (1 to 12, 15, 18, 19, 100, and 116) are presented along with additional objects to be named (e.g., mouse, apple, tree, ball, moon) on a computer display. In the second task, *transcoding* (Cronbach's alpha = .65), three Arabic numerals are visually displayed; the child is subsequently asked to point out a verbally indicated number within this number range (number range: 1 to 150). The *comprehension of quantity* has been assessed with the task *quantity comparison* (Cronbach's alpha = .73). On a screen, two rectangular shapes are displayed, each filled with different amounts of objects (circles, squares, bars). The child is to name the square containing the largest number of objects. The objects vary in size, shape, and arrangement. At the same time, the quantitative proportions differ greatly (e.g., 15:3, respectively 7:6). *Early numerical competence* has been examined in two tasks. For the *addition of two visualized quantities* (Cronbach's alpha = .66), two circles, each containing blocks which indicate quantities, are displayed. The task now is to determine the sum of both quantities. In the task *mental operations involving objects* (Cronbach's alpha = .77), the computer shows either a garage or a rabbit's burrow, into which successively two differing numbers of cars, respectively of rabbits, enter and disappear. Following this presentation, the child is asked to report the number of objects inside the structures. Also, this task assesses primary addition skills, whereby the first subset must be represented mentally in order to add the second subset.

As dependent variables, the children's individual raw points have been divided by the corresponding total of items for every single subtest; resultingly, the values are in the range of 0 to 1.

6.2.2 Working memory

Here, children worked on two tasks from the working memory test battery for children between ages 5 and 12 (AGTB 5–12; Hasselhorn et al., 2012). For the examination of the visuo-static component of the visuo-spatial sketchpad, the *matrix span* has been implicated (retest-reliability $rtt = .51$; split-half-reliability $r = .98$). Patterns of black-and-white sections of a four-by-four matrix are visualized on the display, starting with two sections and increasing up to a maximum of eight black sections. Immediately following this presentation, the children are to reproduce the black sections by pressing the corresponding areas on the empty display matrix. A *corsi-block-span* (retest-reliability $rtt = .60$; split-half-reliability $r = .97$) serves to assess the visuo-spatial sketchpad's spatial-dynamical component. The child is exposed to a touchscreen monitor showing nine arbitrarily arranged white squares, in which for moments little smiley faces appear consecutively. Subsequently, the child is instructed to touch those display areas in which the smiley faces had been visible, in order of their appearance. As the dependent variable, the longest sequence attained in both tasks is being determined.

7 Results

The performances for the entirety of tasks examined of three groups can be obtained from Tab. 2. Initially, the performances in the field of numerical competence have been analyzed via individual univariate analyses of variance (ANOVA). Through this, as presumed, a significant group effect manifested itself for all tasks (Tab. 2). Further post hoc tests (Tukey) for the explanation of group effects illustrated that children of the CA group achieved significantly

Tab. 2: Means (SDs) for numerical competencies and working memory of subgroups.

	SLI	CA	LA	ANOVA			Post hoc comparison
				$F(2,72)$	p	η_p^2	
Numerical competencies							
Counting abilities							
Audibly counting	0.32 (0.14)	0.50 (0.18)	0.33 (0.18)	9.73	.000	.213	SLI = LA < CA
Counting objects	0.53 (0.26)	0.81 (0.22)	0.50 (0.26)	11.70	.000	.245	SLI = LA < CA
Numeral knowledge							
Denomination of Arabic numerals	0.32 (0.20)	0.52 (0.23)	0.20 (0.24)	9.94	.000	.216	LA < SLI < CA
Transcoding	0.64 (0.24)	0.76 (0.20)	0.49 (0.26)	8.48	.000	.191	LA < SLI < CA
Comprehension of quantity							
Quantity comparison	0.72 (0.09)	0.82 (0.10)	0.63 (0.13)	19.21	.000	.348	LA < SLI < CA
Early numerical competence							
Addition of two visualized quantities	0.54 (0.24)	0.77 (0.28)	0.57 (0.26)	5.77	.005	.138	SLI = LA < CA
Mental operations involving objects	0.41 (0.24)	0.57 (0.24)	0.40 (0.22)	4.08	.021	.102	SLI = LA < CA
Visuo-spatial working memory							
Matrix-span	3.28 (0.84)	3.28 (0.74)	2.29 (0.64)	14.40	.000	.289	LA < SLI = CA
Corsi-block-span	3.28 (0.74)	2.96 (0.79)	2.54 (0.59)	6.62	.002	.157	LA < SLI = CA

Note: SLI = specific language disorders; CA = chronological age; LA = language age.

better results in all tasks than children affected by SLI. The comparison with group LA is used to answer the question whether these poor numerical performances of children exhibiting language development anomalies can be attributed to their low level of language development. Here, the results show a heterogeneous pattern. While group SLI yields comparable results in the fields of counting ability and numeracy, they deliver better results than group LA in tasks regarding numeral knowledge and quantity comparison.

As a second step, the performances for working memory's visuo-spatial sketchpad have been examined employing once more univariate analysis of variance (ANOVA) for both tasks separately. Again, substantial group effects appeared (Tab. 2). Yet, subsequent post hoc comparisons uncovered a different result pattern. In both tasks, the SLI and the CA groups yielded similar results, which were significantly higher than those of group LA children.

8 Discussion

The present study closely examined the development of basic numerical preschool competence of SLI-affected children prior to schooling. It became apparent that these individuals showed weaker performances than their peers without language impairments in all fields. In this context, counting ability and first numerical operation skills are comparable to the results of the younger control group. Thus, it appears that the exhibited developmental delay corresponds to the deficit in language competence. However, different result patterns come into view in the examination of quantity comprehension and numeral knowledge. Here, the SLI-affected children's performances rank between those of the two control groups. These findings might be interpreted as evidence for general numerical knowledge of quantity and numerals is not as strongly affected as the specific numerical operations of counting and calculating (cf. Donlan et al., 2007). Presumably, operations of counting and calculating are related to language proficiency more closely than knowledge of quantity and numerals.

Furthermore, it becomes clear that anomalies of numeracy development cannot be associated with a diminished capacity of working memory's visuo-spatial sketchpad. Despite the fundamental role visuo-spatial working memory plays in building up numerical preschool competencies, the children with deficient preschool competence partaking in this survey did not exhibit shortcomings regarding visuo-spatial working memory. Beyond that, functional anomalies of the visuo-spatial sketchpad can be assessed to have a clear connection with a developing dyscalculia (Schuchardt et al., 2008). Asked for a cautious estimate, one can derive from the findings that the SLI-affected children examined here do not possess a significant risk of developing a dyscalculia, since they are unaffected by working memory problems, which frequently are the underlying cause.

Results of the present study rather give evidence for the language deficit hypotheses, according to which language development anomalies represent an obstacle to age-appropriate numeracy development. This obstacle becomes apparent already at preschool age in counting and calculation tasks. Therefore, one can conclude that age-appropriate language competence is not only of central importance in its role as a prerequisite of literacy, but that it is also crucial for the acquisition of preschool numerical competence. For this reason, SLI-affected children after all are particularly prone to developing learning dysfunctions. Since the acquisition of knowledge and the transfer of educational content happen via speech and language, language development deficits constitute a considerable risk for achieving academic success, to which affected children are exposed during preschool and schooling ages.

The results described must be considered in the light of the limitations of the study. It should be noted, for example, that the sample size is rather small. A replication on a larger sample would therefore be desirable. Moreover, only a few working memory tasks were included. A comprehensive battery of tasks that include a wider range of working memory functions would be desirable here. Furthermore, we should be aware of the fact that a diagnosis of SLI is less stable at early ages, and therefore, predictions should be made carefully.

To counteract the emergence and consolidation of numeracy acquisition problems within the context of school education, special tuition is mandatory from early on. As it is the case for all combined developmental deficits (in this case preschool language and numeracy competence), there is a lack of academic studies to recommend an appropriate, evidence-based strategy of support: Would a specific language tuition result in an improvement of numerical competence? Or, as an alternative approach: Is the training of numerical competence the more appropriate intervention, considering the reduced level of speech development?

Possibly, children with language impairments require simplified teaching instructions tailored to their needs, ideally realized via visualization, in order to facilitate the comprehension of numerical concepts. The same problem presents itself during the first years of schooling, even in specialized language learning groups, or in an inclusive learning group offering special language tuition. While both pedagogical environments obviously focus on improving language competence, a methodically and didactically differentiated form of tuition may still be requisite, which explicitly takes into consideration the children's individual learning progress.

References

Alloway, Tracy P., Gathercole, Susan E. & Pickering, Susan J (2006): Verbal and visuospatial short-term and working memory in children: Are they separable? *Child Development* 77, 1698–1716.

Archibald, Lisa M. D. & Gathercole, Susan E (2006): Short-term and working memory in specific language impairment. *International Journal of Language & Communication Disorders* 41, 675–693.

Baddeley, Alan D. (1986): *Working Memory*. Oxford: University Press.

Baddeley, Alan D. (2000): The episodic buffer: A new component of working memory? *Trends in Cognitive Sciences* 4, 417–423.

Beitchman, Joseph H., Brownlie, Elizabeth B., Inglis, Alison, Wild, Jennifer, Ferguson, Bruce, Schachter, Debbie, Lance, William, Wilson, Beth & Mathews, Rachel (1996): Seven-year follow-up of speech/language impaired and control children: Psychiatric outcome. *Journal of Child Psychology and Psychiatry* 37 (8), 961–970.

Beitchman, Joseph H., Wilson, Beth, Johnson, Carla J., Atkinson, Leslie, Young, Arlene, Adlaf, Edward, Escobar, Michaeles & Douglas, Lori (2001): Fourteen-year follow-up of speech/language-impaired and control children: Psychiatric outcome. *Journal of the American Academy of Child & Adolescent Psychiatry* 40 (1), 75–82.

Bulheller, Stephan & Häcker, Hartmut (2002): *Colored Progressive Matrices (CPM)*. Frankfurt/Main: Swets Test Services.

Catts, Hugh, Adolf, Suzanne, Hogan, Tiffany & Ellis Weismer, Susan (2005): Are specific language impairments and dyslexia distinct disorders? *Journal of Speech, Language, and Hearing Research* 48, 1378–1396.

Cowan, Richard (2014): Language disorders, special needs and mathe-matics learning. In Lerman, Steve (Hrsg.): *Encyclopedia of Mathematics Education*. Dordrecht: Springer Netherlands, 336–338.

Cowan, Richard, Donlan, Chris, Newton, Elizabeth J. & Lloyd, Delyth (2005): Number skills and knowledge in children with specific language impairment. *Journal of Educational Psychology* 97, 732–744.

den Bos, Friso-van, Ilona, van der Ven, Sanne, H. G., Kroesbergen, Evelyn H. & van Luit, Johannes E. H (2013): Working memory and mathematics in primary school children: A meta-analysis. *Educational Research Review* 10, 29–44.

Desmarais, Chantal, Sylvestre, Audette, Meyer, Françios, Bairati, Isaballe & Rouleau, Nancie (2008): Systematic review of the literature on characteristics of late-talking toddlers. *International Journal of Langu age and Communication Disorders* 43, 361–389.

Donlan, Chris, Cowan, Richard, Newton, Elizabeth J. & Lloyd, Delyth (2007): The role of language in mathematical development: Evidence from children with specific language impairments. *Cognition* 103, 23–33.

Durkin, Kevin, Mok, Pearl L. & Conti-Ramsden, Gina (2013): Severity of specific language impairment predicts delayed development in number skills. *Frontiers in Psychology* 4. Online unter: https://www.frontiersin.org/articles/10.3389/fpsyg.2013.00581/full.

Durkin, Kevin, Mok, Pearl L. & Conti-Ramsden, Gina (2015): Core subjects at the end of primary school: Identifying and explaining relative strengths of children with specific language impairment (SLI). *International Journal of Language & C ommunication disorders* 50 (2), 226–240.

Fazio, Barbara B. (1996): Mathematical abilities of children with specific language impairment: A 2-year follow-up. *Journal of Speech, Language, and Hearing Research* 39, 839–849.
Grimm, Hannelore (2001): *Sprachentwicklungstest für drei- bis fünfjährige Kinder (SETK 3–5)*. Göttingen: Hogrefe.
Hasselhorn, Marcus, Schumann-Hengsteler, Ruth, Gronauer, Julia, Grube, Dietmar, Mähler, Claudia, Schmid, Inga, Seitz-Stein, Katja & Zoelch, Christof (2012): *Arbeitsgedächtnistestbatterie für Kinder von 5 bis 12 Jahren (AGTB 5–12)*. Göttingen: Hogrefe.
Joye, Nelly, Broc, Lucie, Olive, Thierry & Dockrell, Julie (2018): Spelling performance in children with developmental language disorder: A meta-analysis across european languages. *Scientific Studies of Reading* 23 (2), 129–160.
Kiese-Himmel, Christiane (2005): *Aktiver Wortschatztest für 3- bis 5-jährige Kinder – Revision. (AWST-R)*. Göttingen: Hogrefe.
Klesczewski, Julia, Brandenburg, Janin, Fischbach, Anne, Schuchardt, Kirsten, Grube, Dietmar, Hasselhorn, Marcus & Büttner, Gerhard (2018): Development of working memory from grade 3 to 5. *International Journal of Disability, Development and Education* 65, 509–525.
Krajewski, Kristin (2013): Wie bekommen die Zahlen einen Sinn? Ein entwicklungspsychologisches Modell der zunehmenden Verknüpfung von Zahlen und Größen. In von Aster, Michael, Lorenz, Jens H. (Hrsg.): *Rechenstörungen bei Kindern. Neurowissenschaft, Psychologie, Pädagogik*. Göttingen: Vandenhoeck & Ruprecht, 155–179.
Krajewski, Kristin & Schneider, Wolfgang (2009): Exploring the impact of phonological awareness, visual-spatial working memory, and preschool quantity-number competencies on mathematics achievement in elementary school: Findings from a 3-year-longitudinal study. *Journal of Experimental Child Psychology* 103, 516–531.
Kyttälä, Minna, Aunio, Pirjo, Lehto, Juhani, van Luit, Johannes & Hautamäki, Jarkku (2003): Visospatial memory and early numeracy. *Educational and Child Psychology* 20, 65–76.
Leonard, Laurence B. (2014): *Children with Specific Language Impairment*. Cambridge: MIT press.
Mähler, Claudia, Grube, Dietmar & Schuchardt, Kirsten (2017): Interindividuelle Unterschiede kognitiver Kompetenzen als Herausforderung für die frühkindliche Bildung. *Pädagogische Rundschau* 71, 349–366.
Marton, Klara & Schwartz, Richard G. (2003): Working memory capacity and language processes in children with specific language impairment. *Journal of Speech, Language, and Hearing Research* 46, 1138–1153.
McArthur, Genevieve M., Hogben, John H., Edward, Veronica T., Heath, Steve M. & Mengler, Elise D. (2000): On the "specifics" of specific reading disability and specific language impairment. *Journal of Child Psychological Psychiatry* 41, 869–874.
Michalczyk, Kurt, Malstädt, Nadine, Worgt, Maria, Könen, Tanja & Hasselhorn, Marcus (2013): Age differences and measurement: Invariance of working memory in 5-to 12-year-old children. *European Journal of Psychological Assessment* 29, 220–229.
Montgomery, James W. & Evans, Julia L (2009): Complex sentence comprehension and working memory in children with specific language impairment. *Journal of Speech, Language, and Hearing Research* 52, 269–288.
Nys, Julie, Content, Alain & Leybaerta, Jacqueline (2013): Impact of language abilities on exact and approximate number skills development: evidence from children with specific language impairment. *Journal of Speech, Language, and Hearing Research* 56, 956–970.

Passolunghi, Maria C., Lanfranchi, Silvia, Altoè, Gianmarco & Sollazzo, Nadia (2015): Early numerical abilities and cognitive skills in kindergarten children. *Journal of Experimental Child Psychology* 135, 25–42.
Preßler, Anna-Lena, Krajewski, Kristin & Hasselhorn, Marcus (2013): Working memory capacity in preschool children contributes to the acquisition of school relevant precursor skills. *Learning and Individual Differences* 23, 138–144.
Riccio, Cynthia A., Cash, Deborah L. & Cohen, Morris J. (2007): Learning and memory performance of children with Specific Language Impairment (SLI). *Applied Neuropsychology* 14, 255–261.
Röhm, Alexander, Starke, Anja & Ritterfeld, Ute (2017): Die Rolle von Arbeitsgedächtnis und Sprachkompetenz für den Erwerb mathematischer Basiskompetenzen im Vorschulalter. *Psychologie in Erziehung und Unterricht* 64, 81–93.
Sanjeevan, Teenu, Rosenbaum, David A., Miller, Carol, van Hell, Janet G., Weiss, Daniel J. & Mainela-Arnold, Elina (2015): Motor issues in specific language impairment: A window into the underlying impairment. *Current Developmental Disorders Reports* 2 (3), 228–236.
Schuchardt, Kirsten, Mähler, Claudia & Hasselhorn, Marcus (2008): Working memory deficits in children with specific learning disorders. *Journal of Learning Disabilities* 41, 514–523.
Schuchardt, Kirsten, Piekny, Jeanette, Grube, Dietmar & Mähler, Claudia (2014): Kognitive und häusliche Einflüsse auf den Erwerb früher numerischer Kompetenzen. *Zeitschrift für Entwicklungspsychologie und Pädagogische Psychologie* 46 (1), 24–34.
Schuchardt, Kirsten, Bockmann, Ann-Katrin, Bornemann, Galina & Maehler, Claudia (2013): Working memory functioning in children with learning disorders and SLI. *Topics in Language Disorders* 33, 298–312.
Süß, Heinz-Martin (2001): Prädiktive Validität der Intelligenz im schulischen und außerschulischen Bereich. In Stern, Elsbeth, Guthke, Jürgen (Hrsg.): *Perspektiven der Intelligenzforschung*. Lengerich: Pabst, 109–136.
Tomblin, J. Bruce, Smith, Elaine & Zhanf, Xuyang (1997): Epidemiology of specific language impairment: Prenatal and perinatal risk factors. *Journal of Communication Disorders* 30, 325–342.
Tomblin, J. Bruce, Zhang, Xuyang, Buckwalter, Paula & Catts, Hugh (2000): The association of reading disability, behavioral disorders, and specific language impairment in second grade children. *Journal of Child Psychology and Psychiatry* 41, 473–482.
World Health Organization (2011): ICD: *Classification of mental and behavioural disorders: Clinical descriptions and diagnostic guidelines*, 10th. rev. edn., Geneva, Switzerland: Author.
Yew, Shaun G. K. & O'Kearney, Richard (2013): Emotional and behavioural outcomes later in childhood and adolescence for children with specific language impairments: Meta-analyses of controlled prospective studies. *Journal of Child Psychology and Psychiatry* 54 (5), 516–524.

Elisabeth Moser Opitz and Verena Schindler

Disentangling the relationship between mathematical learning disability and second-language acquisition

1 Introduction

Several studies have established that the mathematical achievement of language minority students (students whose first language differs from the language of instruction) is poorer than that of native speakers (students whose first language is the academic language of the instruction; Haag et al., 2015; Paetsch & Felbrich, 2016; Vukovic & Lesaux, 2013; Warren & Miller, 2015). However, despite the expanding literature on the mathematical learning of language minority students and of native speakers, very little is known about the relationship between mathematical learning disabilities and second-language acquisition. More detailed research on this topic is important for several reasons: Gonzáles and Artiles (2015) report that Latina/o students in the United States who perform below expectations in literacy tests are often diagnosed as having learning difficulties, which, in turn, often leads to their exclusion from mainstream education. Further, language minority students with low mathematical achievement in Switzerland – and probably also in other countries – often receive special second-language support, but they do not receive support for mathematics because it is assumed that their mathematical problems are caused by their language background. Therefore, it is important to investigate the extent to which the problems of language students with mathematical learning disabilities may be caused by math-related, as opposed to language-related, factors.

This study investigates whether the relationship between selected language variables and mathematical achievement gains is similar for native speakers with mathematical learning disabilities and language minority students with mathematical learning disabilities. The research was conducted by evaluating grade 3 students (students who are in the third year of school after attending kindergarten) over the course of a school year.

Note: This project was sponsored by the Swiss National Science Foundation Grant Nr. 134652

https://doi.org/10.1515/9783110661941-013

2 The relationship between mathematical learning and language

The mathematical learning process and therefore the mathematical achievement gain are closely linked to language. According to Morgan et al. (2014: 845), "language has a special role in relation to mathematics because the entities of mathematics are not accessible materially." Language is an important tool that gives access to mathematics. However, the language of instruction, the academic language, differs from everyday language (Cummins, 2000) and has its own characteristics and challenges for all students (Schleppegrell, 2004; Snow & Uccelli, 2009). Prediger et al. (2019) categorized the challenges for students on word, sentence, and text, and on discourse level. The challenges can, individually or in combination, affect the mathematical learning process and thus contribute to learning difficulties in mathematics.

On word level, the linguistic structure of the number words has an influence on the acquisition of numbers (e.g., Klein et al., 2013; Miura et al., 1994). Math vocabulary is also an issue at this level. According to Haag et al. (2015) more difficult lexical features in test items, such as a more specialized vocabulary, increase the item difficulty in math tests.

On a sentence and text level, logical relationships, complex prepositional clauses (Jorgensen, 2011), conditional clauses, and complex issues of cohesion are difficult (e.g., Schleppegrell, 2004). Further, Haag et al. (2015) showed that text length and an increased number of noun phrases made comprehension more challenging for third graders. Koponen et al. (2018) found a link between reading competence and mathematical achievement. Students with a very low performance in reading showed low performance in mathematics across all grades.

Language in mathematics classrooms is also important on a discourse level. Language is both a medium of knowledge transfer and discussion, and a tool for thinking (Morek & Heller, 2012). Moschkovich (2015) and Erath et al. (2018) emphasized the importance of students' participation in discourse for developing conceptual understanding. Moschkovich (2015) points out that it is not the use of formal mathematical words that makes a discussion mathematical, but the use of mathematical concepts. Such concepts may also be expressed using informal words and phrases. Nevertheless, it may be assumed that performance in informal language production, regardless of use of mathematical vocabulary, is an important prerequisite for participation in classroom discourse.

In summary, this short review shows that language factors and mathematical learning are closely related on multiple levels.

3 Language performance and mathematical learning of students with mathematical learning disabilities

Although the scientific community has yet to agree on a formal definition of mathematical learning disability (e.g., Nelson & Powell, 2017), and different countries use different diagnostic criteria, studies have identified some characteristics commonly found in students with mathematical learning disabilities: low competence in counting tasks (Desoete et al., 2009; Stock et al., 2010), problems with understanding different aspects of the base-10 number system (e.g., Herzog et al., 2019; Moeller et al., 2011; Vukovic & Siegel, 2010), and dealing with word problems (e.g., Kingsdorf & Krawec, 2014; Peake et al., 2015; Zhang & Xin, 2012). Students with mathematical learning disabilities also have problems with fact retrieval, which can be related to deficits in working memory (e.g., De Weerdt et al., 2012; Geary et al., 2012). This study uses the term "mathematical learning disabilities" to refer to students with below-average mathematical achievement who have the characteristics described in this section (for cut-off criteria, see instruments).

Little is known about the relationship between language performance and mathematical learning in students with mathematical learning disabilities. Most studies to date have investigated the differences between students with and without comorbid reading disabilities.

On word level, the relationship between mathematics vocabulary and mathematical learning disabilities, with and without reading disabilities, has been investigated (Forsyth & Powell, 2017). Fifth graders with mathematical learning disabilities only or with reading disabilities only demonstrated a significantly weaker grasp of mathematics vocabulary than typically achieving students. Students with both mathematical learning disabilities and reading disabilities scored significantly lower than students who had problems only with either reading or mathematics.

On text and sentence levels, research generally focuses on reading disabilities. Mann Koepke and Miller (2013) conclude that 17–66% of students with mathematical learning disabilities also have reading disabilities. Several studies confirm this relationship (Vukovic, 2012; Vukovic & Siegel, 2010).

Peake et al. (2015) examined another factor on sentence level. They found a relationship between arithmetic problem solving and syntactic awareness in a sample of students with reading disabilities and a group of comorbid disabilities (reading and math), but not with students with mathematical learning disabilities only.

In summary, evidence in the literature supports the hypothesis that students with mathematical disabilities often have problems with mathematical vocabulary and reading.

4 Language performance and mathematical learning of language minority students

The achievement gap in mathematics between native speakers and language minority students (see introduction) is often explained by the difference between everyday language and academic language (Schleppegrell, 2004). This relationship is complex. The research of Martinellio (2008) showed that the poor test results of language minority students in grade 4 were caused by several factors: lack of knowledge of the specific context of a word problem, lack of vocabulary (e.g., likely, unlikely, certain), as well as the linguistic complexity of the items. Bochnik (2017) investigated the relationship between mathematical vocabulary, overall language proficiency in German, and mathematical achievement in a sample of German-speaking native speakers and language minority students in grade 3 who had lower mathematical achievement than their peers. The difference between the samples was predicted by overall proficiency in German. But, proficiency in the technical language of mathematics was the strongest predictor when explaining differences in the mathematical achievement of native speakers and language minority students. Vukovic and Lesaux (2013) investigated the relationship between language ability and mathematical cognition in a sample of language minority and native speakers aged 6 to 9. The authors found that language proficiency predicted gains in data analysis, probability, and geometry, but not in arithmetic, which was assessed with a computation test. The authors concluded that language seems to play a limited role in numerical manipulation but may be necessary for forming mathematical concepts and representations. Prediger et al. (2018) showed in a survey of grade 10 students that language proficiency is the background factor with the strongest connection to mathematics achievement of all the social and linguistic background factors. Language proficiency therefore was more strongly interrelated to mathematics achievement than multilingualism, immigrant status, or socioeconomic status.

Finally, Rodriguez et al. (2001) found that the performance rate of solving word problems by culturally and linguistically diverse learners in special bilingual classrooms was lower than that of students with learning disabilities and of students in general classrooms. This occurred even when reading requirements were minimal and the students were capable of solving computation problems.

The authors assume that the issues with problem solving might be because the student had never seen this kind of problem before.

These studies confirm the strong relationship between language proficiency on different levels (word, sentence, and text, discourse) and mathematical learning for all students. Therefore, it is important to heed Moschkovich (2010), who recommends focusing less on differences between monolinguals and bilinguals, and more on their similarities.

5 Research questions

The literature review reveals that little is known about the relationship between mathematical learning and the contributing language factors of students with mathematical learning disabilities, and as far as we have been able to ascertain, there has been no research into this relationship that looks specifically at language minority students with mathematical learning disabilities. Reported findings from the literature for typically developing students lead to the hypothesis that the mathematical learning gains of language minority students with mathematical learning disabilities and those of native speakers with mathematical learning disabilities may be influenced by similar variables. This would also mean that differences between language minority students with and without mathematical learning disabilities could be explained by math-related factors. This study will examine these assumptions by investigating the following research questions in the framework of a year-long study:

- Are the mathematical learning gains of language minority students with mathematical learning disabilities better explained by math-related or language-related factors?
- Are the mathematical learning gains of native speakers and language minority students with mathematical learning disabilities explained by the same or by different factors?

In order to investigate the relationship between math-related and language-related factors of language minority students with mathematical learning disabilities, it is also important to compare language minority students with mathematical learning disabilities and those without to find an answer to the following question:

- What factors explain the differences between the mathematical learning gains of language minority students with mathematical learning disabilities and those without?

It is assumed that the mathematical learning gains of language minority students with and without mathematical learning disabilities are explained by specific math-related factors.

6 Method

6.1 Participants

The participants were 70 third graders from Switzerland (32 classrooms). These students were selected from an initial sample of 888 students from 58 inclusive classrooms which participated voluntarily in a study on inclusive mathematics instruction (for selection criteria, see section "Measures"). All participants had written parental consent. To reduce the chances of influential variables in a highly selective sample (language minority students, students with mathematical learning disabilities), a matched-sample design was chosen.

Two samples were studied separately. The first sample included matched pairs of native and language minority students with mathematical learning disabilities (sample mathematical learning disabilities, $n = 40$; for criteria see below). The second sample consisted of matched pairs of language minority students with or without mathematical learning disabilities (language minority student sample, $n = 42$). Information on the language background of the students was gathered using a two-step procedure. First, data on languages spoken at home (first, second, third language) were collected using a teacher questionnaire, and students with a first language other than German, or with two first languages, were selected. Then, a language-use telephone interview was conducted with 52 of the 62 parents of these students, following the protocol established by Ritterfeld and Lüke (2011). Based on this information, the first-language variable was dummy-coded (German vs other).

Finding a sufficiently large sample of language minority students with mathematical learning disabilities was challenging due to the small population of such students. Therefore, 12 of the students were part of both samples. Because the analyses for each sample were conducted separately, the problem of dependency can be discounted. Mathematical achievement was measured at the end of grade 2 (t1), then eight months later with a post-test (t2), and at the end of grade 3 with a follow-up (t3). Information on selected language and control variables was collected at t1, after selecting the subsamples.

The mathematical learning disability sample consisted of 20 pairs of matched students ($n = 40$, Tab. 1) with below-average mathematical competence at t1 (for

Tab. 1: Demographic characteristics of the sample mathematical learning disabilities and the sample of language minority learners.

	Mathematical learning disabilities sample			Language minority learner sample		
	First language German	**Other first language**	**Total**	**With math disabilities**	**Without math disabilities**	**Total**
Students	20	20	40	21	21	42
Classes	16	15	24	17	14	23
Gender						
Boys	7	7	14	10	10	20
Girls	13	13	26	11	11	22

criteria, see below). Each pair comprised one student with German as his or her first language and one student with German as a second language. The pairs were matched by mathematical achievement at pre-test (difference math score ≤ 6 points), IQ (difference IQ score ≤ 8 points), age (difference ≤ 6 month), and gender. A Wilcoxon signed-rank test showed no significant difference between the groups on the basis of the matching criteria (math pretest $U = -1.09$, $p > .05$; IQ: $U = -0.13$, $p > .05$; age: $U = -0.11$, $p > .05$). A significant difference was found for socioeconomic status ($U = -3.36$, $p < .01$). The impact of this variable will be controlled in the analysis. The language minority students spoke the following first languages: Albanian (7), Croatian (1), Portuguese (4), Serbian (2), Tamil (2), Tigrinya (1), Turkish (2), and Urdu (1).

The language minority learner sample consisted of 21 matched pairs of students with German as a second language ($n = 42$, Tab. 1). One student from each pair had below-average mathematical achievement (for criteria, see below), and the other had average or good mathematical achievement (difference math score ≥ 10 points). The matching criteria were IQ (difference IQ score ≤ 8 points), age (difference ≤ 6 month), gender, and first language. A Wilcoxon signed-rank test showed no significant difference between the groups on the basis of the matching criteria IQ and age and socio-economic status (IQ: $U = -0.93$, $p > .05$; age: $U = -0.85$, $p > .05$; SES: $U = -0.80$, $p > .05$). As intended, the groups differed significantly in the math pretest ($U = -4.02$, $p < .001$). The language minority students spoke the following first languages: Albanian (9), Chinese (1), English (1), Italian (1), Croatian (2), Macedonian (2), Portuguese (8), Serbian (2), Tamil (6), Tigrinya (1), Turkish (8), and Urdu (1).

6.2 Measures

This study used highly selective samples with reduced variance. This can lead to the measures having low reliability. Therefore, with one exception, standardized tests and scales from standardized instruments were used. Nevertheless, some scales had to be excluded from the analyses due to low reliability scores. All tests were carried out in German. Sum scores were used for all scales.

6.2.1 Mathematics measures

General mathematical achievement: Standardized math tests to diagnose mathematical learning disability with norms for Germany and Switzerland (t1: Moser Opitz et al., 2020; t2 and t3: Moser Opitz, 2019) were conducted. The tests assess basic mathematical competences (e.g., understanding place value, number decomposition, doubling, halving, addition, subtracting, solving simple word problems). Linguistic requirements are minimal as most information is given with tables and pictures. In addition, the test administrators were allowed to read out the short instructions. The cut-off score for determining mathematical learning disability was set in the pre-test on the basis of the test norms (percentile 16) for all students. Average mathematical achievement was defined as scores that were the mean of the initial sample or higher. Table 2 gives an overview of the number of items and Cronbach's alpha.

Solving word problems: Researcher-designed scale with simple and complex comparison, combine, and exchange problems (details see Tab. 2).

Counting competences (counting forward and backwards by twos and tens) was also tested; however, this scale had to be excluded due to low reliability scores.

6.2.2 Language-related variables

In order to answer the research question, different language-related variables on sentence and on text level (reading comprehension, understanding semantic relationships), as well as on discourse level (verbal fluency) were assessed in an in-depth examination. Most of the scales are subtests of the standardized instrument SET5-10 constructed to diagnose language impairment (Petermann, 2010; number of items and Cronbach's alpha; see Tab. 2).

Tab. 2: Overview on the measures.

	Number of items	α sample mathematical learning disabilities	α sample of language minority students
Math pre-test (t1)	30	.62	.84
Math post-test (t2)	43	.81	.93
Math follow-up (t3)	43	.78	.93
Solving word problems	10	.71	.87
Listening comprehension	12	.71	.61
Verbal fluency	7	.75	.63
Reading comprehension	24	.84	.84
Verbal working memory	28	.84	.84

- Listening comprehension: understanding semantic relationships (Petermann, 2010)
- Verbal fluency (forming a sentence from a given word, e.g. lemon and sour; Petermann, 2010)
- Reading comprehension (subtest of K-ABC, Kaufman & Kaufman, 2009)

Mathematical vocabulary, cases, and plural formation were also examined. However, this data had to be excluded due to low reliability scores.

6.2.3 Control variables

Intelligence was tested with CFT 1 at t1 (Weiß & Osterland, 1997). The student's socioeconomic status was determined using the "books-at-home" index (Paulus, 2009), as measures like free meals are not available in Switzerland, and it was not possible to collect data such as the professional qualification of the mother. The books-at-home index involves showing pictures of five bookshelves with different numbers of books, from which the student is asked to choose the one that is most similar to that at home. To improve reliability, students were polled on three occasions (t1, t2, t3). The average of the responses was used in the analyses. Verbal working memory (repeating nonsense words) was tested in-depth using the Mottier-Test (Gamper et al., 2012). Visual processing speed was also assessed, but the reliability score was very low, and therefore this measure could not be used in the analyses.

Table 3 gives an overview on the descriptives of the two samples.

Tab. 3: Descriptives of math-related, language-related, and control variables of the samples.

	Mathematical learning disabilities sample			Language minority student sample		
	First language German	Second language German	All students	With learning disabilities	Without learning disabilities	All students
	M (SD)			M (SD)		
IQ	97.65 (11.70)	97.55 (11.74)	97.70 (11.57)	99.38 (11.66)	98.48 (11.17)	98.39 (11.28)
SES	4.50 (0.76)	2.70 (1.34)	3.6 (1.41)	2.95 (1.40)	3.29 (1.45)	3.12 (1.27)
Age month	102.40 (3.68)	102.30 (3.71)	102.30 (3.6)	103.71 (4.60)	104.76 (4.60)	104.24 (4.70)
Math t1	11.50 (2.19)	11.00 (2.94)	11.25 (2.57)	10.90 (2.55)	24.62 (2.58)	17.76 (7.39)
Math t2	16.85 (6.00)	12.70 (5.44)	14.78 (6.02)	13.52 (5.83)	27.19 (5.31)	20.36 (8.84)
Math t3	18.40 (5.34)	15.25 (5.33)	16.82 (5.50)	16.10 (6.61)	30.95 (4.80)	23.34 (9.45)
Word problems	5.15 (2.62)	3.95 (3.27)	4.55 (2.99)	3.95 (3.04)	6.14 (3.68)	5.05 (3.51)
Reading comprehension	16.50 (2.80)	12.20 (2.75)	15.35 (2.98)	13.67 (2.87)	15.48 (2.06)	14.57 (2.63)
Listening comprehension	9.25 (2.02)	7.40 (2.54)	8.33 (2.45)	7.05 (2.29)	8.14 (2.01)	7.60 (2.20)
Verbal fluency	9.45 (1.43)	6.40 (3.03)	7.93 (2.81)	6.14 (2.71)	6.62 (2.52)	6.38 (2.59)
Verbal working memory	22.90 (4.25)	22.90 (6.55)	22.90 (5.45)	23.67 (5.30)	25.00 (3.58)	24.33 (4.52)

6.3 Data analysis procedures

First, correlations were calculated for both samples. Second, due to the small sample size, which would allow for the inclusion of only few predictors, hierarchical multiple regression analyses were conducted with the dependent variable "mathematical achievement" at the post-test and the follow-up, with separate models. The residuals of these variables were normally distributed for the mathematical learning disabilities sample. This was not the case for the follow-up of the language minority student sample. Because of this and because some students dropped out at t3, the analysis for this sample was carried out with the dependent variable of the post-test only. In order to assess the impact of first language, two models were tested for the language minority student sample. One model put first language in a first step (minimal handicap), followed by math-related variables (math t1, solving word problems), control variables which are known as predictors for mathematical achievement (IQ, socio-economic status, working memory), and finally language-related variables (maximum handicap). In the second model, math t1 was placed in the first step, and first language was put in the last step (maximum handicap). This second model was formulated on the basis of the findings of Vukovic and Lesaux (2013) and Prediger et al. (2018), which show that language proficiency has an impact on mathematical achievement of all students regardless of the language background. For the language minority student sample, two analyses were conducted: one model in which math-related variables were included first, followed by control variables and language-related variables; and a second model with language variables at the first place (minimal handicap). The assumptions of multicollinearity and autocorrelation were not violated.

7 Results

7.1 Correlation analyses

Due to space limitations, only strong correlations are reported (without table). In the students with mathematical learning disabilities sample, a strong correlation ($r = .50^{**}$) was found for math t1 and math t3, and for math t2 and math t3 ($r = .72^{**}$). First language (German/other) was strongly negatively correlated with verbal fluency ($r = -.64^{**}$). In addition, reading comprehension and verbal fluency were strongly correlated ($r = .66^{**}$).

In the language minority student sample, a strong correlation was found between math at t1 and t2 ($r = .78^{**}$). In addition, verbal fluency correlated strongly with reading comprehension ($r = .62^{**}$) and listening comprehension ($r = .54^{**}$).

7.2 Hierarchical multiple regression analysis

7.2.1 Mathematical learning disabilities sample

In the first model for the post-test (Tab. 4), first language was included in step 1. R^2 was .12 ($f = .37$). R^2 rose significantly to $R^2 = .42$ ($\Delta F = 9.40$, $p < .001$) when math-related predictors were inserted with a strong effect size ($f = .85$) according to Cohen (1969). Control variables did not lead to a significant increase of R^2. However, for the language-related variables, R^2 rose significantly to .57 ($\Delta F = 3.02$, $p < .05$, effect size of the increase of R^2 $f = .39$).

Tab. 4: Hierarchical multiple regression analysis summary in sample mathematical learning disabilities (dependent variable post-test and follow-up mathematics) with first language included as first predictor.

Predictor	Post-test				Follow-up			
	R^2	ΔR^2	ΔF	P	R^2	ΔR^2	ΔF	P
Step 1 First language	.12	.12	5.28	.027	.08	.08	3.49	.07
Step 2 Math-related variables	.42	.30	9.40	.001	.54	.46	18.09	.000
Step 3 Control variables	.44	.02	.37	.778	.58	.04	1.01	.40
Step 4 Language-related variables	.57	.3	3.02	.045	.61	.03	.78	.556

The pattern of development over 12 months (dependent variable follow-up math with post-test math as one of the math-related variables) differed (Tab. 4). The proportion of the explained variance, when including first language in step 1, was .08 ($\Delta F = 3.49$, $p < .1$). The significance threshold was missed and the effect size ($f = .29$) is medium, according to Cohen (1969). R^2 increased to .54 when including the math-related variables ($\Delta F = 18.09$, $p < .001$). The effect size for the increase of $R^2 = .46$ was high, with $f = .92$. Neither the control variables nor the language-related variables explained the additional variance.

By including the math-related predictors in step 1 for the post-test (Tab. 5), R^2 was .37, with a very high effect size ($f = .77$). R^2 increased also significantly ($R^2 = .57$, $\Delta F = 4.25$, $p < .05$) when language variables were included. The effect size of the increase was $f = .47$, which is a high effect (Cohen, 1969). The first language variable as well as the control variables did not lead to a significant increase of R^2.

Tab. 5: Hierarchical multiple regression analysis summary in mathematical learning disabilities sample (dependent variable post-test and follow-up mathematics) with math-related variables included as first predictor.

Predictor	Post-test				Follow-up			
	R^2	ΔR^2	ΔF	p	R^2	ΔR^2	ΔF	p
Step 1 Math-related variables	.37	.37	10.83	.000	.54	.54	21.92	.000
Step 2 Control variables	.39	.02	0.44	.730	.56	.02	.46	.710
Step 3 Language-related variables	.57	.18	4.25	.013	.59	.03	.76	.527
Step 4 First language (German/other)	.57	.00	.10	.751	.61	.02	1.67	.207

When math-related variables were included in step 1 for the follow-up (Tab. 5), R^2 was .54 ($\Delta F = 21.92$, $p < .001$). The effect size of $f = 1.08$ was higher than that in the model with the posttest as the dependent variable ($f = .77$). Neither the control variables nor the language variables resulted in a significant change of R^2.

To sum up, math-related variables explained the highest proportion of the variance for mathematical progress, especially over the longer term. First language was found to have a significant impact only for the post-test, when included in the first step of the model.

7.2.2 Language minority student sample

In the language minority student sample, only two models were tested (for depended variable post-test math, see Section 6.3): one model, which put language-related variables in the first step (Tab. 6), and a second model, which inserted math-related variables in the first step (Tab. 7). Putting language-related variables in the first step resulted in $R^2 = .17$ ($\Delta F = 2.57$, $p = .069$. $f = .44$). R^2 increased to .67 ($\Delta F = 27.38$, $p < .001$) when math-related variables were given a heavy weighting with a very high effect size of R^2 ($f = .71$). No impact of control variables was found.

Tab. 6: Hierarchical multiple regression analysis summary in language minority student sample (dependent variable post-test math) with first language included as first predictor.

Predictor	R^2	ΔR^2	ΔF	p
Step 1 Language-related variables	.17	.17	2.57	.069
Step 2 Math-related variables	.67	.50	27.38	.000
Step 3 Control variables	.70	.03	1.12	.360

Tab. 7: Hierarchical multiple regression analysis summary in the language minority student sample (dependent variable post-test math) with math-related variables as first predictor.

Predictor	R^2	ΔR^2	ΔF	p
Step 1 Math-related variables	.66	.66	37.61	.000
Step 2 Control variables	.69	.03	1.15	.344
Step 3 Language-related variables	.7	.01	.44	.725

A very high proportion of the variance (66%) was explained by including the math-related variables ($\Delta F = 37.61$, $p < .001$; $f = 1.40$). No increase of R^2 occurred when control and language variables were added (Tab. 7).

In conclusion, in this sample, only math-related variables led to an increase of R^2.

8 Discussion

This study aimed to investigate which factors contribute to the mathematical learning gains of native speakers with mathematical learning disabilities and language minority students with mathematical learning disabilities. To answer this question, data on the mathematical achievement and on selected language variables (sentence and text, discourse level) as well as control variables were collected in a sample of students with mathematical learning disabilities (native speakers and language minority students) and a sample of language minority students with and without mathematical learning disabilities.

8.1 Correlation analysis

In the sample of students with mathematical learning disabilities, verbal fluency had the highest negative correlation. This is an interesting result which might have an effect on the discourse level. Verbal fluency was assessed with a measure of day-to-day language (forming a sentence with given words). Proficiency in day-to-day language is important for mathematical learning on discourse level, when expressing mathematical concepts. Moschkovic (2015) and Erath et al. (2018) emphasize the importance of students' participation in the classroom discourse in helping them to develop a conceptual understanding of mathematics. However, if students have problems with verbal fluency in day-to-day vocabulary, this may have an impact on their participation in classroom discourse and, therefore, hinder the mathematical learning process. Evidence from Bochnik (2017) supports this assumption. The study found that proficiency in German had an impact on mathematical achievement. However, no causal relationships can be drawn with the aforementioned correlation results.

8.2 Hierarchical multiple regression analysis

The results show that math-related variables had the highest impact on the mathematical learning gains of participants with learning disabilities in mathematics. A significant increase of R^2 for first language was found only when including this variable as a first step. An effect of language-related variables was found for the post-test. However, this effect disappeared completely at the follow-up, four months later. These findings lead to the conclusion that the slower rate of mathematical progress shown by students with mathematical learning disabilities seems to be caused by specific mathematical problems, rather than by language issues. This result is in line with the findings of Vukovic and Lesaux (2013). Their research with typically developing students shows that the relationship between language ability and mathematical cognition seems to be similar for language minority students and native speakers (see Prediger et al., 2018 for older students). This also seems to be true for students with mathematical learning disabilities, regardless of language background. Geary et al. (2017) found, in a sample of typically achieving students, that domain-general effects (intelligence, working memory, reading) on mathematics achievement remained stable across grades, whereas the overall mathematics-specific effects increased across grades. This could also be due to the cumulative nature of mathematics, and might be even more the case for students with mathematical learning disabilities and explains the absence of any effect from the control variables.

In the language minority student sample, only math-related variables had a significant impact on students' mathematical learning gains. This leads to the conclusion that the influences of language and control variables seem to be similar for language minority students, with or without mathematical learning disabilities.

8.3 Limitations

Some limitations of the study should be mentioned. First, the samples were small due to the fundamental challenge of finding a highly selective sample of language minority students with and without mathematical learning disabilities. The highly selective sample led to rather low alpha coefficients in some scales, even when standardized measures were used. The norms of these tests are based on representative samples of students, and might not be suitable for language minority students. Developing suitable measures and testing existing measures for text equity for language minority students are therefore major objectives for future research.

In addition, the choice of language variables and instruments could be queried. The reading test gives only a global score of reading competence, and using an instrument that assesses different components of reading competence could have resulted in different findings. Further, the test for examining working memory was language-based, and a measure without language requirements (e.g., visual processing) could have led to other results. Visual processing was tested, but had to be excluded from the analyses due to a low reliability score. The small sample size also meant that hierarchical regression analyses had to be performed, and the impact of single predictors, such as reading comprehension or working memory, could not be estimated. Therefore, it was not possible to outline effects on word, sentence and text, and discourse level.

Finally, because of the constraints of the budget, it was not possible to compare language minority students with average math achievement with a similar sample of native speakers. However, it is the first longitudinal study disentangling the relationship between language-related variables and mathematical learning gains.

9 Conclusion

The results of the study confirm the assertion by Moschkovich (2010) that it is important for studies to focus not only on differences between monolinguals and bilinguals, but also on their similarities. Our study provides evidence that one of these similarities is the profile of mathematical learning disabilities, irrespective of the student's first language. Our results indicate that these students need support both in second-language acquisition and in mathematics. Future research into this problem would benefit from studies with bigger sample sizes. A very interesting factor, which has rarely been investigated, is verbal fluency and its influence on participation in classroom discourse. This means that, as suggested by Vukovic (2012), research on mathematical learning disabilities and specifically on language minority students with mathematical learning disabilities has to focus on both numerical and language skills.

References

Bochnik, Kathrin (2017): *Sprachbezogene Merkmale als Erklärung für Disparitäten mathematischer Leistung. Differenzierte Analysen im Rahmen einer Längsschnittstudie in der dritten Jahrgangsstufe*. Münster: Waxmann.

Cohen, Jacob (1969): *Statistical Power Analysis for the Behavioral Sciences*. New York: Academic Press.

Cummins, Jim. (2000): *Language Power and Pedagogy*. Clevedon, UK: Multi Lingual Matters.

De Weerdt, Frauke, Desoete, Annemie & Roeyers, Herbert (2012): Working memory in children with reading disabilities and/or mathematical disabilities. *Journal of Learning Disabilities* 46 (5), 461–472.

Desoete, Annemie, Ceulemans, Annelies, Roeyers, Herbert & Huylebroeck, Anne (2009): Subitizing or counting as possible screening variables for learning disabilities in mathematics education or learning? *Educational Research Review* 4 (1), 55–66.

Erath, Kristin, Prediger, Susanne, Heller, Vivien & Quasthoff, Uta (2018): Learning to explain or explaining to learn? Discourse competences as an important facet of academic language proficiency. *Educational Studies in Mathematics* 99 (2), 161–179.
Forsyth, Suzanne R. & Powell, Sarah R. (2017): Differences in the mathematics-vocabulary knowledge of fifth-grade students with and without learning difficulties. *Learning Disabilities Research & Practice* 32 (4), 231–245.
Gamper, Hans, Keller, Ursula, Messerli, Nadine, Moser, Monique & Johannes, Wüst. 2012. *Normen für den Mottier-Test bei 4- bis 12-jährigen Kindern*. Praxisforschung der Erziehungsberatung des Kantons Bern. https://docplayer.org/25529012-Normen-fuer-den-mottier-test.html (accessed 16 October 2019).
Geary, David C., Hoard, Mary K. & Bailey, Drew H. (2012): Fact retrieval deficits in low achieving children and children with mathematical learning disability. *Journal of Learning Disabilities* 45 (4), 291–307.
Geary, David C., Nicholas, Alan, Yaoran, Li & Sun, Jianguo (2017): Developmental change in the influence of domain-general abilities and domain-specific knowledge on mathematics achievement: An eight-year longitudinal study. *Journal of Educational Psychology* 109 (5), 680–693.
Gonzáles, Taucia & Artiles, Alfredo J. (2015): Reframing venerable standpoints about language and learning differences: The need for research on the literate lives of Latina/o language minority students. *Journal of Multilingual Education Research* 6 (3). https://fordham.bepress.com/jmer/vol6/iss1/3/. (accessed 16 October 2019).
Haag, Nicole, Heppt, Birgit, Roppelt, Alexander & Stanat, Petra (2015): Linguistic simplification of mathematics items: Effects for language minority students in Germany. *European Journal of Psychology of Education* 30 (2), 145–167.
Herzog, Moritz, Ehlert, Antje & Fritz, Annemarie (2019): Developement of a sustainable place value understanding. In Annemarie Fritz, Victor G. Haase, Räsänen, Pekka (eds.): *International Handbook of Mathematical Learning Difficulties*. Cham: Springer, 561–579.
Jorgensen, Robyn (2011): Language, culture and learning mathematics: A Bourdieuian analysis of indigenous learning. In Wyatt-Smith, Claire, Elkins, John, Gunn, Stephanie (eds.): *Multiple Perspectives on Difficulties in Learning Literacy and Numeracy*. Dordrecht: Springer, 315–329.
Kaufman, Alan S. & Kaufman, Nadeen L. (2009): *Kaufman Assessment Battery for Children K-ABC*. German Version. Amsterdam: Swets & Zeitlinger.
Kingsdorf, Sheri & Krawec, Jennifer (2014): Error analysis of mathematical word problem solving across students with and without learning disabilities. *Learning Disabilities & Practice* 29 (2), 66–74.
Klein, Elise, Bahnmueller, Julia, Mann, Anne, Pixner, Silvia, Kaufmann, Liane, Nuerk, Hans-Christoph & Moeller, Korbinian (2013): Language influences on numerical development – Inversion effects on multi-digit number processing. *Frontiers in Psychology* 4, 1–6.
Koponen, Tuire, Aro, Mikko, Poikkeus, Anna-Maija, Niemi, Pekka, Lerkkannen, Marja-Kristiina, Ahonen, Timo & Nurmi, Jari-Erik (2018): Comorbid fluency difficulties in reading and math: Longitudinal stability across early grades. *Exceptional Children* 84 (3), 298–311.
Mann Koepke, Kathleen & Miller, Brett (2013): At the intersection of math and reading disabilities: Introduction to the special issue. *Journal of Learning Disabilities* 46 (6), 483–489.
Martiniello, Maria (2008): Language and the performance of English-language learners in math word problems. *Harvard Educational Review* 78 (2), 333–368.

Miura, Irene T., Yukari Okamoto, Chungsoon C., Kim, Marcia Steere & Fayol, Michel (1994): First grader's cognitive representation of number and understanding of place value: Cross-national comparisons – France, Japan, Korea, Sweden, and the United States. *Journal of Educational Psychology* 85 (1), 24–30.

Moeller, Korbinian, Pixner, Silvia, Zuber, Julia, Kaufmann, Liane & Nuerk, Hans-Christoph (2011): Early place-value understanding as a precursor for later arithmetic performance – A longitudinal study on numerical development. *Research in Developmental Disabilities* 32 (5), 1837–1851.

Morek, Miriam & Heller, Vivien (2012): Bildungssprache – Kommunikative, epistemische, soziale und interaktive Aspekte ihres Gebrauchs [Academic language – Communicative, epistemic, social and interactive aspects of its use]. *Zeitschrift für Angewandte Linguistik* 57, 67–101.

Morgan, Candia, Craig, Tracy, Schütte, Marcus & Wagner, David (2014): Language and communication in mathematics education: An overview of research in the field. *ZDM – International Journal on Mathematics Education* 46 (6), 843–853.

Moschkovich, Judit (2010): Recommendations for research on language and mathematics education. In Moschkovich, Judit (ed.): *Language and Mathematics Education*. Charlotte, NC: Information Age, 1–28.

Moschkovich, Judit (2015): Academic literacy in mathematics for English learners. *Journal of Mathematical Behaviour* 40, 43–62.

Moser Opitz, Elisabeth, Meret Stöckli, Grob, Urs, Nührenbörger, Marcus & Reusser, Lis (2019): *BASIS-MATH-G 3⁺. Basisdiagnostik Mathematik für das vierte Quartal der dritten Klasse und das erste Quartal der vierten Klasse*. Bern: Hogrefe.

Moser Opitz, Elisabeth, Meret Stöckli, Grob, Urs, Nührenbörger, Marcus & Reusser, Lis (2020): *BASIS-MATH-G 2⁺. Basisdiagnostik Mathematik für das vierte Quartal der zweiten Klasse und das erste Quartal der dritten Klasse*. Bern: Hogrefe.

Nelson, Gena & Powell, Sarah R. (2017): A systematic review of longitudinal studies of mathematics difficulty. *Journal of Learning Disabilities* 51 (6), 523–539.

Paetsch, Jennifer & Felbrich, Anja (2016): Longitudinale Zusammenhänge zwischen sprachlichen Kompetenzen und elementaren mathematischen Modellierungskompetenzen bei Kindern mit Deutsch als Zweitsprache. *Psychologie in Erziehung und Unterricht* 63 (1), 16–33.

Paulus, Christoph M. 2009. *Die Bücheraufgabe zur Bestimmung des kulturellen Kapitals bei Grundschülern*. http://psydok.psycharchives.de/jspui/handle/20.500.11780/3344 (accessed 16 October 2019).

Peake, Christian, Jiménez, Juan E., Rodriguez, Cristina, Bisschop, Elaine & Villarroel, Rebeca (2015): Syntactic awareness and arithmetic word problem solving in children with and without learning disabilities. *Journal of Learning Disabilities* 48 (6), 596–601.

Petermann, Franz (2010): *Sprachstandserhebungstest für Kinder im Alter zwischen 5 und 10 Jahren (SET 5–10)*. Göttingen: Hogrefe.

Prediger, Susanne, Erath, Kirstin & Moser Opitz, Elisabeth (2019): The language dimension of mathematical difficulties. In Annemarie Fritz, Victor P. Haase, Räsänen, Pekka (eds.): *International Handbook of Mathematical Learning Difficulties*. Cham: Springer, 437–455.

Prediger, Susanne, Wilhelm, Nadine, Büchter, Andreas, Gürsoy, Erkan & Benholz, Claudia (2018): Language proficiency and mathematics achievement – Empirical study of language-induced obstacles in a high stakes test, the central exam ZP10. *Journal für Mathematik-Didaktik* 39 (1), 1–26.

Ritterfeld, Ute & Lüke, Carina (2011). Mehrsprachen-Kontexte. Erfassung der Inputbedingungen von mehrsprachig aufwachsenden Kindern [Multilingual contexts. Recording of the input condition of multilingually raised children]. Dortmund: Technische Universität. Retrieved from http://www.sk.tu-dortmund.de/media/other/Mehrsprachen-Kontexte.pdf. (accessed 18 April 2020).
Rodriguez, Diane, Parmar, Rene & Signer, Barbara (2001): Fourth-grade culturally and linguistically diverse exceptional student's concepts of number lines. *Exceptional Children* 67 (2), 199–210.
Schleppegrell, Mary J. (2004): *The Language of Schooling: A Functional Linguistics Perspective.* Mahwah: Lawrence Erlbaum.
Snow, Catherine E. & Uccelli, Paola (2009): The challenge of academic language. In Olson, David R., Torrance, Nancy (eds.): *The Cambridge Handbook of Literacy.* Cambridge: University Press, 112–133.
Stock, Pieter, Desoete, Annemie & Roeyers, Herbert (2010): Detecting children with arithmetic disabilities from kindergarten: Evidence from a 3-year longitudinal study on the role of preparatory arithmetic abilities. *Journal of Learning Disabilities* 43 (3), 250–268.
Vukovic, Rose K. (2012): Mathematics difficulty with and without reading difficulty: Findings and implications from a four-year longitudinal study. *Exceptional Children* 78 (3), 280–300.
Vukovic, Rose K. & Lesaux, Nonie K. (2013): The language of mathematics: Investigating the ways language counts for children's mathematical development. *Journal of Experimental Child Psychology* 115 (2), 227–244.
Vukovic, Rose K. & Siegel, Linda S. (2010): Academic and cognitive characteristics of persistent mathematics difficulty from first through fourth grade. *Learning Disabilities Research and Practice* 25 (1), 25–38.
Warren, Elizabeth & Miller, Jodie (2015): Supporting English second-language learners in disadvantaged contexts: Learning approaches that promote success in mathematics. *International Journal of Early Years Education* 23 (2), 192–208.
Weiss, Rudolf & Osterland, Jürgen (1997): *Grundintelligenztest Skala 1[Culture fair intelligence test] (CFT 1)*, 5th. Göttingen: Hogrefe.
Zhang, Dake & Xin, Yan Ping (2012): A follow-up meta-analysis for word-problem-solving interventions for students with mathematics difficulties. *The Journal of Educational Research* 105 (5), 303–318.

Margot Buyle, Cathy Marlair, and Virginie Crollen

Blindness and deafness: A window to study the visual and verbal basis of the number sense

1 Introduction

When children acquire numerical skills, they have to learn a variety of specific numerical tools. The most obvious are the numerical codes such as number words (one, two, three, etc.) or Arabic numerals (1, 2, 3, etc.). Other skills will be relatively more abstract: arithmetical facts (i.e., $4 \times 2 = 8$), arithmetical procedures (i.e., borrowing), or arithmetical laws (i.e., $a + b = b + a$). The acquisition of these numerical tools is complex and probably not facilitated by the fact that a numerical expression does not have a single meaning. Indeed, numbers can be used as a kind of label or proper name (i.e., Bus 51). They can also be part of a familiar fixed sequence (i.e., 51 comes immediately after 50 and before 52). They can be used to refer to continuous analogue quantities (i.e., 51,2 grams) (Butterworth, 2005; Fuson, 1988) and, most importantly, they can be used to denote the number of things in a set – the *cardinality* of the set.

Children are able to understand the special meaning of cardinality because they possess a specific and innate capacity for dealing with quantities (Feigenson et al., 2004). Supporting the innate nature of the "number sense," it has been found, for instance, that fetuses in the last trimester are already able to discriminate auditory numerical quantities (Schleger et al., 2014). A large set of behavioral studies using the classic method of habituation has also revealed sensitivity to small numerosities (e.g., Starkey & Cooper, 1980) in young children. In the study of Starkey and Cooper (1980), for example, slides with a fixed number of 2 dots were repeatedly presented to 4- to 6-month-old infants until their looking time decreased, indicating habituation. At that point, a slide with a deviant number of 3 dots was presented and yielded significantly longer looking times, indicating dishabituation and therefore discrimination between the numerosities 2 and 3. This effect was replicated with newborns (Antell & Keating, 1983) and with various stimuli such as sets of realistic objects (Strauss & Curtis, 1981), targets in motion (Van Loosbroek & Smitsman, 1990; Wynn et al., 2002), two- and three-syllable words (Bijeljac-Babic et al., 1991), or puppet making two or three sequential jumps (Wood & Spelke, 2005; Wynn, 1996). However, it was not observed

Note: Margot Buyle and Cathy Marlair are joint first authors on this work.

https://doi.org/10.1515/9783110661941-014

with other numerosities such as 4 and 6 (Starkey & Cooper, 1980). Taken together, these data therefore suggest that infants may possess a concept of small numbers which is dependent on the absolute number of items presented (up to 3 or 4).

Besides their ability to discriminate particularly small numerosities, infants are also able to discriminate large numerosity sets. However, this discrimination ability differs dramatically from that observed with small numerosities: performance no longer depends on the absolute number of items presented but on the numerical ratio that separates the two numerosities to be discriminated. Hence, while 6-month-old infants are able to discriminate numerosities with a 1:2 ratio (4 vs. 8, 8 vs. 16 and 16 vs 32 stimuli) (Brannon et al., 2004; Lipton & Spelke, 2003, 2004; Wood & Spelke, 2005; Xu, 2003; Xu & Spelke, 2000; Xu et al., 2005), they fail to discriminate numerosities with a 2:3 ratio (8 vs. 12 and 16 vs. 24) (Xu & Spelke, 2000). By 9 or 10 months, however, the precision of the representation improves since infants become able to discriminate between 8 and 12 elements (Lipton & Spelke, 2003; Xu & Arriaga, 2007).

Following these two waves of investigations on numerosity discrimination in infancy, it has been suggested that basic numerical abilities could be sustained by two different proto-mathematical systems (e.g., Butterworth, 1999; Carey, 2001; Dehaene, 1992; Feigenson et al., 2004; Xu, 2003; Xu & Spelke, 2000): **(1)** an object-tracking system (e.g., Scholl, 2001; Trick & Pylyshyn, 1994; Uller et al., 1999) which has a clear limit on set size (3 or 4), allows the precise representation of small number of objects, and provides a basis for the verbal and accurate quantification process of counting; **(2)** a large numerosity estimation system, commonly known as the Approximate Number System (ANS), which has no set size limit and allows the approximate representation of large non-symbolic numerosity sets (Xu, 2003; Xu & Spelke, 2000).

Formal mathematics could therefore emerge from these two proto-mathematical systems. A commonly accepted hypothesis is that symbolic number representations acquire their numerical meaning through being mapped onto non-symbolic representations. Accordingly, many visuo-spatial and verbal processes have been linked to the development of complex numerical skills. Among these, we could cite visual working memory (WM; Bull et al., 2011a; LeFevre et al., 2010), visual attention (Anobile et al., 2013), visuo-spatial mental rotation (Reuhkala, 2001), basic visual perception (Lourenco et al., 2012; Tibber et al., 2013), visual movement perception (Sigmundsson et al., 2010), phonological awareness (Alloway et al., 2005; Leather & Henry, 1994; Simmons & Singleton, 2008), and the relative linguistic transparency of the language used (Almoammer et al., 2013; Miller et al., 1995; Miura et al., 1988; Miura & Okamoto, 1989; Miura et al., 1993; Miura et al., 1994).

Interestingly, the relative importance of these visual and verbal processes seems to vary with age and the specific numerical task that is investigated. Several reports, for example, demonstrated that visuo-spatial and verbal processes might be differently engaged in the resolution of specific arithmetic operations. While the resolution of subtraction and addition principally relies on finger-based and visuo-spatial calculation strategies (Siegler & Shrager, 1984), overlearned simple multiplication facts have been associated with verbal memory retrieval (Cooney et al., 1988). Similarly, a recent theoretical framework has developed the idea that different types of spatial information might be engaged in different numerical tasks. In Western populations, people tend to represent number along a left-to-right-oriented mental number line (MNL; Dehaene, 1989, 1992). A compelling demonstration of this strong association between numbers and space resides in the SNARC effect. This effect refers to the observation that responses to small numbers are faster in the left side of space, while responses to large numbers are faster in the right side of space (Dehaene et al., 1990). Because the SNARC effect was observed even when participants crossed their hands, numbers were assumed to be mapped onto an external frame of reference (Dehaene et al., 1993), where small and large numbers facilitate responses in the left and right sides of space irrespective of the hand of response. More importantly, however, while the SNARC effect was assumed to primarily originate from visuo-spatial associations in magnitude comparison tasks, it was assumed to primarily arise from verbal associations in parity judgment tasks (Herrera & Macizo, 2008; van Dijck et al., 2009).

Given the importance of visual and verbal processes in the development of the number concept, it is not surprising to see that visual and verbal deficits can prevent the acquisition of numerical skills. Hence, it has been reported that children presenting non-verbal learning disability frequently show comorbid mathematics learning difficulties (Crollen et al., 2015). Difficulties in arithmetic are also remarkably common in dyslexia, particularly when it comes to retrieving arithmetic facts from semantic long-term memory, as is the case in multiplication (De Smedt & Boets, 2010; Göbel, 2015; Simmons & Singleton, 2008; Träff & Passolunghi, 2015). A possible explanation for this finding is that numerical processing might be influenced by visuo-spatial and phonological processes (Dehaene et al., 2003; De Smedt & Boets, 2010; Geary & Hoard, 2001). However, to date, it is still unknown whether visuo-spatial and verbal processes are mandatory or only co-vary with the development of good numerical skills (Szücs, 2016).

Evidence from sensory-deprived individuals offers a unique opportunity to address this question. If vision and language play a foundational role in the development of the concept of numbers, then blind and deaf individuals should present atypical numerical behavior. Alternatively, if the "number sense" can be acquired without vision and without typical verbal input, then blind and deaf individuals should show typical numerical abilities. In this chapter, we will address this question by reviewing recent empirical evidence examining mathematical reasoning in blind and deaf individuals. We will call into question arguments stating that (1) the development of the number concept is scaffolded mainly by visuo-spatial development, and (2) that language becomes integrated only after the concepts are created.

2 Numerical processing in the blind

Although vision has long been considered as critical in the emergence of numerical representations and skills, a growing set of studies on numerical performances following early visual deprivation nevertheless indicate that blind individuals perform as efficiently as their normally seeing peers in various numerical tasks. It was shown, for instance, that early and congenitally blind participants were as good as sighted individuals in number comparison tasks with both small and large numbers (Castronovo & Seron, 2007a; Szücs & Csépe, 2005). Similar results were found when they had to perform parity judgment task (Castronovo & Seron, 2007a). In addition to that, blind participants have also been shown to perform as accurately as their sighted peers in counting (Crollen et al., 2014) and subitizing tasks (i.e., fast and accurate processing of a small collection of up to three or four elements; Ferrand et al., 2010). Interestingly and more surprisingly, the lack of vision since early age might even have a positive impact on some numerical skills. When submitted to a numerical estimation task, blind individuals indeed demonstrated enhanced abilities as compared to sighted participants, especially when the task involved touch and proprioception (i.e., key press estimation task): they showed greater accuracy in both small (up to 9; Ferrand et al., 2010) and large numerical ranges (up to 64; Castronovo & Seron, 2007b). Moreover, this greater estimation skill in blind individuals was not specific to a particular modality (i.e., tactile), neither to a familiar numerical context (i.e., close to their daily life use of numerical information – a footstep production task). It was indeed found in more unfamiliar contexts requiring verbal, non-tactile processing (i.e., non-word repetition task; Castronovo & Delvenne, 2013). Finally, early blind participants also showed enhanced abilities to perform arithmetic operations: addition, subtraction, and especially multiplication of different complexities (Dormal et al., 2016). In sum, there is plenty of evidence suggesting that early visual deprivation does not prevent the development of good numerical skills but may even induce greater efficiency in estimation and manipulation of numerical quantities.

So far, it seems clear that blind individuals are capable of developing a good number understanding. However, it is still unknown whether the representation of numerical magnitude in visually deprived individuals shares the same "visuo"-spatial properties as the one of the sighted. The close connection between numbers and space is well established in the literature of numerical cognition and illustrated by the metaphor of the logarithmic and left-to-right-oriented mental number line (MNL; Castronovo & Seron, 2007a; de Hevia et al., 2008). This mental spatial organization of numbers has been supported by the recurrent observation of three main effects: (1) the size effect, which refers to the fact that larger numbers are less easy to discriminate than smaller ones (e.g., 2 vs. 4 is easier than 8 vs. 10; Ashcraft & Stazyk, 1981; Brysbaert, 1995; Cantlon et al., 2009; McCloskey et al., 1991); (2) the distance effect, which refers to the fact that it is easier to discriminate numbers that are distant on the MNL than numbers that are close to each other (e.g., 2 vs. 8 is easier than 2 vs. 4; Cantlon et al., 2009; Moyer & Landauer, 1967); and (3) the Spatial Numerical Associations of Response Codes (SNARC) effect, which refers to the observation that responses to small and large numbers are faster when performed in the left and right sides of space respectively (Fias & Fischer, 2005; Wood et al., 2008). Because the SNARC effect was observed even when participants crossed their hands, numbers were assumed to be mapped onto an external frame of reference, where small and large numbers facilitate responses in the left and right sides of space irrespective of the responding hand (Dehaene et al., 1993; Fias & Fischer, 2005). Supporting the idea that the MNL is oriented from left to right, it was also observed that people tend to overestimate the leftward space on the MNL (Loftus et al., 2008). When asked to perform a numerical bisection task (which consists in estimating, without calculating, the number midway between two others), participants indeed present a leftward bias (also called the "pseudo-neglect" effect): they systematically tend to mis-bisect the numerical interval slightly to the left of its objective midpoint. Finally, the spatial organization of numbers was also found in tasks involving arithmetic operations (Masson & Pesenti, 2014; McCrink et al., 2007). It was indeed suggested that additions and subtractions involve attentional shifts along the MNL: toward the right (i.e., larger numbers) for addition and toward the left (i.e., smaller numbers) for subtractions (Knops et al., 2009a, 2009b; Masson & Pesenti, 2014; McCrink et al., 2007; Pinhas & Fischer, 2008).

Does vision root the construction of the relationship between numbers and space? Interestingly, many studies indicate that visually deprived individuals possess a numerical magnitude representation that shares the same spatial characteristics as the one of sighted individuals. Congenitally blind participants indeed show similar distance and size effects as sighted individuals when submitted to number comparison (Castronovo & Seron, 2007a; Szücs & Csépe, 2005) and parity judgment tasks (Castronovo & Seron, 2007a). In addition to that, they both show a pseudo-neglect effect (i.e., a leftward bias) when asked to perform a numerical bisection task (Cattaneo et al., 2011). Both blind and sighted participants also present the SNARC effect when they have to indicate the parity status of a number (odd or even) by means of a manual response with the left or the right hand (Castronovo & Seron, 2007a; Crollen et al., 2013; Szücs & Csépe, 2005). The same SNARC effect was furthermore observed when blind and sighted participants had to judge whether a presented number was smaller or larger than 5 (Crollen et al., 2013). However, unlike sighted individuals, blind individuals were shown to present a reversed SNARC effect when performing the numerical comparison task with the hands crossed over the body midline (Crollen et al., 2013). They indeed produced faster responses to small numbers in the right space (i.e., with the left hand) and to large numbers in the left space (i.e., with the right hand). Consequently, it was proposed that blindness may shape the frame of reference onto which numbers are represented: while sighted individuals rely on a world-centered (external) representation of space, blind individuals rather use a representation that is body-centered (internal). Importantly, blind participants did not present any reversed SNARC effect when performing the parity judgment task with the hands crossed over the body midline. Because this task is thought to involve verbal-spatial processes instead of visuo-spatial processes (van Dijck et al., 2009; but see Huber et al., 2016 for alternative findings), it was therefore suggested that early visual experience shapes the nature of the visual association between numbers and space but does not influence the verbal one (Crollen et al., 2013). Although blind individuals develop a spatial representation of numerical magnitude similar to the one of sighted individuals, their use of this representation might therefore present some specificities.

Another evidence supporting the idea that blindness shapes some qualitative aspects of the numerical processing can be found in the interactions that occur between numbers and fingers (Crollen et al., 2011, 2014). The latter has been illustrated by the use of the finger counting strategy, a procedure that often accompanies the development of basic arithmetic in the sighted population. Although the use of finger counting was thought to play a functional role in the development of a mature numerical system (Butterworth, 2005), it has been shown that blind children less spontaneously use this strategy while learning counting and arithmetic (Crollen et al., 2011, 2014). Fingers were indeed assumed to permit the assimilation of basic numerical skills

(Andres et al., 2008) and the connection between non-symbolic and symbolic numerosities (Fayol & Seron, 2005). However, Crollen and colleagues (2011; 2014) observed that blind children, compared to their sighted peers, used their fingers less spontaneously and in a less canonical way to count and show quantities, despite similar counting performances. Consequently, it was suggested that visual experience drives the establishment of finger-number interactions.

Differences between blind and sighted individuals were also found in more complex numerical domains like arithmetic skills. As already mentioned, early blind individuals show enhanced abilities to perform addition, subtraction, and multiplication of different complexities (Dormal et al., 2016). In addition to these observations, it was recently suggested that blindness may shape the neural foundations of arithmetic reasoning. While it is widely recognized that mathematical skills are supported by a bilateral fronto-parietal network in sighted and blind individuals (Amalric & Dehaene, 2016, 2019; Amalric et al., 2018; Arsalidou & Taylor, 2011; Dehaene et al., 2003), imaging studies have indeed highlighted that blind, but not sighted, participants recruit some early visual areas in addition to this math-responsive network while calculating (Crollen et al., 2019; Kanjlia et al., 2016). Consequently, it was suggested that the occipital cortex – which typically process visual information – might be cognitively pluripotent (i.e., capable of assuming non-visual cognitive functions; Kanjlia et al., 2016). However, this conclusion does not take into account the computational relation between number and visual processes, nor does it consider the variety of strategies that are used while solving arithmetic operations (Campbell & Timm, 2000; Dehaene & Cohen, 1997; Hecht, 1999). Indeed and as mentioned previously, visuo-spatial procedures are principally used to solve subtractions while retrieval is the dominant method for solving easy and overlearned multiplications (Ashcraft, 1992; Campbell & Xue, 2001). It is therefore possible that early visual deprivation selectively affects the brain organization of visuo-spatial arithmetic operations (subtraction) while keeping intact the neural network involved in the arithmetic operations learned by rote verbal memory (multiplication).

This later assumption was recently supported by a study contrasting the brain activity of blind and sighted participants while performing subtraction vs. multiplication arithmetic operations (Crollen et al., 2019). An enhanced activity of the occipital cortex was indeed observed in the blind while they performed subtraction, but was not observed for the multiplication operations (Crollen et al., 2019). This recent study therefore challenges the idea that the brain is cognitively pluripotent and rather suggests that the recruitment of the occipital cortex in the blind actually relates to its intrinsic computational role (Crollen et al., 2019). It is also interesting to note that the brain results obtained by

Crollen et al. (2019) are reminiscent of the behavioral results observed with the SNARC: when crossing their hands over the body midline, blind individuals indeed showed a reversed SNARC effect in tasks relying on visuo-spatial processes (numerical comparison), but not in tasks involving verbal processes (parity judgment; Crollen et al., 2013).

Altogether, numerous studies clearly indicate that early visual experiences are not essential for the development of numerical cognition. Indeed, evidence has shown that (1) blind individuals perform as efficiently as their sighted peers in various numerical tasks; (2) blind individuals even possess enhanced abilities to perform numerosity estimation and calculation tasks; (3) blind individuals develop a numerical magnitude representation that shares the same spatial properties as sighted individuals. However, blindness was found to shape some qualitative aspects of numerical processing. For instance, it was shown that (1) blindness impacts the reference frame in which the associations between numbers and space occur; (2) the lack of vision reduces the use of finger counting strategies; and (3) blindness shapes the neural foundations of arithmetic operations relying on visuo-spatial processes.

Although blindness was found to have a positive impact on numerical abilities such as estimation and calculation, the mechanisms sub-serving these greater abilities are still unknown. One possible explanation lies in the use of enhanced high-level cognitive processes such as WM (Castronovo & Delvenne, 2013). As WM and numerical abilities are linked to each other (De Smedt et al., 2009; Simmons et al., 2012) and as blind individuals present greater WM skills than sighted participants (Crollen et al., 2011; Hull & Mason, 1995; Swanson & Luxenberg, 2009), numerical skills in blind individuals could indeed potentially be accounted by the use of enhanced WM processes.

Finally, although blind individuals present greater skills in some numerical domains, it is still unknown whether they would present some delays in other abilities not tested so far. It could be worth examining basic knowledge of geometry as this mathematical knowledge is intrinsically linked to visuo-spatial representations. Answering these questions in the future may potentially have important implications for mathematics teaching and mathematics rehabilitation programs.

3 Numerical processing in the deaf

Deaf signers show advantages in various visual domains. They, for example, outperform their hearing peers in the speed of shifting visual attention and visual scanning and in the peripheral detection of motion (Bavelier et al., 2000;

Chinello et al., 2012; Proksch & Bavelier, 2002). These advantages are suggested to be primarily due to their experience with sign language since this includes a highly significant spatial component. Accordingly, it is assumed that the primacy of visual cognition in deaf signers may influence numerical skills, and that the spatial components of sign language may have an impact on some visuo-spatial features of the mental number line (Chinello et al., 2012). Despite this common-sense conception, the visuo-spatial advantages of deaf do not appear to support or enhance deaf students' performance compared to hearing students (Ansell & Pagliaro, 2006; Borgna et al., 2018; Marcelino et al., 2019 for review). Numerous studies have consistently showed that deaf children from preschool onward through their school years into higher education, as well as deaf adults tend to be slower and less accurate in numerical processing compared to their hearing counterparts (Ansell & Pagliaro, 2006; Blatto-Vallee et al., 2007; Bull et al., 2006, 2005, 2018, 2011b; Chinello et al., 2012; Korvorst et al., 2007; Marschark et al., 2013, 2015; Rodríguez-Santos et al., 2014; Zarfaty et al., 2004). Several studies have indeed indicated that deaf pupils experience a delay of 2 to 3.5 years in comparison with hearing children on mathematical achievement tests (Bull et al., 2005; Nunes & Moreno, 2002). Growth curves of deaf students are identified to be much flatter than those for hearing learners (Zarfaty et al., 2004) and differences are often noted in: (1) standardized achievement tests; (2) measurement and number concepts; (3) understanding fractions; (4) computation and reasoning; (5) logical thinking; (6) communication about time; and (7) problem solving (Allen, 1995; Ansell & Pagliaro, 2006; Austin, 1975; Bull et al., 2011b; Marschark & Everhart, 1999; Nunes & Moreno, 2002; Pagliaro & Kritzer, 2013; Rodríguez-Santos et al., 2014; Traxler, 2000; Titus, 1995; Zarfaty et al., 2004). Geometry, in contrast, is indicated as an area of strength (Pagliaro & Kritzer, 2013).

While the competences mentioned above are quite complex, some other research has been performed on rather simple abilities such as subitizing. Deaf individuals could have an advantage in performing subitizing tasks because of their enhanced abilities in some aspects of visual and spatial processing (Bull et al., 2006). Nevertheless, the patterns of results are found to be very similar for both deaf and hearing individuals and this is also true for different presentation formats (symbolic and non-symbolic) (Bull et al., 2006). Basic differences in subitizing skills are therefore not believed to be the roots of the mathematical difficulties observed in the deaf (Bull et al., 2006).

Besides subitizing, the accuracy to discriminate quantities is restricted by the ratio difference of the quantities being compared. The closer this ratio is to 1, the more difficult the discrimination of the magnitude will be. Hearing as well as deaf individuals show a distance effect when they are asked to make magnitude judgments, but not when they have to make physical size judgments (Bull et al., 2018). A similar distance effect is found when sign language is used as representation mode in both deaf and hearing adults. This finding implicates that the signed numbers automatically activate information about magnitude for both groups (Bull et al., 2006, 2005). Furthermore, similar size and distance effects are seen in symbolic and non-symbolic tasks for deaf and hearing participants, which demonstrates that deaf individuals have no deficits in building abstract symbolic and non-symbolic numerical representations. Nonetheless, slower reaction times were observed for the symbolic task in deaf individuals. This suggests that both groups have similar quantity representations, but that deaf participants might experience a delay in accessing representations from symbolic codes (Rodriguez-Santos et al., 2014). This conclusion has also been reached in number-to-position tasks, requiring participants to estimate a number's position on a 0–100 number line (Borgna et al., 2018; Bull et al., 2011). Deaf students have consistently made less accurate number-line estimations (Borgna et al., 2018; Bull et al., 2011b) than their hearing peers. This accuracy difference has been found at very young age before much exposure to formal education has taken place (Bull et al., 2018). Deaf individuals indeed appear to be more accurate in arithmetic estimation tasks involving non-symbolic stimuli (Masataka, 2006). In contrast, tasks requiring symbolic processing appear to be more challenging for deaf individuals than for their hearing peers, as this relies more on linguistic skills (Masataka, 2006; Rodriguez-Santos et al., 2014). A reduced accuracy in estimation for deaf participants may therefore be apparent only when number meaning has to be accessed from symbols (Masataka, 2006). This assumption has been called the "access deficit hypothesis" and was first proposed to explain difficulties of children who present mathematical learning disabilities (Rousselle & Noël, 2007). According to Masataka (2006), the difference in performance between deaf and hearing adults might be related to the variability in WM architecture, which is due to the difference of languages both groups acquired. Our WM is traditionally divided into two major domains, namely, a verbal and a visuo-spatial domain. The existence of a sign-based rehearsal loop mechanism that is parallel to the speech-based rehearsal loop is provided in adults who acquire a sign language as their first language, which could thus possibly account for their superior capacity to execute non-symbolic arithmetic (Masataka, 2006).

The ability to compare numbers and to judge whether numbers are odd or even also represents a basic numerical skill. When performing such tasks, both deaf and hearing individuals show faster responses to low numbers with the left hand and to high numbers with the right hand (i.e., SNARC effect). This has been demonstrated with Arabic digits as well as with sign language numerals

(Bull et al., 2006; Chinello et al., 2012) and could therefore indicate that, just like Arabic digits, sign language number signs may be directly mapped into an underlying left-to-right-oriented representation of magnitude (Bull et al., 2006). However, the speed of deaf participants making the SNARC decision in congruent condition (low numbers, left response box) was similar to the speed of hearing participants making the SNARC decision in the incongruent condition (low numbers, right response box) (Bull et al., 2005; Chinello et al., 2012; Iversen et al., 2004). This, again, demonstrates that the processing of numbers may be slower in deaf individuals. In 2007, Korvorst and colleagues presented number triplets (in Arabic digits or in sign language) to deaf and hearing adults who had to determine if the middle number was the numerical mean of the two outer numbers. Hearing individuals appeared to be faster in confirming valid bisection. In the sign language mode, deaf individuals had similar performances as hearing individuals (Korvorst et al., 2007). When asked to estimate as quickly as possible the midpoint of a series of numerical intervals that are presented in ascending and descending order, deaf and hearing participants were equally accurate in their estimations and were significantly biased toward lower numbers (Cattaneo et al., 2014). Nevertheless, the underestimation bias in deaf persons was smaller than in hearing when using a descending order, indicating that the decisions of the hearing individuals fall more systematically to the left (i.e., were more underestimated) than those of deaf participants (Cattaneo et al., 2016).

Finally, Nuerk, Iversen, and Willmes performed a study in 2004 in which they observed that hearing individuals respond faster in an even-right/odd-left condition than for the reverse parity-response box condition. This Markedness Association of Response Codes (MARC) effect has been interpreted as a linguistic markedness congruency effect since "even" and "right" are believed to be the linguistically marked antonyms of "odd" and "left" (Nuerk, H., Iversen, W., & Willmes, K. 2004). The effect also appeared to be stronger for written words than for Arabic numerals, which might reflect a stronger access to verbal-linguistic concepts via verbal stimuli, as suggested by the authors (Hines, 1990). An inversed MARC effect has interestingly been shown in deaf individuals, with native signers responding faster with the left-handed side to even numbers, and responding faster to the odd numbers with the right-handed side (Iversen et al., 2004). This result suggests that the structure of the sign language may influence number representations in a specific way.

To conclude, deaf and hearing individuals show SNARC, distance, and size effects that are normally associated with a representation of magnitude on a visual-analog MNL. However, deaf participants have slower response times when making comparative judgments, which indicates that their numerical representation of magnitude information is not distinct from that of hearing individuals, but that they might process basic numerical information in a less efficient way (Bull et al., 2005; Chinello et al., 2012; Iversen et al., 2004; Rodriguez-Santos et al., 2014). While deaf individuals seem to use a left-to-right-oriented mental number line, it is still not known whether the associations between numbers and space occur in external coordinates or whether deafness, like blindness, shapes the reference frame in which these associations occur.

Therefore, future research should be conducted to: (1) establish a wider base of studies about cognitive abilities among deaf students; (2) determine the specific cognitive mechanisms that are slower in development in deaf individuals and causing a lag in mathematical achievement; (3) assess early representations of number that do not involve counting in younger children to clarify the status of the early abilities in number representation; (4) evaluate if there is an effect of home language, the medium of instruction, and the test language on children's mathematical performance; (5) study the aspects of sign language contributing to mathematical learning; (6) clarify how the human mind spatially represents abstract concepts and the extent to which differences are related to visual characteristics or linguistic values; (7) determine whether poorer acuity of numerical estimation is distinguishable from any language component associated with the task; (8) investigate further the influence of assistive hearing devices on child development and academic functioning; (9) identify differences among deaf individuals and how to accommodate for their needs (Ansell & Pagliaro, 2006; Borgna et al., 2018; Bull et al., 2011b; Cattaneo et al., 2017; Chinello et al., 2012; Gottardis et al., 2011; Korvorst et al., 2007; Marschark et al., 2015; Marcelino et al., 2019; Rinaldi, Merabet, Vecchi & Cattaneo, 2018; Zarfaty et al., 2004).

In summary, future research is necessary to better understand the factors that contribute to the academic achievements for deaf students across various subject areas for both theoretical and practical reasons. This would enhance the scientific understanding of cognitive, social, and linguistic functioning in deaf individuals as well as it would help to develop educational materials, methods, and interventions to support deaf learners in their academic achievement (Marschark et al., 2015).

4 Discussion

The representation of abstract concepts such as numbers has been proposed to originate from sensorimotor interactions within the world around us (Bonato et al., 2012; Winter et al., 2015). Hence, if a normal sensorimotor experience is strictly mandatory in order to represent numbers, we should expect that sensory deprivation would have an impact on the development of this representation. The present chapter examined this question by reviewing experimental data on numerical performances in blind and deaf individuals. From a quantitative point of view, it is interesting to note that blindness does not prevent the emergence of good numerical skills while deafness, in contrast, seems to delay these acquisitions. These observations are at odd with the hypothesis suggesting that mathematical representations are rooted in visuo-spatial thinking and develop through visual experience (Burr & Ross, 2008; Ross & Burr, 2010). They nevertheless support the idea that language plays an important role in learning the meaning of numbers (Spaepen et al., 2011). Recent years have seen a surge in empirical studies examining the role of language in accounting for cross-cultural disparities in children's number understanding and arithmetic competence (Fuson & Kwon, 1992; Göbel et al., 2014; Krinzinger et al., 2011; Wang et al., 2008). It has, for example, been suggested that the superior arithmetic performance of Chinese and other Asian students could be explained by the relative linguistic transparency of the Asian counting systems (Fuson & Kwon, 1992; Miller et al., 2005) which gives a clear and consistent representation of the base-ten system. While comparisons across different auditory languages have been made, examining numerical competences in deaf individuals will additionally allow to compare auditory and visuo-manual languages.

Two hypotheses may account for the existence of good numerical skills in blind individuals. The first one assumes that blind individuals learn mathematics by compensating their visual lack through other modalities. In this case, the same numerical performances in blind and sighted individuals would arise from different neural correlates (e.g., areas involved in auditory or tactile processing). The second hypothesis assumes that mathematical activity is in fact based on highly abstract representations which are amodal rather than primarily visual. In this case, the same mental representation of numbers would be accessed indifferently from visual, auditory, or tactile inputs (Piazza et al., 2006; Riggs et al., 2006; Tokita et al., 2013). In the present chapter, we demonstrated that the reality may probably lie in-between these two main hypotheses. When solving arithmetic operations, congenitally blind adults were indeed shown to activate a number-related network very similar to the one observed in sighted subjects (Crollen et al., 2019; Kanjlia et al., 2016). These findings show that numerical thinking can develop in the absence of visual experience and is rooted in typical number-related brain circuits, therefore lending support to the second hypothesis. However, an additional activity of the occipital cortex was also demonstrated but only when blind participants had to perform subtraction operations (not when they had to perform multiplications) (Crollen et al., 2019). This additional activity was not observed in the sighted and probably reflects the use of compensating strategies to perform a numerical task assumed to primarily rely on visuo-spatial processes.

In the literature, two main visuo-motor functions are assumed to be associated with the representation of numbers. On the one hand, following the recurrent observation that small numbers are preferentially associated with the left side of space while large numbers are preferentially associated with the right side of space (i.e., SNARC effect; Dehaene et al., 1993), numbers were assumed to interact with space. On the other hand, following the observation that children often use their fingers to learn the counting sequence and basic arithmetic operations, numbers were assumed to interact with finger movements (Butterworth, 1999). Interestingly, these two interactions are assumed to take place in the parietal cortex, a brain area which is part of the dorsal visual pathway. Consistently with the idea that blind individuals use alternative strategies to develop their understanding of numbers, we demonstrated that blind participants present a reversed SNARC effect when performing a numerical comparison task with their hands crossed over the body midline (Crollen et al., 2013). Importantly, they did not show a reversed SNARC effect in a parity judgment task (Crollen et al., 2013), suggesting that early visual experience drives the development of the visuo-spatial representation of numbers but do not shape the verbal associations that occur between numbers and space. We also demonstrated that the finger counting strategy was not often used by blind participants while counting and calculating (Crollen et al., 2011, 2014). Together, these observations lend some support to the idea that visual deprivation may promote the development of strategies that allow blind individuals to understand the number concept without relying on visuo-spatial processes.

Several studies already suggested that deaf individuals tend to be slower and less accurate with regard to numerical processing than normally hearing controls (Bull et al., 2011b; Epstein et al., 1994; Rodriguez et al., 2014). Deaf children may also be delayed in developing mathematic skills compared to their normally hearing peers (Gottardis et al., 2011). Interestingly, we showed the opposite dissociation as the one observed in the blind. The delay deaf individuals present in numerical development seems indeed to be more pronounced with symbolic tasks than with non-symbolic tasks. The study of this question should, however, be further investigated in the future. While it has already been demonstrated that deaf individuals represent numerical information along a left-to-right-oriented mental number line (Bull et al., 2005; Chinello et al., 2012; Iversen et al., 2004), the spatial frame of reference they preferentially use to map numbers onto space is still unknown. Moreover, to our knowledge, the spatial frame of reference onto which numbers are represented in deaf has so far never been compared across visuo-spatial and verbal-spatial tasks.

Furthermore, several studies have indicated that WM functioning is correlated with both symbolic and non-symbolic approximation, which points out that the individual variation in our WM could predict the mathematical achievement beyond the effect of approximation skills (Bull et al., 2018). However, symbolic approximation skills appear to correlate with mathematic ability beyond the effect of WM capacity. This might indicate unique contributions from both domain-specific and domain-general abilities (Bull et al., 2018). It has been stated that individuals with hearing loss seem to suffer from difficulties in verbal short-term memory, WM, and executive functioning (Bull et al., 2018; Marcelino et al., 2019 for review). On the other hand, it is suggested that deaf native signers have a better visuospatial WM than hearing individuals (Proksch & Bavelier, 2002).

To be able to better evaluate the respective contribution of visual vs. verbal processes in the development of the number concept, future studies should also examine the brain plasticity phenomenon following deafness. It has already been demonstrated that the temporal "auditory" cortex of deaf individuals changes its functional tuning to support visual or tactile functions (Fine et al., 2005; Finney et al., 2003; Finney & Dobkins, 2001; Nishimura et al., 1999; Petitto et al., 2000; Sadato et al., 2004; Shibata, 2007). However, it is still unknown whether the temporal "auditory" cortex of the deaf can be activated by higher cognitive function such as arithmetic and whether this activation is, as already observed in the blind, operation-specific (observed for multiplication, but not for subtraction in this case). Studying the neural correlates of numerical processes in deaf and comparing this to what has already been observed in the blind will provide a thorough understanding of the development of numerical competencies without vision or audition and give rare insights about the role of experience on the cerebral development of high cognitive functions. This question is really important to understand the principles of brain architecture and its reorganization under sensory deprivation. It will hopefully yield important novel insights into how the brain develops and whether this development is malleable or resistant to atypical sensory experiences.

Beyond this theoretical question, we also argue that a better understanding of the mechanisms underlying number understanding after visual and auditory deprivation plays a critical role in better characterizing what does dyscalculia look like in blind and deaf individuals. An advance in the understanding of this issue is timely since clinicians are currently lacking standardized norms to evaluate the numerical abilities of sensory-deprived individuals. Better understanding number development in these populations will therefore constitute a starting point for elaborating programs that stimulate numerical learning mechanisms in blind and deaf children presenting numerical difficulties. As poor mathematical skills are associated with employment difficulties, developing further such field of research therefore holds the promises to have a substantial fundamental impact, but also some applied, social, and societal implications.

References

Allen, Thomas E. (1995). Demographics and national achievement levels for deaf and hard of hearing students: Implications for mathematics reform. *Moving toward the standards: A national action plan for mathematics education reform for the deaf, 41–49.*

Alloway, Tracy P., Gathercole, Susan E., Adams, Anne-Marie, Willis, Catherine, Eaglen, Rachel & Lamont, Emily (2005): Working memory and phonological awareness as predictors of progress towards early learning goals at school entry. *British Journal of Developmental Psychology* 23, 417–426.

Almoammer, Alhanouf, Sullivan, Jessica, Donlan, Chris, Marušič, Franc, Zaucer, Roc, O'Donnell, Timothy & Barner, David (2013): Grammatical morphology as a source of early number word meanings. *Proceedings of the National Academy of Sciences* 110 (46), 18448–18453.

Amalric, Marie & Dehaene, Stanislas (2016): Origins of the brain networks for advanced mathematics in expert mathematicians. *Proceedings of the National Academy of Sciences* 113 (18), 4909–4917.

Amalric, Marie & Dehaene, Stanislas (2019): A distinct cortical network for mathematical knowledge in the human brain. *Neuroimage* 189, 19–31.

Amalric, Marie, Denghien, Isabelle & Dehaene, Stanislas (2018): On the role of visual experience in mathematical development: Evidence from blind mathematicians. *Developmental Cognitive Neuroscience* 30, 314–323.

Andres, Michael, Di Luca, Samuel & Pesenti, Mauro (2008): Finger counting: The missing tool? *Behavioral and Brain Sciences* 31 (6), 642–643.

Anobile, Giovanni, Stievano, Paolo & Burr, David C. (2013): Visual sustained attention and numerosity sensitivity correlate with math achievement in children. *Journal of Experimental Child Psychology* 116 (2), 380–391.

Ansell, Ellen & Pagliaro, Claudia M. (2006): The relative difficulty of signed arithmetic story problems for primary level deaf and hard-of-hearing students. *Journal of Deaf Studies and Deaf Education* 11 (2), 153–170.

Antell, Sue E. & Keating, Daniel P. (1983): Perception of numerical invariance in neonates. *Child Development* 54, 695–701.

Arsalidou, Marie & Taylor, Margot J. (2011): Is 2+ 2= 4? Meta-analyses of brain areas needed for numbers and calculations. *Neuroimage* 54 (3), 2382–2393.

Ashcraft, Mark H. (1992): Cognitive arithmetic: A review of data and theory. *Cognition* 44 (1–2), 75–106. doi:10.1016/0010-0277(92)90051-i.

Ashcraft, Mark H. & Stazyk, Edmund H. (1981): Menatal addition: A test of three verification models. *Memory & Cognition* 9 (2), 185–196.

Austin, Gary F. (1975): Knowledge of selected concepts obtained by an adolescent deaf population. *American Annals of the Deaf* 360–370.

Bavelier, Daphene, Tomann, Andrea, Hutton, Chloe, Mitchell, Teresa, Corina, David, Liu, Guoying & Neville, Helen (2000): Visual attention to the periphery is enhanced in congenitally deaf individuals. *Journal of Neuroscience* 20 (17), RC93.

Bijeljac-Babic, Ranka, Bertoncini, Josiane & Mehler, Jacques (1991): How do four-day-old infants categorize multisyllabic utterances. *Developmental Psychology* 29, 711–721.

Blatto-Vallee, Gary, Kelly, Ronald R., Gaustad, Martha G., Porter, Jeffrey & Fonzi, Judith (2007): Visual-spatial representation in mathematical problem solving by deaf and hearing students. *Journal of Deaf Studies and Deaf Education* 12 (4), 432–448.

Bonato, Mario, Zorzi, Marco & Umiltà, Carlo (2012): When time is space: Evidence for a mental time line. *Neuroscience & Biobehavioral Reviews* 36 (10), 2257–2273.

Borgna, Georgianna, Walton, Dawn, Convertino, Carol, Marschark, Marc & Trussell, Jessica (2018): Numerical and real-world estimation abilities of deaf and hearing college students. *Deafness & Education International* 20 (2), 59–79.

Brannon, Elizabeth M., Abbott, Sara & Lutz, Donna J. (2004): Number bias for the discrimination of large visual sets in infancy. *Cognition* 93 (2), B59-B68.

Brysbaert, Marc (1995): Arabic number reading: On the nature of the numerical scale and the origin of phonological recoding. *Journal of Experimental Psychology: General* 124 (4), 434–452. doi:10.1037/0096-3445.124.4.434.

Bull, Rebecca, Blatto-Vallee, Gary & Fabich, Megan (2006): Subitizing, magnitude representation, and magnitude retrieval in deaf and hearing adults. *Journal of Deaf Studies and Deaf Education* 11 (3), 289–302.

Bull, Rebecca, Espy, Kimberly A., Wiebe, Sandra A., Sheffield, Tiffany D. & Nelson, Jennifer M. (2011a): Using confirmatory factor analysis to understand executive control in. *Developmental Science* 14 (4), 679–692.

Bull, Rebecca, Marschark, Marc & Blatto-Vallee, Gary (2005): SNARC hunting: Examining number representation in deaf students. *Learning and Individual Differences* 15 (3), 223–236.

Bull, Rebecca, Marschark, Marc, Nordmann, Emily, Sapere, Patricia & Skene, Wendy A. (2018): The approximate number system and domain-general abilities as predictors of math ability in children with normal hearing and hearing loss. *British Journal of Developmental Psychology* 36 (2), 236–254.

Bull, Rebecca, Marschark, Marc, Sapere, Patricia, Davidson, Wendy A., Murphy, Derek & Nordmann, Emily (2011b): Numerical estimation in deaf and hearing adults. *Learning and Individual Differences* 21 (4), 453–457.

Burr, David & Ross, John (2008): A visual sense of number. *Current Biology* 18 (6), 425–428.

Butterworth, Brian (1999): *The Mathematical Brain*. London: Macmillan.

Butterworth, Brian (2005): The development of arithmetical abilities. *Journal of Child Psychology and Psychiatry* 46, 3–18.

Campbell, Jamie I. & Timm, Jennifer C. (2000): Adults' strategy choices for simple addition: Effects of retrieval interference. *Psychonomic Bulletin & Review* 7 (4), 692–699.

Campbell, Jamie I. & Xue, Qilin (2001): Cognitive arithmetic across cultures. *Journal of Experimental Psychology: General* 130 (2), 299–315.

Cantlon, Jessica F., Libertus, Melissa E., Pinel, Philippe, Dehaene, Stanislas, Brannon, Elizabeth M. & Pelphrey, Kevin A. (2009): The neural development of an abstract concept of number. *Journal of Cognitive Neuroscience* 21 (11), 2217–2229.

Carey, Susan (2001): Cognitive foundations of arithmetic: Evolution and ontogenesis. *Mind & Language* 16, 37–55.

Castronovo, Julie & Delvenne, Jean-François (2013): Superior numerical abilities following early visual deprivation. *Cortex* 49 (5), 1435–1440. doi:10.1016/j.cortex.2012.12.018.
Castronovo, Julie & Seron, Xavier (2007a): Semantic numerical representation in blind subjects: The role of vision in the spatial format of the mental number line. *The Quarterly Journal of Experimental Psychology* 60 (1), 101–119. doi:10.1080/17470210600598635.
Castronovo, Julie & Seron, Xavier (2007b): Numerical estimation in blind subjects: Evidence of the impact of blindness and its following experience. *Journal of Experimental Psychology: Human Perception and Performance* 33 (5), 1089–1106. doi:10.1037/0096-1523.33.5.1089.
Cattaneo, Zaira, Cecchetto, Carlo & Papagno, Constanza (2016): Deaf individuals show a leftward bias in numerical bisection. *Perception* 45 (1–2), 156–164.
Cattaneo, Zaira, Fantino, Micaela, Silvanto, Juha, Tinti, Carla & Vecchi, Tomaso (2011): Blind individuals show pseudoneglect in bisecting numerical intervals. *Attention, Perception & Psychophysics* 73 (4), 1021–1028. doi:10.3758/s13414-011-0094-x.
Cattaneo, Zaira, Lega, Carlotta, Cecchetto, Carlo & Papagno, Constanza (2014): Auditory deprivation affects biases of visuospatial attention as measured by line bisection. *Experimental Brain Research* 232 (9), 2767–2773.
Cattaneo, Zaira, Rinaldi, Luca Geraci, Carlo, Cecchetto & Papagno, Constanza (2017): Spatial biases in deaf, blind, and deafblind individuals as revealed by a haptic line bisection task. *Journal of Experimental Psychology*. 71 (11), 2325–2333
Chinello, Alessandro, de Hevia, Maria D., Geraci, Carlo & Girelli, Luisa (2012): Finding the spatial-numerical association of response codes (SNARC) in signed numbers: Notational effects in accessing number representation. *Functional Neurology* 27 (3), 177–185.
Cooney, John B., Swanson, Lee H. & Ladd, Stephen F. (1988): Acquisition of mental multiplication skill: Evidence for the transition between counting and retrieval strategies. *Cognition and Instruction* 5 (4), 323–345.
Crollen, Virginie, Dormal, Giulia, Seron, Xavier, Lepore, Franco & Collignon, Oliver (2013): Embodied numbers: The role of vision in the development of number-space interactions. *Cortex* 49, 276–283. doi:10.1016/j.cortex.2011.11.006.
Crollen, Virginie, Lazzouni, Latifa, Rezk, Mohamed, Bellemare, Antoine, Lepore, F., Noël, Marie-Pascale, Seron, Xavier & Collignon, O. (2019): Recruitment of the occipital cortex by arithmetic processing follows computational bias in the congenitally blind. *Neuroimage* 186, 549–556. doi:10.1016/j.neuroimage.2018.11.034.
Crollen, Virginie, Mahe, Rachel, Collignon, Oliver & Seron, Xavier (2011): The role of vision in the development of finger-number interactions: Finger-counting and finger-montring in blind children. *Journal of Experimental Child Psychology* 109 (4), 525–539. doi:10.1016/j.jecp.2011.03.011.
Crollen, Virginie, NoëL, Marie-Pascale, Seron, Xavier, Mahau, P., Lepore, Franco & Collignon, Oliver (2014): Visual experience influences the interactions between fingers and numbers. *Cognition* 133 (1), 91–96. doi:10.1016/j.cognition.2014.06.002.
Crollen, Virginie, Vanderclausen, Camille, Allaire, Florence, Pollaris, Arnaud & Noël, Marie-Pascale (2015): Spatial and numerical processing in children with non-verbal learning disabilities. *Research in Developmental Disabilities* 47, 61–72.
De Hevia, Maria. D., Vallar, Guiseppe & Girelli, Luisa (2008): Visualizing numbers in the mind's eye: The role of visuo-spatial processes in numerical abilities. *Neuroscience & Biobehavioral Reviews* 32 (8), 1361–1372.
De Smedt, Bert & Boets, Bart (2010): Phonological processing and arithmetic fact retrieval: Evidence from developmental dyslexia. *Neuropsychologica* 48 (14), 3973–3981.

De Smedt, Bert, Janssen, Rianne, Bouwens, Kelly, Verschaffel, Lieven, Boets, Bart & Ghesquière, Pol (2009): Working memory and individual differences in mathematics achievement: A longitudinal study from first grade to second grade. *Journal of Experimental Child Psychology* 103 (2), 186–201.
Dehaene, Stanislas, Piazza, Manuela, Pinel, Philippe & Cohen, Laurent (2003): Three parietal circuits for number processing. *Cognitive Neuropsycholy* 20 (3), 487–506.
Dehaene, Stanislas (1989): The psychophysics of numerical comparison: A reexamination of apparently incompatible data. *Perception & Psychophysics* 45 (6), 557–566.
Dehaene, Stanislas, Dupoux, Emmanuel, & Mehler, Jacques (1990): Is numerical comparison digital? Analogical and symbolic effects in two-digit number comparison. Journal of experimental Psychology: Human Perception and performance 16 (3), 626–641.
Dehaene, Stanislas (1992): Varieties of numerical abilities. *Cognition* 44, 1–42.
Dehaene, Stanislas, Bossini, Serge & Giraux, Pascal (1993): The mental representation of parity and number magnitude. *Journal of Experimental Psychology: General* 122 (3), 371–396.
Dehaene, Stanislas & Cohen, Laurent (1997): Cerebral pathways for calculation: Double dissociation between rote verbal and quantities knowledge of arithmetic. *Cortex* 33 (2), 219–250.
Dormal, Valérie, Crollen, Virginie, Baumans, Christine, Lepore, Franco & Collignon, Oliver (2016): Early but not late blindness leads to enhanced arithmetic and working memory abilities. *Cortex* 83, 212–221. doi:10.1016/j.cortex.2016.07.016.
Epstein, Kenneth I., Hillegeist, Eleanor G. & Grafman, Jordan (1994): Number processing in deaf college students. *American Annals of the Deaf*, 139, 336–347.
Fayol, Michel & Seron, Xavier (2005): About Numerical Representations: Insights from Neuropsychological, Experimental, and Developmental Studies. In Jamie, I. D. Campbell (ed.): *Handbook of Mathematical Cognition*. New York: Psychology Press, 3–22.
Feigenson, Lisa, Dehaene, Stanislas & Spelke, Elisabeth (2004): Core systems of number. *Trends in Cognitive Sciences* 8, 307–314. doi:10.1016/j.tics.2004.05.002.
Ferrand, Ludovic, Riggs, Karen J. & Castronovo, Julie (2010): Subitizing in congenitally blind adults. *Psychonomic Bulletin & Review* 17 (6), 840–845. doi:10.3758/PBR.17.6.840.
Fias, Wim & Fischer, Martin (2005): Spatial Representation of Numbers. In Campbell, J. I. D. (ed.): *Handbook of Mathematical Cognition*. New York: Psychology Press, 43–54.
Fine, Ione, Finney, Eva M., Boynton, Geoffrey M. & Dobkins, Karen R. (2005): Comparing the effects of auditory deprivation and sign language within the auditory and visual cortex. *Journal of Cognitive Neuroscience* 17 (10), 1621–1637.
Finney, Eva M., Clementz, Brett A., Hickok, Gregory & Dobkins, Karen R. (2003): Visual stimuli activate auditory cortex in deaf subjects: Evidence from MEG. *Neuroreport* 14 (11), 1425–1427.
Finney, Eva M. & Dobkins, Karen R. (2001): Visual contrast sensitivity in deaf versus hearing populations: Exploring the perceptual consequences of auditory deprivation and experience with a visual language. *Cognitive Brain Research* 11 (1), 171–183.
Fuson, Karen C. (1988): *Children's Counting and Concepts of Number*. New York: Springer-Verlag.
Fuson, Karen C. & Kwon, Youngshim (1992): Korean children's understanding of multidigit addition and subtraction. *Child Development* 63 (2), 491–506.
Geary, David C. & Hoard, Mary K. (2001): Numerical and arithmetical deficits in learning-disabled children: Relation to dyscalculia and dyslexia. *Aphasiology* 15 (7), 635–647.

Göbel, S. M. (2015). Number processing and arithmetic in children and adults with reading difficulties. In R. C. Kadosh & A. Dowker (Eds.), Oxford library of psychology. *The Oxford handbook of numerical cognition*. Oxford University Press, 696–720.

Göbel, Silke M., Moeller, Korbinian, Pixner, Silvia, Kaufmann, Liane & Nuerk, Hans-Christiph (2014): Language affects symbolic arithmetic in children: The case of number word inversion. Journal of Experimental Child Psychology 119, 17–25.

Gottardis, Laura, Nunes, Terezinha & Lunt, Ingrid (2011): A Synthesis of research on deaf and hearing children's mathematical achievement. *Deafness & Education International* 13 (3), 131–150.

Hecht, Steven A. (1999): Individual solution processes while solving addition and multiplication math facts in adults. *Memory & Cognition* 27 (6), 1097–1107.

Herrera, Amparo & Macizo, Pedro (2008): Cross-notational semantic priming between symbolic and nonsymbolic numerosity. *The Quarterly Journal of Experimental Psychology* 61 (10), 1538–1552.

Hines, Terence M. (1990): An odd effect: Lengthened reaction times for judgments about odd digits. *Memory & Cognition*. 18 (1), 40–46

Huber, Stefan, Klein, Elise, Moeller, Korbinian & Willmes, Klaus (2016): Spatial-numerical and ordinal positional associations coexist in parallel. *Frontiers in Psychology* 7 (438). doi:10.3389/fpsyg.2016.00438.

Hull, Tim & Mason, Heather (1995): Performance of blind children on digit-span tests. *Journal of Visual Impairment & Blindness*. 89 (2), 166–169

Iversen, Wiebke, Nuerk, Hans-Christoph & Willmes, Klaus (2004): Do signers think differently? The processing of number parity in deaf participants. *Cortex* 40 (1), 176–178.

Kanjlia, Shipra, Lane, Connor, Feigenson, Lisa & Bedny, Marina (2016): Absence of visual experience modifies the neural basis of numerical thinking. *Proceedings of the National Academy of Sciences* 113 (40), 11172–11177.

Knops, André, Thirion, Bertrand, Hubbard, Edward M., Michel, Vincent & Dehaene, Stanislas (2009a): Recruitment of an area involved in eye movements during mental arithmetic. *Science* 324 (5934), 1583–1585. doi:10.1126/science.1171599.

Knops, André, Viarouge, Arnaud & Dehaene, Stanislas (2009b): Dynamic representations underlying symbolic and nonsymbolic calculation: Evidence from the operational momentum effect. *Attention, Perception & Psychophysics* 71 (4), 803–821. doi:10.3758/APP.71.4.803.

Korvorst, Marjolein, Nuerk, Hans-Christoph & Willmes, Klaus (2007): The hands have it: Number representations in adult deaf signers. *Journal of Deaf Studies and Deaf Education* 12 (3), 362–372.

Krinzinger, Helga, Gregoire, Jacques, Desoete, Annemie, Kaufmann, Liane, Nuerk, Hans-Christoph & Willmes, Klaus (2011): Differential language effects on numerical skills in second grade. *Journal of Cross-Cultural Psychology* 42 (4), 614–629.

Leather, Cathy V. & Henry, Lucy A. (1994): Working memory span and phonological awareness tasks as predictors of early reading ability. *Journal of Experimental Child Psychology* 58, 88–111.

LeFevre, Jo A., Fast, Lisa, Skwarchuk, Sheri-Lynn, Smith-Chant, Brenda L., Bisanz, Jeffrey, Kamawar, Deepthi & Penner-Wilger, Marcie (2010): Pathways to mathematics: Longitudinal predictors of performance. *Child Development* 81 (6), 1753–1767.

Lipton, Jennifer S. & Spelke, Elizabeth (2003): Origins of number sense: Large number discrimination in human infants. *Psychological Science* 14, 396–401.

Lipton, Jennifer S. & Spelke, Elizabeth S. (2004): Discrimination of large and small numerosities by human infants. *Infancy* 5, 271–290.
Loftus, Andrea M., Nicholls, Micheal E., Mattingley, Jason B. & Bradshaw, John L. (2008): Left to right: Representational biases for numbers and the effect of visuomotor adaptation. *Cognition* 107 (3), 1048–1058.
Lourenco, Stella F., Bonny, Justin W., Fernandez, Edmund P. & Rao, Sonia (2012): Nonsymbolic number and cumulative area representations contribute shared and unique variance to symbolic math competence. *Proceedings of the National Academy of Sciences* 109 (46), 1837–1842.
Marcelino, Lilia, Sousa, Carla & Costa, Conceição (2019). Cognitive foundations of mathematics learning in deaf students: A systematic literature review. *Proceedings of EDULEARN19 Conference, Palma, Mallorca, Spain.*
Marschark, M., Morrison, Carolyn, Lukomski, Jennifer, Borgna, Georgianna & Convertino, Carol (2013): Are deaf students visual learners? *Learning and Individual Differences* 25, 156–162.
Marschark, Marc & Everhart, Victoria S. (1999): Problem-solving by deaf and hearing students: Twenty questions. *Deafness & Education International* 1 (2), 65–82.
Marschark, Marc, Shaver, Debra M., Nagle, Katherine M. & Newman, Lynn A. (2015): Predicting the academic achievement of deaf and hard-of-hearing students from individual, household, communication, and educational factors. Exceptional Children 81 (3), 350–369.
Masataka, Nabuo (2006): Differences in arithmetic subtraction of nonsymbolic numerosities by deaf and hearing adults. *Journal of Deaf Studies and Deaf Education* 11 (2), 139–143.
Masson, Nicolas & Pesenti, Mauro (2014): Attentional bias induced by solving simple and complex addition and subtraction problems. *The Quarterly Journal of Experimental Psychology* 67 (8), 1514–1526. doi:10.1080/17470218.2014.903985.
McCloskey, Michael, Harley, Walter & Sokol, Scott M. (1991): Models of arithmetic fact retrieval: An evaluation in light of findings from normal and brain-damaged subjects. *Journal of Experimental Psychology: Learning, Memory, and Cognition* 17 (3), 377–397. doi:10.1037/0278-7393.17.3.377.
McCrink, Koleen, Dehaene, Stanislas & Dehaene-Lambertz, Ghislaine (2007): Moving along the mental number line: Operational momentum in nonsymbolic arithmetic. *Perception & Psychophysics* 69 (8), 1324–1333.
Miller Kevin, F., Melissa, Kelly & Xiaobin, Zhou (2005): Learning Mathematics in China and the United States: Cross-Cultural Insights into the Nature and Course of Preschool Mathematical Development. In Jamie, I. D. Campbell (ed.): *Handbook of Mathematical Cognition*. New York: Psychology Press, 163–178.
Miller, Kevin F., Smith, Catherine M., Zhu, Jinjuan & Zhang, Houcan (1995): Preschool origins of cross-national differences in mathematical competence: The role of number-naming systems. *Psychological Science* 6 (1), 56–60.
Miura, Irene T., Kim, Chungsoon C., Chang, Chih-Mei & Okamoto, Yukari (1988): Effects of language characteristics on children's representation of number: Cross-national comparisons. *Child Development* 59, 1445–1450.
Miura, Irene T. & Okamoto, Yukari (1989): Comparisons of US and Japanese first graders' cognitive representation of number and understanding of place value. *Journal of Educational Psychology* 81, 109–113.

Miura, Irene T., Okamoto, Yukari, Kim, Chungsoon C., Chang, Chih-Mei, Steere, Marcia & Fayol, Michel (1993): First graders cognitive representation of number and understanding of place-value: Cross-national comparisons – France, Japan, Korea, Sweden and the Unites States. *Journal of Educational Psychology* 85 (1), 24–30.

Miura, Irene T., Okamoto, Yukari, Kim, Chungsoon C., Chang, Chih-Mei, Steere, Marcia & Fayol, Michel (1994): Comparison of childrens' cognitive representation of number: China, France, Japan, Korea, Sweden, and the United States. *International Journal of Behavioural Development* 17, 401–411.

Moyer, Robert S. & Landauer, Thomas K. (1967): Time required for judgments of numerical inequality. *Nature* 215, 1519–1520.

Nishimura, Hiroshi, Hashikawa, Kazuo, Doi, Katsumi, Iwaki, Takako, Watanabe, Yoshiyuki, Kusuoka, Hideo, Nishimura, Tsunehiko & Kubo, Takeshi (1999): Sign language 'heard' in the auditory cortex. *Nature* 397 (6715), 116.

Nuerk, H., Iversen, W., & Willmes, K. (2004). Notational Modulation of the SNARC and the MARC (Linguistic Markedness of Response Codes) Effect. The Quarterly Journal of Experimental Psychology Section A, 57(5), 835–863. https://doi.org/10.1080/02724980343000512

Nunes, Terezinha & Moreno, Constanza (2002): An intervention program for promoting deaf pupils' achievement in mathematics. *Journal of Deaf Studies and Deaf Education* 7 (2), 120–133.

Pagliaro, Claudia M. & Kritzer, Karen L. (2013): The math gap: A description of the mathematics performance of preschool-aged deaf/hard-of-hearing children. *Journal of Deaf Studies and Deaf Education* 18 (2), 139–160.

Petitto, Laura A., Zatorre, Robert J., Gauna, Kristine, Nikelski, Erwin J., Dostie, Deanna & Evans, Alan C. (2000): Speech-like cerebral activity in profoundly deaf people processing signed languages: Implications for the neural basis of human language. *Proceedings of the National Academy of Sciences* 97 (25), 13961–13966.

Piazza, Manuela, Mechelli, Andrea, Price, Cathy J. & Butterworth, Brian (2006): Exact and approximate judgements of visual and auditory numerosity: An fMRI study. *Brain Research* 1106 (1), 177–188.

Pinhas, Michal & Fischer, Martin (2008): Mental movements without magnitude? A study of spatial biases in symbolic arithmetic. *Cognition* 109 (3), 408–415.

Proksch, Jason & Bavelier, Daphne (2002): Changes in the spatial distribution of visual attention after early deafness. *Journal of Cognitive Neuroscience* 14 (5), 687–701.

Reuhkala, Minna (2001): Mathematical skills in ninth-graders: Relationship with visuo-spatial abilities and working memory. *Educational Psychology* 21, 387–399.

Riggs, Kevin J., Ferrand, Ludovic, Lancelin, Denis, Fryziel, Laurent, Dumur, Gérard & Simpson, Andrew (2006): Subitizing in tactile perception. *Psychological Science* 17 (4), 271–272.

Rinaldi, Luca, Lotfi, Merabet, B., Vecchi, Tomaso & Cattaneo, Zaira (2018): The spatial representation of number, time, and serial order following sensory deprivation: A systematic review. *Neuroscience and Biobehavioral Reviews* 90, 371–380.

Rodriguez-Santos, José M., Calleja, Marina, Garcia-Orza, Javier, Iza, Maurizio & Damas, Jesús (2014): Quantity processing in deaf and hard of hearing children: Evidence from symbolic and nonsymbolic comparison tasks. *American Annals of the Deaf* 159 (1), 34–44.

Ross, John & Burr, David C. (2010): Vision senses number directly. *Journal of Vision* 10 (2). doi:10.1167/10.2.10.

Rousselle, Laurence & Noël, Marie-Pascale (2007): Basic numerical skills in children with mathematics learning disabilities: A comparison of symbolic vs non-symbolic number magnitude processing. *Cognition* 102 (3), 361–395.
Sadato, Norihiro, Yamada, Hiroki, Okada, Tomohisa, Yoshida, Masaki, Hasegawa, Takehiro, Matsuki, Ken-Ichi, Yonekura, Yoshiharu & Itoh, Harumi (2004): Age-dependent plasticity in the superior temporal sulcus in deaf humans: A functional MRI study. *BMC Neuroscience* 5 (1), 56.
Schleger, Franziska, Landerl, Karin, Muenssinger, Jana, Draganova, Rossitza, Reinl, Maren, Kiefer-Schmidt, Isabelle, Weiss, Magdalene, Wacker-Gußmann, Annette, Huotilainen, Minna & Preissl, Hubert (2014): Magnetoencephalographic signatures of numerosity discrimination in fetuses and neonates. *Developmental Neuropsychology* 39 (4), 316–329. doi:10.1080/87565641.2014.914212.
Scholl, Brian J. (2001): Objects and attention: The state of the art. *Cognition* 80, 1–46.
Shibata, Darryl K. (2007): Differences in brain structure in deaf persons on MR imaging studied with voxel-based morphometry. *American Journal of Neuroradiology* 28 (2), 243–249.
Siegler, Robert S. & Shrager, J. (1984): Strategy Choices in Addition and Subtraction: How do Children know What to do? In Sophian, Catherine (ed.): *The origins of cognitive skills*. Hillsdale: Erlbaum, 229–293.
Sigmundsson, Hermundur, Anholt, Synne K. & Talcott, Joel B. (2010): Are poor mathematics skills associated with visual deficits in temporal processing? *Neuroscience letters* 469 (2), 248–250.
Simmons, Fiona R. & Singleton, Chris (2008): Do weak phonological representations impact on arithmetic development? A review of research into arithmetic and dyslexia. *Dyslexia* 14 (2), 77–94.
Simmons, Fiona R., Willis, Catherine & Adams, Anne-Marie (2012): Different components of working memory have different relationships with different mathematical skills. *Journal of Experimental Child Psychology* 111 (2), 139–155.
Spaepen, Elizabet, Coppola, Marie, Spelke, Elizabeth S., Carey, Susan E. & Goldin-Meadow, Susan (2011): Number without a language model. *Proceedings of the National Academy of Sciences* 108 (8), 3163–3168.
Starkey, Paul & Cooper, R. G. Jr. (1980): Perception of numbers by human infants. *Science* 210 (4473), 1033–1035.
Strauss, Mark S. & Curtis, Lynne E. (1981): Infant perception of numerosity. *Child Development* 52, 1146–1152.
Swanson, H. Lee & Luxenberg, Diana (2009): Short-term memory and working memory in children with blindness: Support for a domain general or domain specific system? *Child Neuropsychology* 15 (3), 280–294.
Szücs, Denes (2016): Subtypes and comorbidity in mathematical learning disabilities: Multidimensional study of verbal and visual memory processes is key to understanding. *Progress in Brain Research* 227, 277–304.
Szücs, Denes & Csépe, Valéria (2005): The parietal distance effect appears in both the congenitally blind and matched sighted controls in an acoustic number comparison task. *Neuroscience letters* 384 (1–2), 11–16.
Tibber, Marc S., Manasseh, Gemma S., Clarke, Robert C., Gagin, Galina, Swanbeck, Sonja N., Butterworth, Brian, Beau, Lotto, R. & Dakin, Steven C. (2013): Sensitivity to numerosity is

not a unique visuospatial psychophysical predictor of mathematical ability. *Vision Research* 89, 1–9.

Titus, Janet C. (1995): The concept of fractional number among deaf and hard of hearing students. *American Annals of the Deaf* 140, 255–263.

Tokita, Midori, Ashitani, Yui & Ishiguchi, Akira (2013): Is approximate numerical judgment truly modality-independent? Visual, auditory, and cross-modal comparisons. *Attention, Perception, & Psychophysics* 75 (8), 1852–1861.

Träff, Ulf & Passolunghi, Maria C. (2015): Mathematical skills in children with dyslexia. *Learning and Individual Differences* 40, 108–114.

Traxler, Carol B. (2000): The Stanford Achievement Test: National norming and performance standards for deaf and hard-of-hearing students. *Journal of Deaf Studies and Deaf Education* 5 (4), 337–348.

Trick, Lana & Pylyshyn, Zenon W. (1994): Why are small and large numbers enumerated differently? A limited capacity preattentive stage in vision. *Psychological Review* 101, 80–102.

Uller, Claudia, Huntley-Fenner, Gavin, Carey, Susan & Klatt, Laura (1999): What representations might underlie infant numerical knowledge? *Cognitive Development* 14, 1–36.

van Dijck, Jean-Philippe, Gevers, Wim & Fias, Wim (2009): Numbers are associated with different types of spatial information depending on the task. *Cognition* 113, 248–253.

Van Loosbroek, Erik & Smitsman, Ad W. (1990): Visual perception of numerosity in infancy. *Developmental Psychology* 26, 916–922.

Wang, Jian, Lin, Emily, Tanase, Madalina & Sas, Midena (2008): Revisiting the influence of numerical language characteristics on mathematics achievement: Comparison among China, Romania, and US. *International Electronic Journal of Mathematics Education* 3 (1), 24–46.

Winter, Bodo, Matlock, Teenie, Shaki, Samuel & Fischer, Martin H. (2015): Mental number space in three dimensions. *Neuroscience & Biobehavioral Reviews* 57, 209–219.

Wood, Guilherme, Willmes, Klaus, Nuerk, Hans-Christoph & Fischer, Martin H. (2008): On the cognitive link between space and number: A meta-analysis of the SNARC effect. *Psychology Science Quarterly* 50 (4), 489–525.

Wood, Justin N. & Spelke, Elizabeth S. (2005): Infant's enumeration of actions: Numerical discrimination and its signature limits. *Developmental Science* 8, 173–181.

Wynn, Karen (1996): Infants' individuation and enumeration of actions. *Psychological Science* 7 (3), 164–169.

Wynn, Karen, Bloom, Paul & Chiang, Wen-Chi (2002): Enumeration of collective entities by 5-month-old infants. *Cognition* 83 (3), B55-B62.

Xu, Fei (2003): Numerosity discrimination in infants: Evidence for two systems of representations. *Cognition* 89 (1), B15-B25.

Xu, Fei & Arriaga, Rosa I. (2007): Number discrimination in 10-month-old infants. *British Journal of Developmental Psychology* 25 (1), 103–108.

Xu, Fei, Spelke, Elisabeth S. & Goddard, Sydney (2005): Number sense in human infants. *Developmental Science* 8 (1), 88–101.

Xu, Fei & Spelke, Elizabeth S. (2000): Large number discrimination in 6-month-old infants. *Cognition* 74 (1), B1-B11.

Zarfaty, Y., Nunes, T., & Bryant, P. (2004). The performance of young deaf children in spatial and temporal number tasks. Journal of deaf studies and deaf education, 9(3), 315–326.

V Language as learning resource in school

Ricardo Moura, Vitor Geraldi Haase, Júlia Beatriz Lopes-Silva, Luana Teixeira Batista, Fernanda Rocha de Freitas, Julia Bahnmueller, and Korbinian Moeller

Reading and writing words and numbers: Similarities, differences, and implications

1 Introduction

Literacy and numeracy are culturally acquired abilities that are well established as crucial for educational and vocational prospects (Parsons & Bynner, 1997; Ritchie & Bates, 2013; Romano et al., 2010). When investigating these abilities in children, researchers from educational and cognitive sciences often focus on the writing and reading of either words or numbers. Accordingly, these usually represent two independent lines of research. Nevertheless, in recent years there is increasing research interest into relevant commonalities between learning to read and write words as well as numbers (e.g., Lopes-Silva et al., 2016).

It has been argued that efficient processing of words and numbers requires a partially overlapping cognitive architecture including basic perceptual abilities, attention, working memory (WM), verbal, visuo-spatial and visuo-constructional processing as well as graphomotor sequencing, among others (e.g., Collins & Laski, 2019; Geary, 2005). Over the last decades, researchers have mostly been focusing on either phonological processing as a cognitive precursor of reading and writing words (Castles & Coltheart, 2004) or on numerical magnitude understanding as the most important precursor of number processing (Siegler & Braithwaite, 2017). In this chapter, we aim at bringing together both lines of research by discussing the role of phonological and magnitude processing for the understanding of words and numbers, as well as interactions between these processes in more detail. In particular, we will address aspects of the structure and the acquisition of symbolic (both verbal and Arabic) codes in young children. Moreover, we will discuss similarities and specificities of both codes and how they acquire semantic meaning in early stages of human development. Furthermore, we will elaborate on the comorbidity between math and reading difficulties in light of the interaction between the development of symbolic codes for words and numbers. Finally, we will integrate these lines of argument

https://doi.org/10.1515/9783110661941-015

by exemplarily reviewing the cognitive underpinnings of number transcoding (a numerical task with clear verbal aspects), focusing on the role played by different subcomponents of phonological processing.

2 Words and numbers: Common developmental footprints?

The process of mastering the representational codes for words and numbers is marked by a change from an early period when children learn the primitives and begin to construct a lexicon, to a later period in which this lexicon is fully and readily available and can be operated on. This can be observed in number transcoding tasks that demand converting numbers from different notations, such as reading Arabic digits aloud or writing Arabic numbers from dictation. Previous research investigating number transcoding performance observed relatively high frequencies of lexical errors in younger children (up to the second grade), and of syntactic errors in older children (Moura et al., 2013, 2015; Power & Dal Martello, 1990; Seron et al., 1992; Seron & Fayol, 1994). During the first years of schooling, processing of words and specifically also number words is usually more procedural and serial in nature (i.e., starting to read letter by letter and counting-based strategies to assign cardinality to sets). At this point, processing of words poses high demands on WM based on the segmentation of words into smaller units (i.e., phonemes) and their recoding (Share, 1999). Additionally, the processing of number words highly depends on the actual task at hand. For instance, in young children the precise numerical magnitude meaning of a number word is often accessed by counting-based strategies which, later on, may also be employed to solve simple calculations (Fritz et al., 2013). Additionally, both (multi-digit) number words and numbers in the form of Arabic digits are segmented in order to be processed (Bahnmüller et al., 2016; Barrouillet et al., 2004). All these processes represent a considerable challenge for children at the respective age of acquisition and depend heavily on working memory resources (Camos, 2008; Hecht, 2002; Noël, 2009).

Commonalities between the acquisition of the verbal and numerical codes are reflected at the theoretical level. Brysbaert (2005) called attention to the similarities between the process of word reading, as described by the dual-route model (Coltheart et al., 2001), and the processing of single-digit numbers. In particular, Brysbaert (2005) suggested that learning of both verbal and numerical codes proceeds from initial sequential processing based on phonological and working memory resources to later more holistic/parallel and automatized forms of processing.

According to the dual-route model of single word reading (Coltheart et al., 2001, see Fig. 1a for an illustration) reading starts by the visual orthographic analysis of the written word, with identification and grouping of its graphic components in parallel, followed by serial processing of the word following different routes. Along the sublexical or phonological route, processing occurs by rules for converting written units into sound units (i.e., grapheme-phoneme conversion). Along the lexical route, familiar words, stored in a lexicon that combines contextual, visual, phonological, and orthographic information, are recognized directly, bypassing grapheme-phoneme conversions. These two routes work simultaneously and in a horse-race manner so that the more efficient route results in reading or speaking a target word out loud first. As such, reading unfamiliar words is usually associated with the phonological route, while familiar words are more likely read via the lexical route primarily. While less proficient readers might have access only to the phonological, more sequentially operating route, proficient readers can flexibly draw from both routes in parallel.

Barrouillet and colleagues (2004) also explicitly explored similarities between verbal and numerical processing in the ADAPT (A Developmental, Asemantic, and Procedural Transcoding) model of writing numbers in digital-Arabic notation – a dual-route model of number dictation (see Fig. 1b). The ADAPT model explains transcoding of verbally spoken number words to digital-Arabic numbers through the interplay of recovering content from long-term memory and applying algorithm-based conversion rules. The model suggests a first step in which verbal input is temporarily stored in a phonological buffer. In case this content matches a lexical unit stored in long-term memory, the digital form can be retrieved directly (cf. the lexical route in dual-route model of single word reading; Coltheart et al., 2001). When this is not possible, a parsing process divides the respective content into units that can be processed. At this stage, a set of procedural rules are applied sequentially processing the content held in the phonological buffer and deriving a syntactic frame which is then filled with the respective digital forms.

In general, dual-route models assume that words and number words are initially processed in a laborious sequential way at the phonemic level. As children become more experienced, lexical entries gradually develop and processing of words and some number words and digital-Arabic numbers becomes less WM demanding and increasingly based on parallel processing (Barrouillet et al., 2004). Practice in word reading allows for applying more holistic or parallel visual word processing based on recurring grapheme ensembles and their progressive

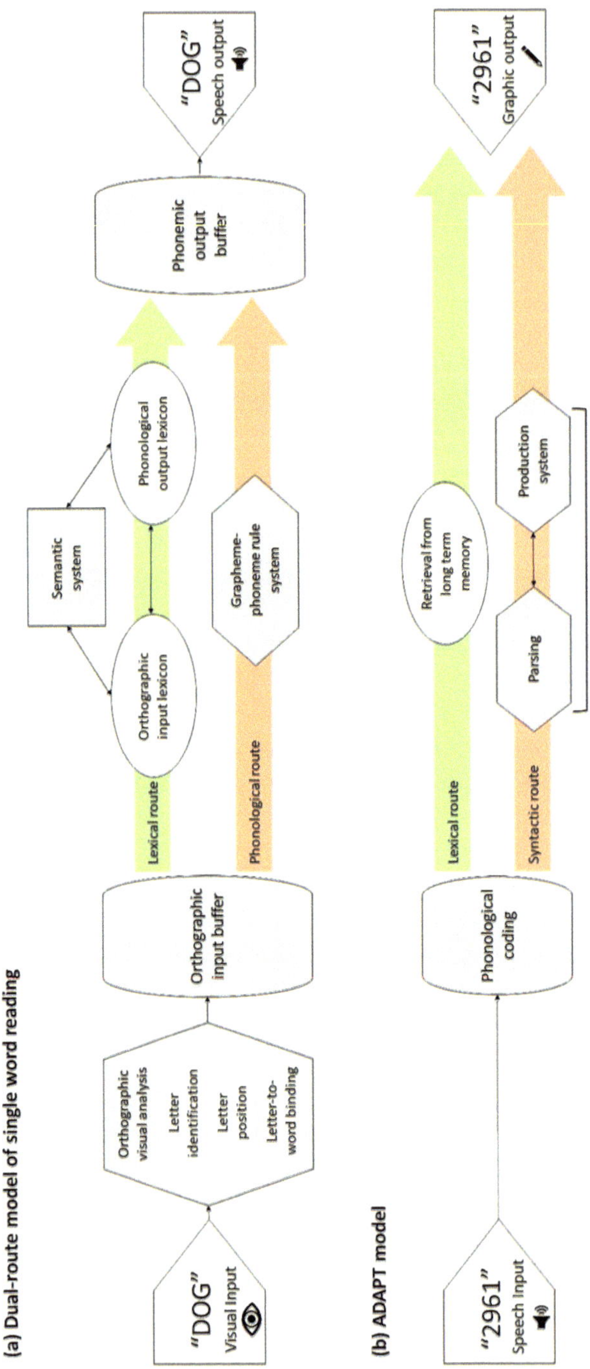

Fig. 1: Dual-Route models of (a) single word reading and (b) numbers writing in digital-Arabic notation.

association with pronunciation and meaning. Dehaene (2009) suggested that these holistic strategies are not acquired at the lexical level, eventually building a "sight lexicon," but at the sublexical level, consisting of recurring patterns of associations among graphemes digrams, such as "ll" (Treiman et al., 2018), that are processed preferentially.

Something similar to the lexicalization of word processing can occur with respect to the processing of single-digit number words and Arabic numbers. Growing experience with smaller and more frequent numbers in this range can facilitate more direct processing of these symbols, allowing fast access to the represented numerical magnitudes (Brysbaert, 2005). Empirical evidence also indicates that more frequent numerals with two or more digits with associated verbal lexical-semantic meanings may be accessed more efficiently (747, 1945, etc. See, e.g., Delazer & Girelli, 1997). However, access to and processing of the quantitative meaning of number words and Arabic numbers with two or more digits remains dependent on more laborious serial processing strategies (Bahnmueller et al., 2016).

Primary units of symbolic representations are then used to build more elaborate representations, with words leading to lexical-semantic access, and multi-digit Arabic numbers leading to the ability to represent and manipulate increasingly larger quantities in an abstract way. Figure 2 illustrates this

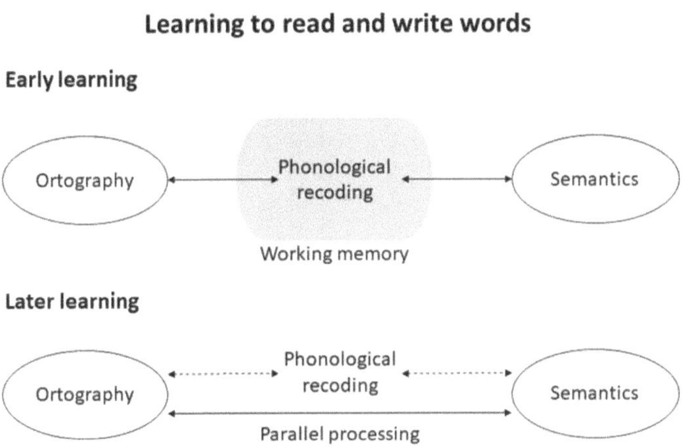

Fig. 2: In early phases of reading acquisition, associations between orthography and semantics primarily rely on sequential phonological recoding. In later phases, lexical representations are gradually built and access to semantics from orthography becomes more direct through parallel processing of sublexical subcomponents such as digrams and trigrams.

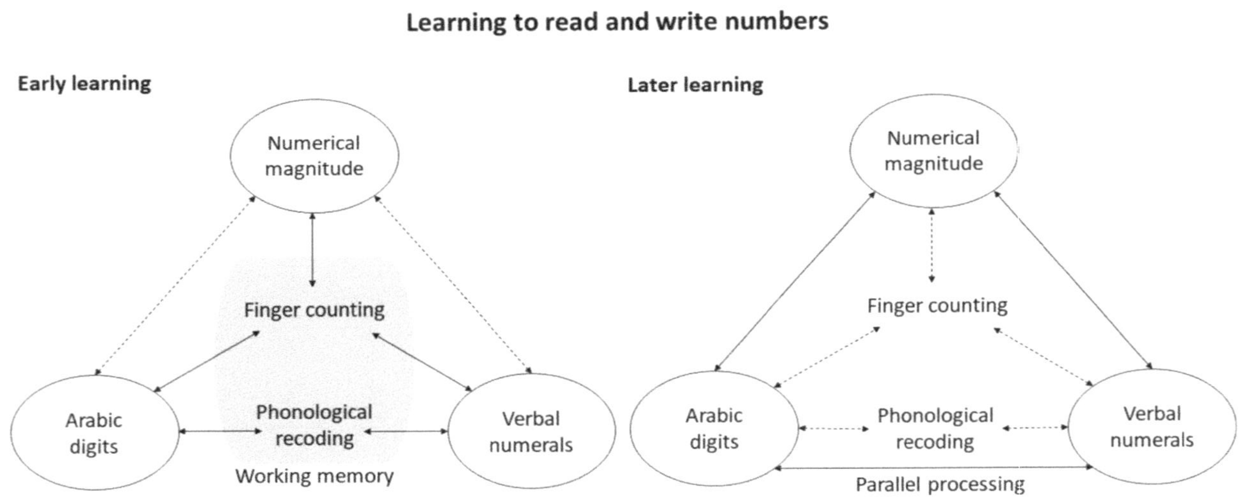

Fig. 3: In early phases, associations of numerical magnitude and their symbolic representations (i.e., Arabic digits and number words) also seem to depend on sequential processing in WM. Phonological recoding of, for instance, the counting sequence, is crucial for associating verbal and digital-Arabic representational codes. Finger counting may help offloading WM through concrete representations. With age and experience some number words and digital-Arabic numbers, at least in the single-digit range, become increasingly lexicalized. This means that in later phases of numerical processing their direct access to the corresponding quantity information depends less on sequential processing in WM. Accordingly, the importance of finger counting as a link between numerical magnitudes and number words fades out. In addition, transcoding between digital-Arabic numbers and number words also depends less on phonological recoding, at least for more frequently encountered numbers charged with verbal lexical-semantic meaning.

assumed development of associations between graphemes and lexical entries in word reading. Corresponding associations of numerical magnitude and number words and Arabic digits/multi-digit numbers are illustrated in Fig. 3.

As indicated in Fig. 2 and 3, an important difference during the acquisition of word, number word, and Arabic digit knowledge is the role of bodily experiences of fingers (for counting). Finger-based numerical representations (e.g., thumb, index, and middle finger representing three) and finger counting are extremely common (Crollen et al., 2011; Wasner et al., 2014). As finger-based representations and finger counting provide concrete representations of number magnitude, they may play an important role in offloading working memory. Thereby, resources that facilitate the acquisition of more abstract symbolic representations and calculation procedures may be set free (Alibali & DiRusso, 1999; Costa et al., 2011).

3 Shared deep structural features: Symbolic mapping and relational reasoning

The detailed mechanisms by which phonological processing mediates the development of literacy and numeracy are not yet clear. Collins and Laski (2019) proposed an analytical framework intending to foster our understanding of interactions between word and number processing during developmental progression. According to the authors, early literacy and numeracy skills differ in surface features such as the physical signs (letters, words, digits, arithmetic symbols, etc.). On the other hand, literacy and numeracy skills share some deep structural features, which rely on common processes (i.e., processing rules, principles, or schemas). These common processes may, in part, explain the observed associations between both domains. Importantly, the authors called attention to specific similarities in the deep structure of literacy and numeracy, mainly pertaining to symbolic mapping and relational reasoning.

Symbolic mapping reflects the establishing of connections between symbols and labels (i.e., identification of letters and digits as relevant codes) as well as symbols and referents (i.e., mapping of letters onto sounds and digits onto magnitudes). Relational reasoning is defined as the ability to discern meaningful patterns within otherwise unconnected information (Dumas et al., 2013). As such, relational reasoning abilities allow for making comparisons and recognizing similarities and differences between sets of information to infer meaningful relationships, structures, and patterns.

An important subcomponent of relational reasoning similarly involved in literacy and numeracy is part-whole thinking. Part-whole thinking is defined as understanding how units of information (parts) combine into larger units of meaning (wholes, cf. Fritz et al., 2013). With respect to literacy, phonemic awareness, the prime cognitive correlate of literacy, allows singling out specific phonemic segments from words to create new words (e.g., What is cup without the /c/?). As regards numeracy, part-whole thinking plays a role in recognizing that several parts can make up a whole (i.e., composing numbers of other numbers, e.g., $2+4=6$), and wholes can be divided into parts (i.e., decomposing numbers), which also underlies children's basic understanding of first arithmetic procedures (i.e., addition and subtraction, e.g., Krajewski & Schneider, 2009), but also fractions and proportions later on, e.g., Siegler et al. (2011).

Despite numbers and words sharing some features, numerical symbols are unique for many reasons. The special status of numerical symbols is attributable to the syntactic structure of number words and Arabic numbers, which imposes specific hurdles during development. In the next section the literature on the acquisition of the numerical symbols will be addressed in more detail.

4 The numerical Arabic system

Numerical representations develop side-by-side with language in children (e.g., Le Corre & Carey, 2007). Almost as early in their cognitive development as children begin to speak, they start using the first oral number words. However, it takes years of informal learning but also formal instruction until children master the use of symbolic numerical notations (Moura et al., 2013, 2015). At first, children learn to count by reciting a sequence of number words. However, these number words are still devoid of any quantitative meaning (Sarnecka & Lee, 2009). Gradually, these number words become associated with non-symbolic numerical representations (Krajewski & Schneider, 2009; Le Corre & Carey, 2007). As the mapping between the sequence of number words and their respective numerical magnitude meaning is established, children become able to successfully perform several new tasks. For instance, they may then use these number words to indicate the quantity reflected by a set (i.e., say "six" when they quickly look at a set of six objects). Additionally, they can now also produce quantities of a certain magnitude (e.g., delivering two toys requested by a caregiver). These activities are only completely developed around the age of five, when children have mastered the so-called cardinality principle, according to which the last number word recited when counting corresponds to the magnitude of the set (Le Corre & Carey, 2007). From then on, children possess a list of number words which is progressively associated with specific numerical magnitudes, which still need to be automatized and associated with other numerical codes, in particular the digital-Arabic code.

Mastering symbolic numerical codes is one of the first challenges faced by young children in math instruction at school (McLean & Rusconi, 2014). At this point in their numerical development, most children already acquired lexical entries necessary for reciting number words and recognizing single-digit Arabic numbers (Moura et al., 2013, 2015; Power & Dal Martello, 1990, 1997; Seron et al., 1992). However, they usually still struggle with larger numbers and with switching between numerical notations, this means transcoding from number words to digital-Arabic notation and vice versa. In order to successfully acquire these skills, children have to master not only the lexical and syntactic structure of number words and Arabic numbers, but also be aware of similarities and specificities of the two codes.

Learning the digital-Arabic code is, in fact, an important landmark in the development of children's numerical abilities, and one of the first important difficulties they have to deal with (McLean & Rusconi, 2014). But why is it considered and experienced as difficult? The main reason why understanding the structure of the Arabic number system is difficult may be because it is fully symbolic, and not based on any previously acquired numerical ability (e.g., counting) or acquired intuitively. In fact, the learning of the Arabic number system demands explicit and systematic instruction and it usually requires several years until children have mastered its structure (Gervasoni & Sullivan, 2007; Moura et al., 2015).

From an evolutionary perspective, representing numbers in symbolic notations was a big challenge for human civilizations. The origins of the first symbolic numerical codes go back to the time when humans developed written language and may have originated from the necessity to store and share the results of enumeration (Chrisomalis, 2004; Ifrah, 2000; Zhang & Norman, 1995). Early tally-like notations, mostly based on one-to-one correspondence, failed when larger numerical magnitudes needed to be represented (Coolidge & Overmann, 2012). Symbolic codes were then proposed, but initially they did not take advantage of a compositional place-value structure to reduce the complexity imposed by larger sequences of symbols as numerical magnitudes increased (e.g., MCMXLVIII for 1948 in the Roman code; Bender & Beller, 2018). As in other numerical notations (e.g., the Babylonian), in the Arabic number system, numbers are represented by sequences of lexical primitives (i.e., the digits 1 to 9) in accordance with a so-called place-value structure. The latter allowed for an economic representation of large numbers by only using a small set of digits. In particular, in the place-value structure of the Arabic number system, the numerical value of a digit is indicated by its position in the digit string, with the relative magnitude of a digit increasing from right to left by powers of ten. The relative magnitude of a single digit in the digit string is given by the multiplication of its absolute value and its base (following a multiplicative composition principle). The overall magnitude of a multi-digit number is given by the sum of the relative values of all digits (following an additive composition principle). For example, the overall value of 291 is equal to $2 \times 10^2 + 9 \times 10^1 + 1 \times 10^0$ (i.e., $200 + 90 + 1$). Finally, "0" (zero) is an indispensable placeholder that indicates the absence of a given power of ten in a multi-digit Arabic number.

Despite being of clear symbolic nature, the Arabic number system is also influenced by language characteristics such as, for example, the transparency of the respective number word system. Asian languages, such as Mandarin, Korean, and Japanese, are known for having highly transparent number words as they clearly reflect the place-value structure of the Arabic number system (Fuson, 1990; Miura et al., 1993). For example, numbers between 11 and 19 are spoken as "ten one" ($1 \times 10^1 + 1 \times 10^0$), "ten two" ($1 \times 10^1 + 2 \times 10^0$), and so on, until 20, which is spoken as "two tens" (2×10^1) (Fuson, 1990). Contrarily, some languages such as German and Dutch are rather in-transparent as the order of number words is inverted compared to the digital-Arabic notation. For example, the German number word for 24 is "vierundzwanzig" (literally "four and twenty"). Interestingly, previous studies indicate that children speaking languages with transparent number words seem to encounter fewer difficulties in learning number transcoding when compared to speakers of languages with less transparent number words such as English (e.g., thirteen instead of ten three; Miura et al., 1993) and German (e.g., Moeller et al., 2015). For instance, one consistent finding is that a large portion of transcoding errors observed in children speaking languages with in-transparent number words like German, Dutch, or Czech are related to the inversion property of the verbal number system (e.g., Zuber et al., 2009; Moeller et al., 2015; Pixner et al., 2011a; Pixner et al., 2011b).

The complexity of the Arabic number system for young students becomes evident when we consider how performance in number transcoding (i.e., Arabic number writing and reading) increases with age. When investigating Italian speakers, Power and Dal Martello (1990) observed that typically developing first graders were well able to write two-digit numbers flawlessly but experienced problems when writing down three- and four-digit numbers. Interestingly, these difficulties were more pronounced for Arabic numbers with internal zeros (e.g., 1007). This is well in line with more recent findings by Camos (2008). When investigating the performance of French second graders, Camos (2008) also found that these children were perfect in writing Arabic numbers up to 100 and committed fewer errors in three-digit, as compared to four-digit, Arabic numbers. Using a longitudinal design, Seron, Deloche and Nöel (1992; see also Seron & Fayol, 1994) assessed number transcoding skills of second and third graders three times within one school year and reported performance improvements over time with overall better performance for the Arabic number reading as compared to the Arabic number writing condition. In particular, second graders showed an improvement in performance from the beginning to the end of the school year. On the other hand, third graders showed only a small improvement due to ceiling effects from the middle of the school year on. Finally, Moura et al. (2015) studied writing of one- to four-digit Arabic numbers in Brazilian children from first to fourth grades and observed significant improvements from first to third grades, but not from third to fourth grades, substantiating the idea of a plateau or ceiling effect from the third grade onward.

Mastery of the place-value structure of the Arabic number system by children has received increasing research interest recently. This is mostly due to its importance for succeeding in school but also everyday life in general (e.g., Gervasconi & Sullivan, 2007). In the educational context, mastery of the place-value structure of the Arabic number system allows children to represent larger (multi-digit) numbers, and to apply more sophisticated calculation strategies. Not surprisingly, early mastery of the place-value structure of the Arabic number system was found to be predictive of later mathematics achievement. Moeller et al. (2011) administered several numerical tasks to first graders and observed that place-value understanding, assessed by multi-digit Arabic number transcoding and two-digit number magnitude comparison, was highly predictive of performance in multi-digit addition but also math grades two years later. More recently, Lambert and Moeller (2019) showed that difficulties in two-digit addition (in particular in problems requiring a carry over, e.g., 15 + 17 = __) in children with mathematics learning difficulties (MLD) were driven by deficits in their place-value understanding.

Moreover, employing an Arabic number writing task, Moura et al. (2013) showed that children with MLD experienced pronounced difficulties when required to write more complex Arabic numbers (i.e., three- and four-digit Arabic numbers, and Arabic numbers with internal zeros, e.g., 405). Importantly, the most frequently observed errors were due to insufficient syntactical understanding of the place-value structure of the Arabic number system (e.g., writing three hundred forty-five as 300405). Interestingly, these errors were even more common in children with MLD.

Besides requiring specific understanding of the place-value structure of the Arabic number system, processing multi-digit Arabic numbers is also demanding with respect to WM resources. Camos (2008) studied number transcoding in 7-year-old children and found a strong positive association between transcoding performance and WM capacities. More specifically, a critical role for visuo-spatial and central executive components of WM in number transcoding was reported by Zuber et al. (2009) when investigating syntactic errors (such as unit-decade inversion) produced by typically developing German-speaking children in number transcoding.

Importantly, the significant role of WM for number transcoding may reflect one of the underlying factors associated with specific learning difficulties in the domains of both mathematics (Salvador et al., 2019) and reading (Peterson & Pennington, 2015). These respective developmental disabilities frequently co-occur and, in the next sections, we discuss hypotheses put forward to explain this high comorbidity. Moreover, we specifically focus on hypotheses relating phonological WM, as well as other subcomponents of phonological processing (namely phonemic awareness and lexical access) to developmental disabilities in both domains. Afterward, we discuss studies that investigated the role of phonological processing in number transcoding, suggesting that, besides WM, phonemic awareness and lexical access should also be taken into account when it comes to the evaluation of subjacent factors to Arabic number processing.

5 The association between math and reading disabilities

In a meta-analysis, Joyner and Wagner (2019) found that students suffering from MLD are over two times more likely to also present a reading disability compared to children that do not have MLD. According to Moll et al. (2019), basic linguistic skills such as phonemic awareness may be precursors not only for later reading skills but also for verbal numerical skills, such as counting and transcoding, which in turn were found to underlie later arithmetic skills (see above).

The high comorbidity rate for math and reading difficulties may be explained by the double deficit hypothesis (Landerl et al., 2004), according to which children who present both learning difficulties suffer from simultaneous deficits in phonological processing and the processing of number magnitude. In contrast, Simmons and Singleton (2008) suggested a common deficit account to describe cognitive impairments associated with difficulties in both reading and mathematics. According to these authors, MLD may be caused by the phonological deficits commonly associated with dyslexia. It is assumed that phonological representations of dyslexic children are weak, which leads to an impairment in cognitive processes that demand and build on phonological codes. In particular, Simmons and Singleton (2008) proposed the weak phonological representation hypothesis, according to which the poorly specified nature of phonological representations would lead to poor performance in tasks that involve the retention, retrieval, or manipulation of phonological codes. Because Arabic number writing requires access to the verbal representation of number words, it seems sensible to assume that children's phonological processing abilities should also influence their numerical (i.e., transcoding) attainment.

With respect to the high comorbidity rate between math and reading difficulties, Moll et al. (2014) proposed that mathematics has both verbal and nonverbal components and poor performance may be due to different patterns of deficits in verbal and/or nonverbal number processing. The authors assessed children from 6 to 12 years and concluded that children with both reading and math difficulties presented an additive profile of deficits. In line with this argument, Jordan (2007) suggested that reading deficits aggravate – but not necessarily cause – math difficulties, because children with both difficulties would also struggle in using language-based compensatory mechanisms.

For instance, when children have to write Arabic numbers to dictation, the respective input is verbal. Hence, children must be able to differentiate between speech sounds to correctly comprehend the verbal number word that should be transcoded into the digital-Arabic notation. De Clercq-Quaegebeur et al., (2018) assessed arithmetic and number processing abilities of 47 dyslexic French children and found their performance to be lower than one standard deviation below the mean on number transcoding tasks. This result supports the claim that, independently of math learning difficulties, impairments in phonological processing may impact number transcoding performance.

However, most studies on children with reading difficulties, who present phonological processing deficits, have focused on their general arithmetic performance, and did not explore their performance in basic number processing in a differential way (e.g., De Smedt, 2018; Simmons & Singleton, 2008). As such, it is still not clear whether number transcoding may be consistently impaired in these children because of its verbal processing components when transcoding from verbal number words to digital-Arabic notation. Despite this potential impact of phonological processing on basic number processing skills, to the best of our knowledge, there are no studies so far that systematically investigated the association between phonological processing and number transcoding in more depth.

6 Words and numbers: The role of phonological skills and WM

As outlined in the first section, both the dual-route model of single-word reading and the ADAPT model of Arabic number writing assume an important role of phonological processes for the acquisition of the respective symbolic codes. The term “phonological processing” was proposed to refer to a set of cognitive abilities associated with literacy acquisition such as (i) the speed of phonological recoding in lexical access (referring to the recoding of a written stimulus into a sound-based representation to get from the written word to its lexical referent) (e.g., assessed by rapid automatized naming tasks), (ii) processes associated with maintaining sound-based representation in working memory (e.g., measured using verbal span tasks), and (iii) phonemic awareness, reflecting awareness of the sound structure of language (e.g., assessed by phoneme deletion tasks; Wagner & Torgesen, 1987). This set of abilities seems also relevant to number processing.

Lopes-Silva et al. (2014) assessed children’s general cognitive abilities, verbal and visuo-spatial WM, non-symbolic magnitude comparison, phonemic awareness, and verbal to Arabic number transcoding in a sample of 172 children from second to fourth grades. At first glance, a hierarchical regression model showed that verbal WM was a significant predictor of transcoding after considering effects of age and general cognitive abilities. However, adding phonemic awareness in a third step of the regression analyses led to the exclusion of verbal WM. Therefore, the authors conducted path analyses including all of the previous measures to determine possible mediation effects on number transcoding. When phonemic awareness was not included as a mediator of the influence of verbal WM on number transcoding, model fit indices were not acceptable. The model in which the effect of WM was partially mediated by phonemic awareness was the one fitting the empirical data best indicating that this phonological skill is associated specifically with number transcoding. Both phonemic awareness and phonological working memory can thus be interpreted as indexes of the quality of children’s phonological representation which may influence performance on numerical tasks requiring number words representations, such as number transcoding.

Phonemic awareness has also been consistently associated with reading performance (Peterson & Pennington, 2015; Vellutino et al., 2004). To extend this result, possible shared associations between phonemic awareness and digital-Arabic as well as word writing and reading skills were investigated. Lopes-Silva et al. (2016) aimed at disentangling the role of phonemic awareness and its impact on verbal to Arabic transcoding tasks as well as on single-word reading and spelling, controlling for other cognitive variables such as WM. The authors conducted a series of hierarchical regression models with scores of reading and writing of single words and Arabic numbers as dependent variables. They observed that performance on each numerical task (i.e., reading or writing Arabic numbers) was predicted by the corresponding verbal tasks (i.e., reading or spelling words) and vice versa as well as by phonemic awareness – even beyond the influence of general cognitive abilities. Phonological WM was also significantly associated with word reading, but to a smaller extent as compared to the influence of phonemic awareness. Interestingly, phonological WM was not associated with number transcoding. Potentially, this was due to possible shared variance with phonemic awareness. In addition, Teixeira and Moura (2020) observed that children with reading difficulties also present difficulties in writing Arabic numbers, committing both syntactic and lexical errors, whereas lexical errors were hardly observed in typically developing children. These difficulties may be explained by differences in phonological processing abilities, mainly with respect to phonemic awareness, but also regarding speed of lexical access and phonological memory.

Adding to the studies mentioned above, Batista et al. (in preparation) investigated the association between phonological processes, WM and Arabic number transcoding more thoroughly by considering different aspects of phonological processing as well as different WM aspects in the same study. In particular, in a sample of third and fourth graders they assessed variables including phonemic awareness, speed of lexical access as well as verbal and visuo-spatial WM. Hierarchical regressions controlling for influences of general cognitive abilities showed that Arabic number writing performance was predicted by visuo-spatial WM and lexical access. Interestingly, considering lexical access in the regression models led to the exclusion of phonemic awareness.

These findings are in line with the weak phonological representation hypothesis by Simmons and Singleton (2008), according to which phonological processing deficits impair aspects of numerical processing that require the manipulation of verbal codes (transcoding but also counting, arithmetic fact retrieval, etc.), while other nonverbal aspects of number processing that rely less on verbal codes (e.g., magnitude manipulations, estimation, subitizing) should remain unimpaired. In number transcoding, the input is verbal; hence, the child must be able to differentiate between speech sounds to correctly comprehend the verbal number word that needs to be transcoded into Arabic notation. The results reviewed above suggest that the poor phonological representation hypothesis may also hold for numerical transcoding tasks in the sense that number transcoding should also be interpreted as a verbally mediated numerical ability, at least partially relying on phonological processing.

However, the actual working mechanisms underlying the association between phonological processing and numerical abilities more broadly remains unclear so far, even though above-described results may allow for a preliminary conclusion with respect to the interplay of phonological processing and number transcoding abilities. To further substantiate our suggestions, future studies should simultaneously consider different subcomponents of phonological processing (i.e., lexical access and phonemic awareness) to investigate their specific influences. To illustrate this, lexical access has so far been associated with arithmetic fact retrieval (De Smedt, 2018) and fluency in reading words (Papadopoulos et al., 2016). In a similar vein, Geary (1993) suggested that the comorbidity between math and reading difficulties may be associated with deficits in lexical access. Moreover, a meta-analysis by Koponen et al. (2017) investigated the association of lexical access with a range of numerical abilities. Results indicated that rapid automatized naming was more strongly associated with simple numerical tasks (e.g., arithmetic fluency) than with more complex ones (e.g., multi-digit calculations). Also, lexical access is required when processing numerical or operational symbols in simple tasks, while in more complex calculations, multiple cognitive skills are involved (e.g., Koponen et al., 2017). Regarding number transcoding, deficits in lexical access may lead children to commit more lexical errors due to incorrect access to the digital-Arabic representation corresponding to the dictated verbal number word (Barrouillet, 2004).

However, phonemic awareness may be strongly associated with lexical access to numerical symbols. In many languages investigated so far, there are phonologically similar number words that one may confuse – especially children – and specific strategies may be required to differentiate them orally. For instance, in German "zwei" (two) and "drei" (three) sound quite similar. When dictating a phone number people could say "zwo" instead of "zwei," to avoid errors. This is also observed in Portuguese for "três" (three) and "seis" (six), on which people often say "meia" (half a dozen) instead of "seis" (six) to avoid misunderstandings. Furthermore, it is obvious that an accurate understanding of the phonological structure of verbal number words is crucial to derive the corresponding Arabic symbols correctly. Thus, it is especially important to investigate the role of lexical access *and* phonemic awareness as subcomponents of phonological processing because most studies only considered influences of phonological WM so far (see Camos, 2008; Moura et al., 2013; Zuber et al., 2009) – even though the ADAPT model suggests that the first step of transcoding from verbal number words to digital-Arabic notation is the phonological encoding of the respective number word.

Taken together, we reviewed evidence suggesting that phonological processes are important not only for acquiring word reading and writing but also for reading and writing (multi-digit) numbers. However, it is not clear from previous research which role different components of phonological processing may play – in particular in reading and writing (multi-digit) numbers. From a developmental point of view, it seems that phonological processing might be important in early stages of the acquisition of basic numerical and arithmetic abilities whereas the exact role of phonological processes in numerical cognition in adults is controversial (De Rammelaere et al., 2001; De Rammelaere & Vandierendonck, 2001; DeStefano & LeFevre, 2004; Seitz & Schumann-Hengsteler, 2000). In this context, initial evidence also indicates that the relevance of phonological and visuospatial working memory may vary considerably with age and experience (Krajewski & Schneider, 2009; McKenzie et al., 2003). This may suggest that there might be different paths to acquire symbolic numerical and arithmetic abilities in the transition from kindergarten to primary school, one verbal (phonological) and the other visuospatial (LeFevre et al., 2010).

7 Conclusions

Reading and writing words as well as numbers are core challenges elementary students face in their first years of schooling. Despite considerable bodies of research dedicated to each of these tasks, there is still a lack of research on potential overlaps between the cognitive mechanisms underlying word and symbolic number processing, and how they interact during children's cognitive development. Recent results indicated a prominent role for phonological skills for the development of both reading and numerical abilities. Moreover, evaluating performance in tasks that simultaneously draw on phonological as well as numerical aspects, such as number transcoding, seems to be particularly informative. Laborious and sequential phonological processing may be crucial for the initial processing of both words and symbolic numbers in children's development. Practice allows for more efficient forms of processing of words and smaller and more frequent Arabic digits. These may then form the building blocks for reading comprehension and processing of more complex multi-digit symbolic numbers. Better understanding of how representations of words and numbers are associated may foster our understanding of the cognitive underpinnings of learning to read and write words and numbers and, as a consequence, the diagnosis of specific learning difficulties.

References

Alibali, Martha W. & Di Russo, Alyssa A. (1999): The function of gesture in learning to count: More than keeping track. *Cognitive Development* 14, 37–56. ISSN 0885–2014.

Bahnmueller, Julia, Huber, Stefan, Nuerk, Hans-Cristoph, Göbel, Silke M. & Moeller, Korbinian (2016): Processing multi-digit numbers: A translingual eye-tracking study. *Psychological Research* 80, 422–433. doi:10.1007/s00426-015-0729-y.

Barrouillet, Pierre, Camos, Valerie, Perruchet, Pierre & Seron, Xavier (2004): ADAPT: A developmental, asemantic, and procedural model for transcoding from verbal to arabic numerals. *Psychological Review* 111 (2), 368–394. doi:10.1037/0033-295X.111.2.368.

Batista, Luana T., Gomides, Mariuche, Bahnmueller, Julia, Moeller, Korbinian, Salles, Jerusa, Haase, Vitor G. & Lopes-Silva, Julia B. Differential impact of phonological processing components on Arabic number writing. (in preparation)

Bender, Andrea & Beller, Sieghart (2018): Numeration systems as cultural tools for numerical cognition. In *Language and Culture in Mathematical Cognition*. Academic Press, 297–320. doi:https://doi.org/10.1016/b978-0-12-812574-8.00013-4.

Brysbaert, Marc (2005): Number recognition in different formats. In Jamie, I. D. Campbell (ed.): *The Handbook of Mathematical Cognition*. New York: Psychology Press, 23–42.

Camos, Valerie (2008): Low WM capacity impedes both efficiency and learning of number transcoding in children. *Journal of Experimental Child Psychology* 99, 37–57. doi:10.1016/j.jecp.2007.06.006.

Castles, Anne & Coltheart, Max (2004): Is there a causal link from phonological awareness to success in learning to read? *Cognition* 91 (1), 77–111. doi:10.1016/S0010-0277(03)00164-1.

Chrisomalis, Stephen (2004): A cognitive typology for numerical notation. *Cambridge Archaeological Journal* 14 (1), 37.

Collins, Melissa A. & Laski, Elida V (2019): Digging deeper: Shared deep structures of early literacy and mathematics involve symbolic mapping and relational reasoning. *Early Childhood Research Quarterly* 46, 201–212. doi:10.1016/j.ecresq.2018.02.008.

Coltheart, Max, Rastle, Kathleen, Perry, Conrad, Langdon, Robyn & Ziegler, Johannes (2001): DRC: A dual route cascaded model of visual word recognition and reading aloud. *Psychological Review* 108 (1), 204. Inc. doi:I0.1037//0033-295X.108.1.204.

Coolidge, Frederick L. & Overmann, Karenleigh A. (2012): Numerosity, abstraction, and the emergence of symbolic thinking. *Current Anthropology*. doi:https://doi.org/10.1086/664818.

Costa, Anneliese J., Lopes-Silva, J. G., Pinheiro-Chagas, Pedro, Krinzinger, Helga, Lonnemann, Jan, Willmes, Klaus, Wood, Guilherme & Haase, Vitor G. (2011): A hand full of numbers: A role for offloading in arithmetics learning. *Frontiers in Psychology* 2, 368. doi:10.3389/fpsyg.2011.00368.

Crollen, Virginie, Seron, Xavier & Noël, Marie-Pascale (2011): Is finger-counting necessary for the development of arithmetic abilities? *Frontiers in Psychology* 2, 242. doi:10.3389/fpsyg.2011.00242.

De Clercq-Quaegebeur, Maryse, Casalis, Severine, Vilette, Bruno, Lemaitre, Marie-Pierre & Vallée, Louis (2018): Arithmetic abilities in children with developmental dyslexia: Performance on French ZAREKI-R test. *Journal of Learning Disabilities* 51 (3), 236–249. doi:10.1177/0022219417690355.

De Rammelaere, Stijn, Stuyven, Els & Vandierendonck, André (2001): Verifying simple arithmetic sums and products: Are the phonological loop and the central executive involved? *Memory & Cognition* 29 (2), 267–273. doi:https://doi.org/10.3758/BF03194920.

De Rammelaere, Stijn & Vandierendonck, André (2001): Are executive processes used to solve simple arithmetic production tasks. Current Psychology Letters. *Behaviour, Brain & Cognition* 5, 79–89, URL:: http://journals.openedition.org/cpl/231.

De Smedt, Bert (2018): Language and arithmetic: The potential role of phonological processing. In *Heterogeneity of Function in Numerical Cognition*. Academic Press, 51–74. doi:https://doi.org/10.1016/B978-0-12-811529-9.00003-0.

Dehaene, Stanislas (2009): *Reading in the Brain: The New Science of How We Read*. New York Penguin.

Delazer, Margarete & Girelli, Luisa (1997): When 'Alfa Romeo' facilitates 164: Semantic effects in verbal number production. *Neurocase* 3 (6), 461–475.

DeStefano, Diana & LeFevre, Jo-Anne (2004): The role of working memory in mental arithmetic. *European Journal of Cognitive Psychology* 16 (3), 353–386. doi:https://doi.org/10.1080/09541440244000328.

Dumas, Dennis, Alexander, Patricia A. & Grossnickle, Emily M. (2013): Relational reasoning and its manifestations in the educational context: A systematic review of the literature. *Educational Psychology Review* 25 (3), 391–427. doi:10.1007/s10648-013-9224-4.

Fritz, Annemarie, Ehlert, Antje & Balzer, Lars (2013): Development of mathematical concepts as basis for an elaborated mathematical understanding. *South African Journal of Childhood Education* 3 (1), 38–67. http://www.scielo.org.za/scielo.php?script=sci_art text&pid=S2223-76822013000100004&lng=en&tlng=en.

Fuson, Karen C. (1990): Conceptual structures for multiunit numbers: Implications for learning and teaching multidigit addition, subtraction, and place value. *Cognition and Instruction* 7 (4), 343–403.

Geary, David C. (1993): Mathematical disabilities: Cognitive, neuropsychological, and genetic components. *Psychological Bulletin* 114 (2), 345–362. doi:10.1037/0033-2909.114.2.345.

Geary, David C. (2005): *The Origin of Mind*. Washington, DC: American Psychological Association.

Gervasoni, Ann & Sullivan, Peter (2007): Assessing and teaching children who have difficulty learning arithmetic. *Educational & Child Psychology* 24 (2), 40–53.

Hecht, Steve A. (2002): Counting on working memory in simple arithmetic when counting is used for problem solving. *Memory & Cognition* 30 (3), 447–455. doi:https://doi.org/10.3758/BF03194945.

Ifrah, Georges (2000): *The Universal History of Numbers*. London: Harvill.

Jordan, Nancy C. (2007): Do words count? Connections between reading and mathematics difficulties. In Berch, D. B., Mazzocco, M. M. M. (eds.): *Why is Math So Hard for Some Children*. Baltimore, MD: Brooks, 107–120.

Joyner, Rachel E. & Wagner, Richard K. (2019): Co-Occurrence of reading disabilities and math disabilities: A meta-analysis. *Scientific Studies of Reading*, 1–9. doi:10.1080/10888438.2019.1593420.

Koponen, Tuine, Georgiou, George, Salmi, Paula, Leskinen, Markku & Aro, Mikko (2017): A meta-analysis of the relation between RAN and mathematics. *Journal of Educational Psychology* 109 (7), 977. doi:10.1037/edu0000182.

Krajewski, Kristin & Schneider, Wolfgang (2009): Early development of quantity to number-word linkage as a precursor of mathematical school achievement and mathematical difficulties: Findings from a four-year longitudinal study. *Learning and Instruction* 19 (6), 513–526. doi:10.1016/j.learninstruc.2008.10.002.

Lambert, Katharina & Moeller, Korbinian (2019): Place-value computation in children with mathematics difficulties. *Journal of Experimental Child Psychology* 178, 214–225. doi:10.1016/j.jecp.2018.09.008.

Landerl, Karin, Bevan, Anna & Butterworth, Brian (2004): Developmental dyscalculia and basic numerical capacities: A study of 8–9-year old students. *Cognition* 93 (2), 99–125. doi:10.1016/j.cognition.2003.11.004.

Le Corre, Mathieu & Carey, Susan (2007): One, two, three, four, nothing more: An investigation of the conceptual sources of the verbal counting principles. *Cognition* 105 (2), 395–438. doi:https://doi.org/10.1016/j.cognition.2006.10.005.

LeFevre, Jo-Anne, Fast, Lisa, Skwarchuk, Sheri-Lynn L., Smith-Chant, Brenda L., Bisanz, Jeffery, Kamawar, Deepthi & Penner-Wilger, Marcie (2010): Pathways to mathematics: Longitudinal predictors of performance. *Child Development* 81 (6), 1753–1767. doi:https://doi.org/10.1111/j.1467-8624.2010.01508.x.

Lopes-Silva, Julia B., Moura, Ricardo, Júlio-Costa, Anneliese, Haase, Vitor G. & Wood, Guilherme (2014): Phonemic awareness as a pathway to number transcoding. *Frontiers in Psychology* 5, 13. doi:10.3389/fpsyg.2014.00013.

Lopes-Silva, Julia B., Moura, Ricardo, Júlio-Costa, Anneliese, Wood, Guilherme, Salles, Jerusa F. & Haase, Vitor G. (2016): What is specific and what is shared between numbers and words? *Frontiers in Psychology* 7, 22. doi:10.3389/fpsyg.2016.00022.

McKenzie, Bruce, Bull, Rebecca & Gray, Colin (2003): The effects of phonological and visual-spatial interference on children's arithmetical performance. *Educational and Child Psychology* 20, 93–108. https://www.researchgate.net/profile/Rebecca_Bull/publication/285025652_The_effects_of_phonological_and_visual-spatial_interference_on_children's_arithmetic_performance/links/5853552308ae7d33e01ab741/The-effects-of-phonological-and-visual-spatial-interference-on-childrens-arithmetic-performance.pdf.

McLean, Janet F. & Rusconi, Elena (2014): Mathematical difficulties as decoupling of expectation and developmental trajectories. *Frontiers in Human Neuroscience* 8, 44. doi:10.3389/fnhum.2014.00044.

Miura, Irene T., Okamoto, Yukari, Kim, Chungsoon C., Steere, Marcia & Fayol, Michel (1993): First graders' cognitive representation of number and understanding of place value: Cross-national comparisons – France, Japan, Korea, Sweden, and the United States. *Journal of Educational Psychology* 85, 24–30. doi:10.1037/0022-0663.85.1.24.

Moeller, Korbinian, Pixner, Silvia, Zuber, Julia, Kaufmann, Liane & Nuerk, Hans-Christoph (2011): Early place-value understanding as a precursor for later arithmetic performance – A longitudinal study on numerical development. *Research in Developmental Disabilities* 32 (5), 1837–1851. doi:10.1016/j.ridd.2011.03.012.

Moeller, Korbinian, Zuber, Julia, Olsen, Naoko, Nuerk, Hans-Christoph & Willmes, Klaus (2015): Intransparent German number words complicate transcoding – A translingual comparison with Japanese. *Frontiers in Psychology* 6 (JUN). doi:https://doi.org/10.3389/fpsyg.2015.00740.

Moll, Kristina, Göbel, Silke & Snowling, Margaret (2014): Basic number processing in children with specific learning disorders: Comorbidity of reading and mathematics disorders. *Child Neuropsychology* 21 (3), 399–417. doi:10.1080/09297049.2014.899570.
Moll, Kristina, Landerl, Karin, Snowling, Margaret J. & Schulte-Körne, Gerd (2019): Understanding comorbidity of learning disorders: Task-dependent estimates of prevalence. *Journal of Child Psychology and Psychiatry* 60 (3), 286–294. doi:10.1111/jcpp.12965.
Moura, Ricardo, Lopes-Silva, Julia B., Vieira, Laura R., Paiva, Giulia M., Prado, Ana C. D. A., Wood, Guilherme & Haase, Vitor G (2015): From "five" to 5 for 5 minutes: Arabic number transcoding as a short, specific, and sensitive screening tool for mathematics learning difficulties. *Archives of Clinical Neuropsychology* 30 (1), 88–98. doi:10.1093/arclin/acu071.
Moura, Ricardo, Wood, Guilherme, Pinheiro-Chagas, Pedro, Lonnemann, Jan, Krinzinger, Helga, Willmes, Klaus & Haase, Vitor G. (2013): Transcoding abilities in typical and atypical mathematics achievers: The role of WM and procedural and lexical competencies. *Journal of Experimental Child Psychology* 116 (3), 707–727. doi:10.1016/j.jecp.2013.07.008.
Noël, Marie-Pascale (2009): Counting on working memory when learning to count and to add: A preschool study. *Developmental Psychology* 45 (6), 1630–1643. doi:10.1037/a0016224.
Papadopoulos, Timothy C., Spanoudis, George C. & Georgiou, George K. (2016): How is RAN related to reading fluency? A comprehensive examination of the prominent theoretical accounts. *Frontiers in Psychology* 7, 1217. doi:10.3389/fpsyg.2016.01217.
Parsons, Samantha & Bynner, John (1997): Numeracy and employment. Education + Training, 39 (2), 43–51. https://doi.org/10.1108/00400919710164125
Peterson, Robin L. & Pennington, Bruce F (2015): Developmental Dyslexia. *Annual Review of Clinical Psychology* 11 (9), 1–25.
Pixner, Silvia, Moeller, Korbinian, Hermanová, Veronika, Nuerk, Hans-Christoph & Kaufmann, Liane (2011a): Whorf reloaded: Language effects on nonverbal number processing in first grade: A trilingual study. *Journal of Experimental Child Psychology* 108, 371–382.
Pixner, Silvia, Zuber, Julia, Hermanová, Veronika, Kaufmann, Liane, Nuerk, Hans-Christoph & Moeller, Korbinian (2011b): One language, two number-word systems, and many problems: Numerical cognition in the Czech language. *Research in Developmental Disabilities* 32, 2683–2689.
Power, Richard & Dal Martello, Maria F. (1997): From 834 to eighty thirty four: The reading of arabic numerals by seven-year-old children. *Mathematical Cognition* 3 (1), 63–85. doi: https://doi.org/10.1080/135467997387489.
Power, Richard J. D & Dal Martello, Maria F (1990): The dictation of Italian numerals. *Language and Cognitive processes* 5 (3), 237–254. doi:10.1080/01690969008402106.
Ritchie, Stuart J. & Bates, Timothy C. (2013): Enduring links from childhood mathematics and reading achievement to adult socioeconomic status. *Psychological Science* 24 (7), 1301–1308. doi:https://doi.org/10.1177/0956797612466268.
Romano, Elisa, Babchishin, Lyzon, Pagani, Linda S. & Kohen, Dafna (2010): School readiness and later achievement: Replication and extension using a nationwide Canadian survey. *Developmental Psychology* 46 (5), 995–1007. doi:10.1037/a0018880.
Salvador, Larissa S., Moura, Ricardo, Wood, Guilherme & Haase, Vitor G. (2019): Cognitive heterogeneity of math difficulties: A bottom-up classification approach. *Journal of Numerical Cognition* 5 (1), 55–85. doi:10.5964/jnc.v5i1.60.

Sarnecka, Barbara W. & Lee, Michael D. (2009): Levels of number knowledge during early childhood. *Journal of Experimental Child Psychology* 103 (3), 325–337. doi:10.1016/j.jecp.2009.02.007.

Seitz, Katja & Schumann-Hengsteler, Ruth (2000): Mental multiplication and working memory. *European Journal of Cognitive Psychology* 12 (4), 552–570. doi:https://doi.org/10.1080/095414400750050231.

Seron, Xavier, Deloche, Gerard & Noël, Marie-Pascale (1992): Number transcribing by children: Writing Arabic numbers under dictation. In Bideaud, Jacqueline, Meljac, Claire, Fischer, Jean-Paul (eds.): *Pathways to Number: Children's Developing Numerical Abilities*. Lawrence Erlbaum Associates, Inc, 245–264. https://psycnet.apa.org/record/1992-97947-013.

Seron, Xavier & Fayol, Michele (1994): Number transcoding in children: A functional analysis. *British Journal of Developmental Psychology* 12 (3), 281–300. doi:https://doi.org/10.1111/j.2044-835X.1994.tb00635.x.

Share, David L. (1999): Phonological recoding and orthographic learning: A direct test of the self-teaching hypothesis. *Journal of Experimental Child Psychology* 72, 95–129. doi:https://doi.org/10.1006/jecp.1998.2481.

Siegler, Robert S. & Braithwaite, David W. (2017): Numerical development. *Annual Review of Psychology* 68, 187–213. doi:10.1146/annurev-psych-010416-044101.

Siegler, Robert S., Thompson, Clarissa A. & Schneider, Martin (2011): An integrated theory of whole number and fractions development. *Cognitive Psychology* 62, 273–296. doi:https://doi.org/10.1016/j.cogpsych.2011.03.001.

Simmons, Fiona R. & Singleton, Chris (2008): Do weak phonological representations impact on arithmetic development? A review of research into arithmetic and dyslexia. *Dyslexia* 14 (2), 77–94. doi:10.1002/dys.341.

Teixeira, Renata M. & Moura, Ricardo (2020): Arabic number writing in children with developmental dyslexia. *Estudos de Psicologia (Campinas)* 37 (e180179). doi:http://dx.doi.org/10.1590/1982-0275202037e180179.

Treiman, Rebecca, Kessler, Brett, Boland, Kelly, Clocksin, Hayley & Chen, Zhengdao (2018): Statistical learning and spelling: Older prephonological spellers produce more wordlike spellings than younger prephonological spellers. *Child Development* 89 (4), e431-e443. doi:https://doi.org/10.1111/cdev.12893.

Vellutino, Frank R., Fletcher, Jack M., Snowling, Margaret J. & Scanlon, Donna M. (2004): Specific reading disability (dyslexia): What have we learned in the past four decades? *Journal of Child Psychology and Psychiatry* 45 (1), 2–40. doi:10.1046/j.0021-9630.2003.00305.x.

Wagner, Richard K. & Torgesen, Joseph K. (1987): The nature of phonological processing and its causal role in the acquisition of reading skills. *Psychological Bulletin* 101 (2), 192–212. doi:https://doi.org/10.1037/0033-2909.101.2.192.

Wasner, Miram, Moeller, Korbinian, Fischer, Martin H. & Nuerk, Hans-Christoph (2014): Aspects of situated cognition in embodied numerosity: The case of finger counting. *Cognitive Processing* 15 (3), 317–328. doi:10.1007/s10339-014-0599-z.

Zhang, Jiajie & Norman, Donald A. (1995): A representational analysis of numeration systems. *Cognition* 57 (3), 271–295. doi:https://doi.org/10.1016/0010-0277(95)00674-3.

Zuber, Julia, Pixner, Silvia, Moeller, Korbinian & Nuerk, Hans-Christoph (2009): On the language specificity of basic number processing: Transcoding in a language with inversion and its relation to WM capacity. Journal of Experimental Child Psychology 102, 60–77. doi:10.1016/j.jecp.2008.04.003.

Sarah R. Powell, Samantha E. Bos, and Xin Lin

The assessment of mathematics vocabulary in the elementary and middle school grades

1 Introduction

Students use academic language, which involves vocabulary, grammatical structures, and linguistic functions, to learn knowledge and perform tasks in a specific discipline (e.g., mathematics; Cummins, 2000). Understanding these discipline-specific ways of using language requires deep knowledge of discipline-specific content and a keen understanding connecting academic language to learning (Fang, 2012). Therefore, not surprisingly, academic language has been shown to be closely related to academic performance (Kleemans et al., 2018) and a significant predictor of academic achievement (Townsend et al., 2012). Mathematics, a challenging discipline for many students (Berch & Mazzocco, 2007), also develops academic language specific to the discipline, which is often referred to as mathematics language. Mathematics language is used to express mathematical ideas and to define mathematical concepts, and it can facilitate connections among different representations of mathematical ideas (Bruner, 1966).

In this Introduction, we provide a definition of mathematics vocabulary and discuss the importance of understanding mathematics vocabulary. Then, we review why and how students experience difficulty with mathematics vocabulary. In the rest of the chapter, we describe the development and testing of several measures of mathematics vocabulary. These measures could be used by educators to understand which mathematics vocabulary cause difficulty for students and could be a focus of mathematics instruction.

1.1 Definition of mathematics vocabulary

Mathematics vocabulary, a key component of mathematics language (Moschkovich, 2015; Simpson & Cole, 2015), includes terms routinely used in mathematics instruction, textbooks, and assessments (Monroe & Orme, 2002; Moschkovich, 2013). In this chapter, we define *mathematics vocabulary* as terms used to describe specific mathematical concepts or procedures. We use the word "term" because a large proportion of mathematics vocabulary includes more than one word (e.g., *greater than*,

https://doi.org/10.1515/9783110661941-016

minus sign, quarter past, rectangular prism, unequal shares). In the mathematics learning progression (Browning & Beauford, 2011; Purpura & Logan, 2015), mathematics vocabulary first emerges as number terms (e.g., *one, three*), terms representing quantities (e.g., *more, less*), and terms representing spatial relations (e.g., *above, below*). Some of the earliest learning of mathematics occurs through learning mathematics vocabulary (Purpura et al., 2017).

As the complexity of mathematics skills increases by grade level, mathematics vocabulary becomes accumulatively complex with students expected to understand hundreds of different mathematics vocabulary terms by middle school (Powell et al., 2017). Mastering foundational mathematics vocabulary may be necessary for understanding advanced mathematics vocabulary and concepts. For example, students need to master the term *multiple* to understand the term *least common multiple*, which describes part of the procedure for identifying common denominators when adding or subtracting fractions. Given the complexity and accumulative nature of mathematics vocabulary, an explicit focus on vocabulary has become a point of interest in mathematics education (Browning & Beauford, 2011; Riccomini et al., 2015). In the next section, we discuss mathematics standards and research practices that highlight the importance of mathematics vocabulary.

1.2 Importance of mathematics vocabulary

Mathematics practice standards in the United States (U.S.) highlight the use of mathematics vocabulary as a medium to learn and perform mathematics. For example, within the Curriculum Focal Points of National Council of Teachers of Mathematics in the U.S. (2006), students are expected to "develop vocabulary to describe" various attributes of shapes (p. 31) or use "language" to compare quantities (p. 11). Similarly, use of mathematics vocabulary as a medium to learn mathematics is also shown in the mathematics standards used in the U.S. (National Governors Association Center for Best Practices & Council of Chief State School Officers, 2010). Specifically, mathematics standards suggest that students be able to use clear vocabulary to communicate precisely to others, explain how to solve problems, construct viable arguments, and critique the mathematic reasoning of others. For example, it is outlined that students should use "language to describe" (p. 42) or "describe their physical world using . . . vocabulary" (p. 9).

In addition to a focus in mathematical standards in the U.S., mathematics vocabulary is important because of its association to mathematics performance. For example, Powell and Nelson (2017) noted a significant correlation between the mathematics vocabulary and mathematics fluency (i.e., fluency with mathematics facts such as $4 + 7$ or $36 \div 6$) scores of U.S. first-grade students.

Similarly, Powell et al. (2017) identified significant correlations between a test of mathematics vocabulary and mathematics computation (i.e., addition, subtraction, multiplication, or division or multi-digit numbers involving algorithms) for both third- and fifth-grade U.S. students. Peng and Lin (2019) noted an analogous pattern with Chinese fourth-grade students with significant correlations on measures of mathematics vocabulary and mathematics fluency, computation, and word problems. Besides correlational findings, Fuchs et al. (2015) further demonstrated that mathematics vocabulary may partially or fully explain the relation between general cognitive skills and students' word-problem solving performance.

Consistent with the focus of mathematics vocabulary in U.S. practice standards for mathematics and more recent research demonstrating an association between mathematics-vocabulary knowledge and mathematics performance, an increasing number of researchers have started to focus on the importance of instruction related to mathematics vocabulary (Harmon et al., 2005; Livers & Elmore, 2018; Monroe & Orme, 2002; Riccomini et al., 2015). Monroe and Panchyshyn (1995) explained instruction needs to occur for four categories of mathematics vocabulary. First, students need to learn technical terms that include terms specific to mathematics (e.g., *decagon*). Second, students need instruction on subtechnical terms. Subtechnical terms have multiple meanings, one of which is mathematics related (e.g., *cube*). Third, students require instruction on symbolic terms (e.g., *minus sign*). Fourth, educators need to ensure students understand general terms. These are non-mathematics terms (e.g., *measure*) used in the mathematics classroom.

Many researchers and educators have provided suggestions for teaching mathematics vocabulary including using explicit instruction (Bay-Williams & Livers, 2009; Monroe & Orme, 2002), mnemonic strategies (Riccomini et al., 2015), and graphic organizers with the definitions, characteristics, examples, and nonexamples of a mathematics vocabulary term (Bruun et al., 2015). However, in classrooms, mathematics vocabulary instruction is not often prioritized. Specifically, except for the instruction of definitions in textbook, educators provide students with few opportunities to explicitly learn mathematics vocabulary (Monroe & Orme, 2002). In addition, educators of mathematics often use informal language in the classroom – for example, *diamond* for *rhombus*, *bottom number* for *denominator*, and *line* for *fraction bar* (Karp et al., 2014; Rubenstein & Thompson, 2002). Because students do not receive enough mathematics vocabulary instruction, understanding mathematics vocabulary can be difficult.

1.3 Difficulty of understanding mathematics vocabulary

As described by Barrow (2014), mathematics is not a universal language, and all students should be considered learners of mathematics vocabulary and language. Not understanding academic language may prohibit students from engaging fully in the mathematics classroom (Ernst-Slavit & Mason, 2011; Schleppegrell, 2012). Mathematics vocabulary may be difficult for many students because of the complexity of the vocabulary. Rubenstein and Thompson (2002) listed 11 difficulties students may encounter when learning mathematics vocabulary terms: (1) when used in mathematics context, some common English terms have alternative meanings (e.g., *expression, face*); (2) some terms have similar but more precise meanings (e.g., *area, average*); (3) some terms involve technical terms specific to mathematics (e.g., *parallelogram, integer*); (4) some terms have more than one mathematical definition, such as *cube* as a solid figure versus to *cube* a number; (5) some terms used in mathematics have different technical meanings when used in other disciplines (e.g., *prism* is a solid figure in mathematics versus *prism* is an object that refracts light in science); (6) some mathematical terms have homophones or homographs, such as *pi* versus *pie*; (7) some related mathematical terms have distinct meanings, but are easily confused (e.g., *divisor* and *dividend*); (8) the translation of a single mathematical term into another language may have multiple ways, which may cause confusion – for example, the Spanish word *tabla* can be translated to the data *table*, but not the *table* we eat from (this would be *mesa*); (9) the spelling of terms is not regular (e.g., *half* vs. *halves*); (10) some mathematical terms can be verbalized in more than one way (e.g., *one-quarter* and *one-fourth*); and (11) the use of informal language in many classrooms, as discussed earlier, make the learning of mathematics vocabulary more difficult.

Given the difficulty of understanding mathematics vocabulary, research on mathematics-vocabulary instruction is in need. Only one experimental study (Petersen-Brown et al., 2019), according to our knowledge, has specifically focused on the instruction of mathematics vocabulary. Although their study showed the effectiveness of mathematics-vocabulary instruction for third and fourth graders, their study only included instruction about eight mathematics-vocabulary terms. Future experimental research involving more grade-level important mathematics vocabulary is in need. However, before conducting experimental work, it is necessary to understand how to develop a measure of mathematics vocabulary. Understanding such process could help determine important mathematics vocabulary terms to be involved in mathematics-vocabulary intervention. Such measures should assist in determining the efficacy of any mathematics-vocabulary intervention in addition to providing an understanding of the baseline levels of mathematics vocabulary of students and variability in mathematics-vocabulary performance. In the next section, we describe our efforts at designing several mathematics-vocabulary measures at different grade levels.

2 Development of mathematics-vocabulary measures

Across the last few years, we developed a series of mathematics-vocabulary measures for use at grade 1 (ages 6–7), 3 (ages 8–9), 5 (ages 10–11), 7 (ages 12–13), and 8 (ages 13–14). We utilized a similar development process across the measures, and we describe this process in the following paragraphs to demonstrate our development framework and to aid any researchers or educators who want to develop their own mathematics-vocabulary measures in English or another language.

2.1 Determine grade-level mathematics vocabulary

First, we compiled lists of grade-specific mathematics vocabulary. Unfortunately, educators in the U.S. do not have access to a common list of mathematics vocabulary at each grade level, so we accessed several mathematics textbooks at a single grade level (e.g., first-grade mathematics textbooks) and created our own grade-level lists of mathematics vocabulary. Our list of textbooks included *enVision MATH*, *Everyday Mathematics*, *Go Math!*, and *Houghton Mifflin Math*. We reviewed glossaries in the textbooks and created a database of mathematics-vocabulary terms and definitions. We accessed two or three textbooks at each grade level from kindergarten through eighth grade. Our complete database contained 1,220 mathematics-vocabulary terms with many terms appearing at several grade levels. Fig. 1 displays the counts of mathematics-vocabulary terms within each grade level.

In the early elementary grades (i.e., kindergarten through second grade), glossaries featured approximately 150 different terms. In the third grade, we noted a large increase in mathematics-vocabulary expectations, and we attributed this to the introduction of multiplication, division, and fractions in U.S. classrooms in the third grade (National Governors Association Center for Best Practices & Council of Chief State School Officers, 2010). We identified another substantial increase from fifth grade to sixth grade as U.S. students enter middle school, and instruction about algebra and rational numbers becomes a core focus of mathematics instruction.

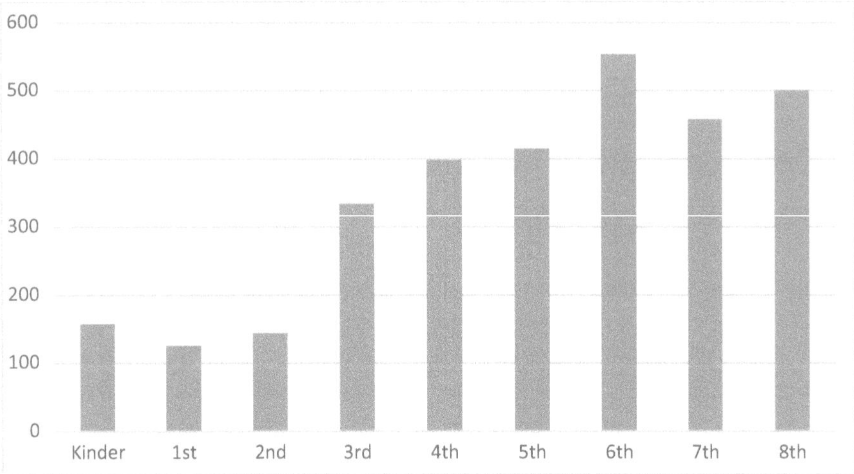

Fig. 1: Mathematics terms featured within grade-level textbooks.

2.2 Focus on important mathematics vocabulary

After we compiled the vocabulary database, we determined it was important to streamline the mathematics vocabulary that would be included in any measure. That is, we could not possibly create a test with 553 separate mathematics-vocabulary terms identified from sixth-grade glossaries; no student would take that test, and no educator would want to grade that test.

Our process for streamlining involved the following. We awarded 1 or 2 points if a term appeared in one or two additional glossaries within a specific grade level. We believed a term gained importance if multiple textbooks written by different author teams used the same term. We identified many textbook-specific terms (e.g., *break apart a ten*, *count back*, *in all*) that did not appear in more than one textbook. Then, we awarded 1 to 3 points if the term appeared in glossaries at other grade levels. For example, during the first development of the first-grade measure, a term earned points for also appearing in kindergarten, second grade, and third grade glossaries. We noted the use of some terms (e.g., *half past* or *T-chart*) occurred only at a single grade level. Other terms (e.g., *greater than*, *length*, *number line*, *octagon*) were utilized across several grade levels, which indicated such terms had carried greater weight because students would hear, see, and use these terms across multiple years. Third, we consulted mathematics standards used across the U.S. (National Governors Association Center for Best Practices & Council of

Chief State School Officers, 2010) and determined whether standards explicitly used terms. For example, at first grade, *addend*, *quarters*, and *trapezoid* are each explicitly named within the first-grade mathematics standards. A term earned 1 point if it is specifically mentioned within standards. In sum, we included the terms with the highest point values.

For the development of a middle school measure for use in seventh and eighth grades, we further tightened our method for streamlining vocabulary terms (Hughes et al., 2020). After confirming terms were featured across glossaries and grade levels and within standards, we had approximately 200 terms remaining. We placed the terms on an online survey and asked middle school educators to select the 50 most important mathematics-vocabulary terms for their grade level. For our measure, we included approximately 70 terms that multiple educators categorized as important. We noted consistency across teachers; for example, 53 of 53 educators agreed *expression* was important with 52 teachers coming to agreement on *distributive property* and 50 educators stating *equation* was an important middle school mathematics vocabulary term.

To avoid basement or ceiling effects, we included terms introduced before and beyond the target grade levels. We did this based on grade level of introduction of the term in the textbook glossaries. For example, at third grade, we included many terms (e.g., *circle*, *odd*) introduced before the third grade, some as early as kindergarten, to ensure students who experienced difficulty with third-grade terms could answer some questions about more familiar terms. At fifth grade, we included several terms introduced beyond the fifth grade (e.g., *positive integer*, *slope*) to ensure a distribution of mathematics-vocabulary scores and limit ceiling effects. For this reason, it was important to collect terms from textbook glossaries across the elementary and middle school grades to be able to understand the grade level of introduction of a term, in how many grades the term appeared, and the grade level of disappearance of a term.

2.3 Develop questions to assess mathematics-vocabulary knowledge

We designed each of our measures for educators to use in the general education classroom. Therefore, we planned for a paper-and-pencil task in which students would read prompts and respond via writing. For each selected term, we created three levels of questions: recall, comprehension, and use in complex tasks (Haladyna & Rodriguez, 2013). Figure 2 displays sample questions. Often, our recall questions involved matching a letter to a term. Comprehension questions

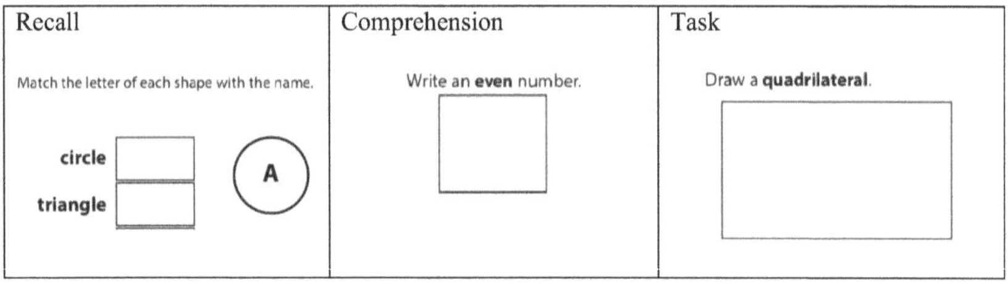

Fig. 2: Examples of different questions on third-grade measure.

encouraged the students to provide a quick response. Task questions asked students to draw something. After generating the levels of questions, we selected the level of question that would be easiest for student response, but we also considered balancing levels so that students answered a variety of questions.

When creating the test forms, we started each test with an easier question as the first problem. We then grouped similar items together (e.g., questions about *even* and *odd* numbers or *numerator* and *denominator* appeared next to one another), but we distributed terms by mathematical domain. For all grade levels except the first grade, examiners read a set of directions and then provided time (e.g., 20 min or 30 min) for students to work. Examiners read no questions or terms aloud. Our decision to not read the test aloud meant that students had to use reading to interpret the mathematics-vocabulary term, just as students would do in a textbook, on a test, or on a computer screen. At the first grade, because of the limited reading experiences of the students, examiners read each prompt and answer choices aloud and permitted students to respond to the question before moving on to read the next prompt aloud.

3 Student performance on mathematics-vocabulary measures

In the U.S., we administered mathematics-vocabulary measures to students at different grade levels. In the following sections, we describe five different studies. In the first three studies, we investigated the use of mathematics-vocabulary measures at understanding the mathematics-vocabulary knowledge of students in grade 1 (Powell & Nelson, 2017), in grades 3 and 5 (Powell et al., 2017), and grades 7 and 8 (Hughes et al., 2020). In the fourth study (Forsyth & Powell, 2017), we compared the mathematics-vocabulary scores of fifth-grade students with and without mathematics or reading difficulty. In the fifth study

(Powell et al., 2020), we compared mathematics-vocabulary scores of third-grade students with and without mathematics difficulty who also categorized as dual-language learners or native English speakers. Tab. 1 presents an overview of the data from each study.

3.1 Elementary students' mathematics vocabulary knowledge

3.1.1 Grade 1

Powell and Nelson (2017) investigated the mathematics-vocabulary knowledge of first-grade students and also explored the relationship between general English vocabulary, mathematics fluency, and mathematics vocabulary. Students showed wide variability in mathematics-vocabulary performance. Findings suggested a significant and positive relationship between both general English vocabulary and mathematics vocabulary as well as mathematics fluency and mathematics vocabulary.

Powell and Nelson (2017) explored whether students struggled with a particular type of mathematics vocabulary. Few clear patterns emerged regarding which mathematics-vocabulary terms caused the most difficulty for students. According to categories suggested by Monroe and Panchyshyn (1995), students had higher accuracy rates in identifying general mathematics-vocabulary terms (91.1%) than symbolic terms (54.5%), technical terms (42.0%), or subtechnical terms (56.4%). Student accuracy varied based on whether the vocabulary term was introduced in kindergarten, first or second grade. That is, students demonstrated a 67.1% accuracy rate with terms introduced in kindergarten textbooks, followed by an accuracy of 48.8% of terms introduced in the first grade. As expected, accuracy on terms introduced in the second grade was 29.2%. Powell and Nelson (2017) tested the measure near the final weeks of the first grade when students should have mastered kindergarten and first-grade mathematics vocabulary. This finding suggested students may already struggle with mathematics-vocabulary terms as early as the first grade.

3.1.2 Grades 3 and 5

Powell et al. (2017) continued to explore the mathematics vocabulary knowledge of elementary-aged students, examining trends in the knowledge base of third- and fifth-grade students. The authors analyzed the mathematics vocabulary knowledge, general English vocabulary, and mathematics computation knowledge

Tab. 1: Descriptions of Mathematics-Vocabulary Assessments.

Study	Grade	n	Comparison	Measure	Maximum terms	M	SD	Range	Correlations or comparison
Powell & Nelson (2017)	1	104	All students	Mathematics Vocabulary – Grade 1	64	36.3	8.1	15–55	r = .697 with general English vocabulary r = .586 with mathematics fluency
Powell et al. (2017)	3	65	All students	Mathematics Vocabulary – Grades 3 and 5	129	35.6	14.0	6–68	r = .606 with general English vocabulary r = .669 with mathematics computation
	5	128	All students			57.5	20.6	5–100	r = .659 with general English vocabulary r = .626 with mathematics computation
Hughes et al. 2020	7	338	All students	Mathematics Vocabulary – Grades 7 and 8	69	45.0	12.6	NR	d = 0.22 (8th grade > 7th grade)
	8	153	All students			48.7	11.2	NR	
Forsyth & Powell (2017)	5	70	No difficulty	Mathematics Vocabulary – Grades 3 and 5	129	68.1	15.1	NR	g = 1.37 (typical > MD) g = 1.40 (typical > RD) g = 3.21 (typical > MD + RD)
		16	MD			46.0	19.4	NR	g = -0.03 (MD = RD) g = 1.41 (MD > MD + RD)
		18	RD			46.6	16.1	NR	g = 1.67 (RD > MD + RD)
		10	MD + RD			21.9	10.4	NR	

Powell et al. (2020)	3	242	DLL; no difficulty	Mathematics Vocabulary – Grade 3 (revised)	45	14.1	5.7	0–42	$d = 0.78$ (no difficulty non-DLL > DLL)
		545	Non-DLL; no difficulty			19.1	7.0	0–44	
		36	DLL; EQD			10.6	5.3	3–28	$d = 0.54$ (EQD non-DLL > DLL)
		89	Non-DLL; EQD			13.6	5.9	3–40	
		101	DLL; WPD			9.0	4.7	0–23	$d = 0.32$ (WPD non-DLL = DLL)
		34	Non- DLL; WPD			10.4	3.5	4–20	
		136	DLL; EQ + WPD			7.0	4.3	0–21	$d = 0.24$ (EQ + WPD non-DLL = DLL)
		75	Non- DLL; EQ + WPD			8.0	4.5	0–26	

of both third- and fifth-grade students. Similar to the findings of Powell and Nelson (2017), Powell et al. (2017) determined general English vocabulary and mathematics computation knowledge were both significant predictors of mathematics-vocabulary knowledge in third- and fifth-grade students, although the strength of the relationships varied depending on the students' level of mathematics vocabulary. The relation was significantly stronger in third-grade students. That is, third-grade students demonstrated greater dependence on general English vocabulary and computation compared to fifth-grade students.

Furthermore, the authors examined whether the influence of general vocabulary and mathematics computation on mathematics vocabulary varied for students with different levels of mathematics vocabulary. The findings showed that general English vocabulary was a more accurate predictor of mathematics vocabulary for students with lower mathematics-vocabulary scores, but mathematics computation served as a stronger predictor for students with higher mathematics-vocabulary scores in the third-grade sample. In the fifth-grade sample, however, mathematics computation was a stronger predictor for students with lower mathematics-vocabulary performance but not for students with the higher mathematics-vocabulary performance; general English vocabulary was a significant predictor across all performance levels of mathematics vocabulary. This reversal in the trend lines between grades could reflect the growing development of a mathematical lexicon in the third grade and therefore a greater dependence on general English vocabulary compared to fifth-grade students.

3.2 Secondary students' mathematics knowledge

3.2.1 Grades 7 and 8

Hughes et al. (2020) measured mathematics vocabulary of both seventh- and eighth-grade students. The authors determined the mathematics-vocabulary measure was well targeted for middle school students and measured mathematics vocabulary with high validity and high reliability. The differential performance between seventh- and eighth-grade students provided evidence to suggest that a mathematics-vocabulary measure can detect differences in the growth of mathematics vocabulary from one grade level to the next. Also notable was the wide variability in students' scores. The average score of both versions of the mathematics-vocabulary measure showed that students knew only two-thirds of vocabulary deemed essential by middle school educators, textbooks, and standards.

3.3 Mathematical vocabulary knowledge of students with difficulties

3.3.1 Students experiencing mathematics and reading difficulty

Forsyth and Powell (2017) examined the impact of mathematics and reading difficulties on the mathematics-vocabulary knowledge of fifth-grade students. Using the same measure of mathematics vocabulary as Powell et al. (2017), Forsyth and Powell (2017) examined the mathematics-vocabulary scores of students who experienced a reading-only difficulty (RD-only), a mathematics-only difficulty (MD-only), or comorbid reading and mathematics difficulty (MDRD), determined by performing below cut-off benchmarks on a test of general English vocabulary, a test of mathematics computation, or both assessments, respectively. The authors compared scores to typically developing students. Typically developing students demonstrated significantly higher performance over students with RD-only, MD-only, and MDRD, with the largest effects in comparison to students with MDRD. Students with RD-only and MD-only did not differ significantly on mathematics-vocabulary scores, but both groups of students had significantly higher mathematical-vocabulary scores than students with MDRD.

When Forsyth and Powell (2017) examined the impact of the year the term was introduced to students upon student knowledge of that term, the same pattern held, in which typically developing students outperformed students with MD-only, RD-only, and MDRD; students with MD-only and RD-only did not significantly differ; and students with MD-only and students with RD-only outperformed students with MDRD, with the greatest contrast between typically developing students and students with MDRD. This pattern did not hold for mathematics-vocabulary introduced in fifth-grade and sixth-grade textbooks and not included in textbook glossaries. Specifically, when mathematics-vocabulary were terms introduced in fifth-grade or not included in textbook glossaries, typical students significantly outperformed all other difficulty groups, with no significant differences among MD-only, RD-only, and MDRD groups. Given that mathematics-vocabulary introduced in sixth-grade was difficult for all groups, there was no significant group difference among typically developing and other disability groups.

3.3.2 Dual-language learners

Powell et al. (2020) assessed the mathematics-vocabulary performance of third-grade students to determine if performance differences existed among dual-language learners and native English speakers with and without MD. Powell et al. (2020) categorized students based on performance on measures of equation solving and word problems. This categorization included: equation-only difficulty (EQD; i.e., performing below 27th percentile on equation solving), word-problem-only difficulty (WPD; i.e., performing below 28th percentile on word-problem solving), or word-problem and equation difficulty (EQ + WPD). In addition, the study included students without equation or word-problem difficulty.

The Grade 3 measure of mathematics vocabulary was a revised version of the mathematics-vocabulary measure used in grades 3 and 5 (Powell et al., 2017). The revised assessment did not contain terms introduced in grades 4, 5, or 6. Students with no MD who were native English speakers outperformed students with no MD who were dual-language learners. The same was true for students with equation-only difficulty in which native English speakers outperformed dual-language learners. This pattern, however, did not hold for students with word-problem difficulty or combined difficulty. Instead, Powell et al. (2020) identified no significant differences between native English speakers and dual-language learners with word-problem-only difficulty or combined difficulty.

In addition, Powell et al. (2020) noted differences in students' mathematics vocabulary knowledge between different types of mathematics difficulty. For dual-language learners, students with no MD outperformed students with any type of MD (i.e., equation-only difficulty, word-problem-only difficulty, and combined difficulty). Although there was not a significant difference in the mathematics vocabulary between dual-language learners with equation-only difficulty and word-problem-only difficulty, both groups significantly outperformed dual-language learners with the comorbid difficulty. Across native English speakers, students without a MD outperformed their peers with any form of MD. Native English speakers with equation-only difficulty significantly outperformed native English speakers with combined difficulty. Similar to the findings of Forsyth and Powell (2017), students with comorbid difficulties demonstrated the weakest mathematics vocabulary. It is noteworthy that students without any mathematics difficulty, on average, did not score above 50% on the third-grade mathematics-vocabulary measure, suggesting many students struggle answering questions about mathematics vocabulary.

4 Conclusion

In this chapter, we described five studies in which students in the U.S. answered questions about mathematics vocabulary. All five studies above demonstrated wide variability in students' understanding of mathematics vocabulary. That is, within a grade level, the mathematics-vocabulary scores of students ranged from very low – in some cases, zero – to well above average. Across studies, the average mathematics-vocabulary score was at or below 67% of all terms on a measure. This indicated that all students have room for improvement on measures of mathematics vocabulary. Some students answered fewer than 10% of items on a mathematics-vocabulary measure even when the terms were introduced in previous grades and students should have experienced multiple opportunities to interact with such terms.

In a few studies, we noted significant correlations between mathematics-vocabulary scores and general English vocabulary as well as mathematics performance on measures of fluency or computation. The connection between mathematics and general English vocabulary is important, but our analyses did not explore whether greater mathematics vocabulary led to greater English vocabulary or vice versa. Future research should investigate this relationship. Similarly, we would ask researchers to conduct more research on the connection between mathematics-vocabulary knowledge and the understanding of mathematics concepts. We noted several significant correlations between mathematics vocabulary and fluency or computation, but it would be important for both researchers and educators to understand whether increased mathematics-vocabulary knowledge contributes to improved understanding of a concept or procedure. That is, how important is mathematics vocabulary on the pathway to learning mathematics?

Furthermore, students experiencing mathematics difficulty scored significantly below peers without mathematics difficulty. In the study by Forsyth and Powell (2017), the average score for students with both mathematics and reading difficulty was markedly below-average scores of students experiencing difficulty in only mathematics or reading. Such results should help educators understand which students in a classroom require more or less mathematics-vocabulary support. In the study by Powell et al. (2020), we observed differing average scores for dual-language learners and non-dual-language learners when the students experienced no mathematics difficulty or only equations difficulty. When students experienced word-problem difficulty, the difference between non-dual- and dual-language learners faded. These results should inform discussions about mathematics language support for dual-language learners.

Even with the need to conduct future research, we suggest all educators should use mathematics-vocabulary measures to understand the mathematics-vocabulary profiles of their students. This knowledge could inform mathematics instruction. Based on our findings that many students demonstrated low mathematics-vocabulary scores, we also suggest that educators provide explicit instruction on mathematics vocabulary to help all students develop a deep lexicon related to mathematics vocabulary. Such explicit instruction may be essential for students experiencing mathematics difficulty to ensure these students have access to the mathematics curriculum throughout their education.

References

Barrow, Melissa (2014): Even math requires learning academic language. *Phi Delta Kappan Magazine* 95 (6), 35–38.

Bay-Williams, Jennifer & Livers, Stefanie (2009): Supporting math vocabulary acquisition. *Teaching Children Mathematics* 16 (4), 238–245.

Berch, Dan & Mazzocco, Michelle (eds.) (2007): *Why is Math so Hard for some Children? The Nature and Origins of Mathematical Learning Difficulties and Disabilities*. Baltimore, MD: Brookes.

Browning, Sandra & Beauford, Judith (2011): Language and number values: The influence of number names on children's understanding of place values. *Investigations in Mathematics Learning* 4 (2), 1–24.

Bruner, Jerome (1966): *Toward a Theory of Instruction*. Cambridge, MA: Harvard University Press.

Bruun, Faye, Diaz, Joan & Dykes, Valerie (2015): The language of mathematics. *Teaching Children Mathematics* 21 (9), 530–536.

Cummins, Jim. (2000): Academic language learning, transformative pedagogy, and information technology: Towards a critical balance. *TESOL Quarterly* 34 (3), 537–548.

Ernst-Slavit, Gisela & Mason, Michele (2011): Words that hold us up:" Teacher talk and academic language in five upper elementary classrooms. *Linguistics and Education* 22 (4), 430–440.

Fang, Zhihui (2012): Language correlates of disciplinary literacy. *Topics in Language Disorders* 32 (1), 19–34.

Forsyth, Suzanne & Powell, Sarah (2017): Differences in the mathematics-vocabulary knowledge of fifth-grade students with and without learning difficulties. *Learning Disabilities Research and Practice* 32 (4), 231–245.

Fuchs, Lynn, Fuchs, Douglas, Compton, Donald, Hamlett, Carol & Wang, Amber (2015): Is word-problem solving a form of text comprehension?. *Scientific Studies of Reading* 19 (3), 204–223.

Haladyna, Thomas & Rodriguez, Michael (2013): *Developing and Validating Test Items*. New York, NY: Routledge.

Harmon, Janis, Hedrick, Wanda & Wood, Karen (2005): Research on vocabulary instruction in the content areas: Implications for struggling readers. *Reading & Writing Quarterly* 21 (3), 261–280.
Hughes, Elizabeth, Powell, Sarah & Lee, Joo-Young (2020) – this is no longer in press: Development and psychometric report of a middle school mathematics vocabulary measure. *Assessment for Effective Intervention*. 34 (2), 417–447.
Karp, Karen, Bush, Sarah & Dougherty, Barbara (2014): 13 rules that expire. *Teaching Children Mathematics* 21 (1), 18–25.
Kleemans, Tijs, Segers, Eliane & Verhoeven, Ludo (2018): Role of linguistic skills in fifth-grade mathematics. *Journal of Experimental Child Psychology* 167, 404–413.
Livers, Stefanie & Elmore, Patricia (2018): Attending to precision: Vocabulary support in middle school mathematics classrooms. *Reading & Writing Quarterly* 34 (2), 160–173.
Monroe, Eula & Orme, Michelle (2002): Developing mathematical vocabulary. *Preventing School Failure: Alternative Education for Children and Youth* 46 (3), 139–142.
Monroe, Eula & Panchyshyn, Robert (1995): Vocabulary considerations for teaching mathematics. *Childhood Education* 72 (2), 80–83.
Moschkovich, Judit (2013): Principles and guidelines for equitable mathematics teaching practices and materials for English language learners. *Journal of Urban Mathematics Education* 6 (1), 45–57.
Moschkovich, Judit (2015): Scaffolding student participation in mathematical practices. *ZDM: Mathematics Education* 47 (7), 1067–1078.
National Council of Teachers of Mathematics (2006): *Curriculum Focal Points for Pre-Kindergarten through Grade 8 Mathematics*. Reston, VA: Author.
National Governors Association Center for Best Practices & Council of Chief State School Officers (2010): *Common Core State Standards Mathematics*. Washington, DC: Author.
Peng, Peng & Lin, Xin (2019): The relation between mathematics vocabulary and mathematics performance among fourth graders. *Learning and Individual Differences* 69, 11–21.
Petersen-Brown, Shawna, Lundberg, Ashlee, Ray, Jannine, Paz, Iwalani Dela, Riss, Carrington & Panahon, Carlos (2019): Applying spaced practice in the schools to teach math vocabulary. *Psychology in the Schools* 56 (6), 977–991.
Powell, Sarah, Berry, Katherine & Tran, Le (2020): Performance differences on a measure of mathematics vocabulary for English learners and non-English learners with and without mathematics difficulty. *Reading and Writing Quarterly: Overcoming Learning Difficulties* 36 (2), 124–141.
Powell, Sarah, Driver, Melissa, Roberts, Greg & Fall, Anna-Marla (2017): An analysis of the mathematics vocabulary knowledge of third-and fifth-grade students: Connections to general vocabulary and mathematics computation. *Learning and Individual Differences* 57, 22–32.
Powell, Sarah & Nelson, Gena (2017): An investigation of the mathematics-vocabulary knowledge of first-grade students. *The Elementary School Journal* 117 (4), 664–686.
Purpura, David & Logan, Jessica (2015): The nonlinear relations of the approximate number system and mathematical language to early mathematics development. *Developmental Psychology* 51 (12), 1717–1724.
Purpura, David, Napoli, Amy, Wehrspann, Elizabeth & Gold, Zachary (2017): Causal connections between mathematical language and mathematical knowledge: A dialogic reading intervention. *Journal of Research on Educational Effectiveness* 10 (1), 116–137.

Riccomini, Paul, Smith, Gregory, Hughes, Elizabeth & Fries, Karen (2015): The language of mathematics: The importance of teaching and learning mathematical vocabulary. *Reading and Writing Quarterly* 31 (3), 235–253.
Rubenstein, Rheta & Thompson, Denisse (2002): Understanding and supporting children's mathematical vocabulary development. *Teaching Children Mathematics* 9 (2), 107–113.
Schleppegrell, Mary (2012): Academic language in teaching and learning: Introduction to the special issue. *The Elementary School Journal* 112 (3), 409–418.
Simpson, Amber & Cole, Mikel (2015): More than words: A literature review of language of mathematics research. *Educational Review* 67 (3), 369–384.
Townsend, Dianna, Filippini, Alexis, Collins, Penelope & Biancarosa, Gina (2012): Evidence for the importance of academic word knowledge for the academic achievement of diverse middle school students. *The Elementary School Journal* 112 (3), 497–518.

Judit Moschkovich and Judith Scott

Language issues in mathematics word problems for English learners

1 Introduction

This paper describes language issues in mathematics word problems for English language learners (ELLs). We first summarize research relevant to the linguistic complexity of mathematics word problems from studies in mathematics education, reading comprehension, and vocabulary. Based on that research, we make recommendations for addressing language complexity and vocabulary in designing word problems for instruction, curriculum, or assessment. We then use examples of word problems[1] to illustrate how to apply those recommendations to designing or revising word problems and creating supports for students to work with word problems.

2 Summary of relevant research on language issues in mathematics

This section contextualizes recommendations for the design of word problems and support for word problem instruction for ELLs, using research on language issues in mathematics. In reviewing the relevant research on language and mathematics, we focused on research specific to the domain of mathematics and word problems in particular.

1 These word problems are sample released items for the U.S. assessment created by the Smarter Balanced Assessment Consortium. The Smarter Balanced Assessment Consortium (SBAC) is a standardized test consortium that created Common Core State Standards–aligned tests ("adaptive online exams") to be used in several states in the United States. The Common Core State Standards Initiative is an educational initiative from 2010 that details what K–12 students throughout the United States should know in English language arts and mathematics at the conclusion of each grade.

https://doi.org/10.1515/9783110661941-017

2.1 Research on language issues in mathematics word problems

There are multiple uses of the terms *language* in mathematics education research. Some interpretations of the phrases "language in mathematics" or "mathematical language" reduce their meaning to single words or the proper use of technical vocabulary. In contrast, we ground this chapter in research in mathematics education that provides a more complex view of mathematical language. Such work (e.g., Pimm, 1987) provides a view of mathematical language as not only specialized vocabulary—new words and new meanings for familiar words—but also as extended discourse that includes syntax and organization (Crowhurst, 1994), the mathematics register (Halliday, 1975), and discourse practices (Moschkovich, 2007).

Researchers in mathematics education have examined many topics related to mathematics and language, some of them relevant to designing word problems used with ELLs, for example, mathematical texts (O'Halloran, 2005), words with multiple meanings or polysemy (Pimm, 1987), and differences between mathematical registers at school and at home (Walkerdine, 1988). One contribution that is especially relevant to word problems is a shift from seeing the mathematics register as merely technical mathematical language. The mathematics register should not be interpreted as merely a set of words and phrases particular to mathematics. The mathematics register includes styles of meaning, modes of argument, and mathematical practices. It also has several levels of complexity that go beyond the word or phrase level to include background knowledge level and complexity at the sentence or paragraph level.

The following word problem illustrates how the mathematics register is not simply about vocabulary specific to mathematics and involves more than only the word level:

A boat in a river with a current of 3 mph can travel 16 miles downstream in the same amount of time it can go 10 miles upstream. Find the speed of the boat in still water.

The complexity involved in making sense of this word problem is not at the level of technical mathematical vocabulary, but lies principally in the background knowledge (Martiniello & Wolf, 2012) for understanding and imagining the context or situation for the problem. In this case, the reader needs to imagine and understand that there is a boat traveling up and down a river, that the speed was measured in still water (presumably a lake), and that the speed of the boat increases (by the speed of the current) when going downstream, and decreases (by the speed of the current) when going upstream. The language complexity lies not in understanding mathematical terms, but having the background knowledge to imagine the situation. A glossary for non-mathematical words such as *upstream, downstream*, and the phrase *in still water* would certainly help. However, also notice that much of the language complexity is not at the word level, but at the sentence and paragraph levels, in the use of the passive voice without an agent and in the multiple subordinate clauses and nested constructions (Cook & MacDonald, 2013).

Another contribution from mathematics education work relevant to mathematics word problems is that we know there are international differences in the meaning of some mathematical terms. For example, the definitions of *trapezium* and *trapezoid* (a quadrilateral with no sides parallel) are often interchanged. In Spanish, "The word *trapezoid* is reserved for a quadrilateral without any parallel sides, whereas *trapezium* is used when there is one pair of parallel sides (This is opposite to American English usage)" (Hirigoyen, 1997: 167).

2.2 Research on language complexity of mathematics word problems in assessments

The research on word problems on assessments is relevant to considering the language complexity in word problems. According to Abedi (2002), linguistic complexity of assessment word problems unrelated to the content being assessed may at least be partly responsible for the performance gap between ELLs and non-ELLs, and linguistic complexity of assessment word problems may invalidate achievement on tests. In particular, Shaftel et al. (2006) found that fourth graders had difficulty with vague words, complex verbs (verbs with three or more words, e.g., *had been going*), pronouns, prepositions, and mathematical vocabulary. The greater the number of linguistic elements, the more difficult the word problem proved to be. Grade 7 students found it hard to understand mathematical vocabulary and comparative terms (*greater than, less than*). Overall, unfamiliar words, rarely used vocabulary, and passive voice hinder comprehension (Abedi & Lord, 2001).

Abedi (2009) studied several supports (computerized tests, a pop-up glossary, a customized English dictionary, extra testing time, and small-group testing). The dictionary was customized and did not include any content-related vocabulary. He found that all the supports made a significant difference for ELL students for the more linguistically complex word problems. Another study (Sato et al., 2010) used an original and modified version[2] and found that linguistically

2 This study used the Grade 8 National Assessment of Educational Progress (NAEP) and the Grade 7 California mathematics assessment.

modified versions more reliably measured mathematical proficiency of students labeled as ELL or Non-English Proficient (NEP) than the original format and linguistic modification did not alter the targeted math constructs being assessed.

Martiniello (2008) found that understanding word problems that involve polysemous words can be challenging for ELLs. Polysemous words are words with different meanings or connotations, depending on the context provided by the text or discourse. Martiniello gave the following example: "Find the amount of money each fourth-grade class raised for an animal shelter using the table below." The word *raised* here refers to collecting funds. Other meanings are "raise your hands," "raise the volume," "raising the rent," or "receiving a raise." Martiniello found that ELLs tended to interpret the word *raise* as "increase" and did not understand the connotation of *raise* in fund raising.

Martiniello and Shaftel both found that fourth-grade students struggled with specific categories of vocabulary. These included words with multiple meanings, slang or conversation words, and words learned in an English-speaking home (Martiniello, 2008; Shaftel et al., 2006). Martiniello concluded, "It is important to distinguish between school and home related vocabulary as a potential source of differential difficulty for ELLs." She suggests that since ELLs learn English primarily at school, school-related words (*student, pencil, ruler, school, day, book*, etc.) are likely to be more familiar than words related to the home (her examples included *raking leaves, chore, washing dishes, vacuum, dust, rake*, and *weed*). Martiniello's general recommendations for word problems include avoiding unnecessary linguistic complexity not relevant to mathematics, refining linguistic complexity measures so they include issues that are specific to ELLS (e.g. home vocabulary, polysemy, familiarity), and, for assessments, including thorough review by experts on ELLs.

2.3 Summary of research on the language of mathematics word problems

The following key understandings from research contribute to recommendations for how to design, revise, or support mathematics word problems when working with ELLs. The language features of word problems documented as problematic for ELLs are at three levels: Cultural (background level), syntactic (sentence and paragraph level), and lexical (word and phrase level). All three levels should be considered in designing mathematics word problems for curriculum materials, in designing assessment problems, and when considering supports such as glossaries.

A crucial aspect of word problems is the background knowledge and cultural references necessary for understanding the setting of a word problem. Since it is not possible to predict what settings, context, or background knowledge all students share, it is important to provide some support for any setting described in an item. The syntactic level involves the way sentences are put together. Several challenging aspects of mathematics word problems include the use of the passive voice without an agent, multiple subordinate clauses, nested constructions, and long noun phrases. The lexical level involves unfamiliar words, unfamiliar phrases, and unfamiliar connotations of words with multiple meanings (polysemy).

Vocabulary that is specific to mathematics is not the only source of difficulty. Since the language complexity of mathematics word problems and the language complexity issues for ELLs are not all necessarily at the word level, the overall recommendation is that the design of word problems and the guidelines for glossaries should focus on cultural and syntactic levels, in addition to the lexical level.

3 Summary of relevant research on reading comprehension

This section contextualizes recommendations for the design of word problems and support for word problem instruction for ELLs. In reviewing the relevant research on vocabulary and reading, we focused on a broad view of reading comprehension and word knowledge, not constrained by the domain of mathematics. The following key understandings from research on reading comprehension and vocabulary can contribute to recommendations.

3.1 Background knowledge

Background knowledge plays a large role in understanding text and making inferences about a word's meaning. Research in the last 40 years has shown that language comprehension requires knowledge of the world as well as knowledge of the language (McNamara et al., 1991). Reading comprehension depends on interaction between the reader (e.g., background, knowledge, abilities, and experience), the activity (e.g., instruction, grouping, purpose), and the text (e.g., genre, structure, words), embedded within a sociocultural context (RAND reading study group; Snow, 2002). "Text can be difficult or easy depending on factors inherent in the text, on the relationship between the text and the knowledge and abilities of the reader, and on the activities in which the reader is engaged" (RAND: 14). Within an individual text, such as a word problem, the vocabulary load and linguistic structure interact with the readers' knowledge during the comprehension process. Comprehension is affected when there is a mismatch between the text and student's knowledge and experience. A synthesis of research on background knowledge and assessment with ELLs revealed that underdeveloped background knowledge or lack of background knowledge hinders performance on all types of assessments, including word problems (August & Shanahan, 2006).

3.2 Words have multiple meanings

Multiple meanings for words are a common source of confusion for students. Words are the cornerstone of effective communication, but work knowledge is complex and multifaceted (Anderson & Nagy, 1991; Beck & McKeown, 1991; Nagy & Scott, 2000). In addition to multiple meanings, words are often abstract, are used in idioms, vary according to register, and differ according to context.

Anderson et al. (1976) propose, "A word does not have *a* meaning, but has, rather, a *family* of potential meanings" (p. 667). The more frequent a word is in English, the more likely it is to have multiple meanings (Nagy, 2009). For instance, a *round* of golf, singing in a *round*, *round*ing up numbers, and a *round* shape are all acceptable uses for the word *round*. The meaning of individual words (and phrases) must be inferred from context and is nuanced by that context, even if it is a familiar word (Nagy & Scott, 2000).

3.3 Word frequency

Most of the words in English, particularly academic words, occur infrequently. In a 17.5-million-word corpus from K–12 schoolbooks (Zeno et al., 1995), the frequency distribution shows that a core set of 5,600 words accounts for almost 80% of the words. The other 20% are over 150,000 unique words, most of which are seen less than once per million words of text (Nagy & Hiebert, 2011). Given this distribution, it is likely that many ELLs will not know or recognize infrequent words. On average, ELLs at fourth- and fifth-grade score more than two grade levels below their native English-speaking peers in English vocabulary knowledge and tend to use high-frequency words for communication rather than low-frequency vocabulary (Manyak, 2012).

3.4 Verbs and nouns

Verbs are harder to learn than nouns. Young children learn concrete nouns before verbs in part because verbs have a less transparent relationship to the perceptual world (Gentner, 1982). For concrete nouns, the mapping between a word and its referent is tangible. For verbs, this mapping depends on the language used, and the way that information is captured by verbs varies across languages (Gentner, 2006). For instance, English verbs include manner of motion (e.g., *the little bird hops out of the cage*) while Spanish typically includes path of motion, and not manner (e.g. *El pajarito sale de la jaula dando saltitos*—The little bird leaves from the cage giving hops) (Negueruela et al., 2004: 118). Note that the word *exit* could have been used instead of *leaves from* to avoid the propositional verb.

3.5 Using high-frequency words and shorter sentences is not the solution

The use of high-frequency words and shorter sentences may make a text more difficult to read rather than easier. Abedi et al. (2004), in a thorough review of assessment supports for ELLs, established that minor changes in the wording or the syntactic complexity of mathematics problems can affect student performance and recommend using straightforward, uncomplicated language when developing word problems for assessments. While this makes sense as a general rule, substituting more familiar words for more precise, less frequently seen, words might also be problematic. Since high-frequency words are more likely to have multiple meanings (Nagy, 2009), this needs to be factored into decisions about word use.

Syntactical simplification is also problematic, as shortening sentences by eliminating words that establish connective relationships (*because, therefore*, etc.) can make text harder to read rather than easier (Davidson & Kantor, 1982). In a study with Puerto Rican students learning English as a second language, researchers found that eighth-grade students' comprehension benefited from longer sentences that showed relationships rather than from choppy sentences with simple syntax. Thus, sentences like, "If the manufacturer and the market are a long distance apart, then it can be a big expense for the manufacturer to get goods to market" were easier to understand than several shorter sentences like, "Manufacturers must get goods to market. Suppose that the manufacturer and the market are a long distance apart. This can be a big expense" (Blau, 1982: 518).

3.6 Other factors relating to word recognition

Two different studies indicate that both word frequency and the age of acquisition of a word are features that influence the ease or difficulty of word knowledge, and the second study also reported word length, number of syllables, and concreteness or imageability as significant factors (Hiebert et al., 2019). Age of acquisition captures when children can typically understand or use a word in oral language, and thus relates to familiarity with the word. Concrete words, words that can be easily thought of as a picture or an image, are easier than abstract words, and imageability interacts with both the complexity of sentences and word familiarity to influence word recognition (Mesmer et al., 2012). These additional factors should also be considered in choosing words for word problems.

Some words are more easily decoded because their spelling pattern maps onto recognizable phonetic patterns (Adams, 1990). For Spanish speakers, the use of cognates with a similar meaning may facilitate word recognition (Nagy et al., 1993). For instance, *rotation/ rotación* are cognates. Using *rotation* in a word problem instead of the word *spin* or *flip* would help those Spanish-speaking students who recognize this relationship. However, the use of complex or unfamiliar orthographic patterns (e.g., *bouquet, answer, souvenir*) can prove challenging, for non-ELLs as well as ELLs (Peregoy & Boyle, 2000).

3.7 Definitions

Definitions are often not student friendly. Using typical definitions to understand the meaning of unknown words is a difficult metalinguistic task (McKeown, 1993; Miller & Gildea, 1987; Scott & Nagy, 1997). The conventions of standard definitions are largely the result of the need to conserve space in their printed form (Landau, 1984). Definitions are "even more decontextualized, more terse, and less like oral language than most of the written language to which children have been exposed" (Scott & Nagy, 1997: 187). Graves et al. (2012) suggest that student-friendly definitions are "longer, often written in complete sentences, phrased in ways that are as helpful as possible to second-language learners, and do not include words more difficult than the word being defined. Also, sentences that give an example of the thing named can be a useful add-on to a student friendly definition" (pg. 58). They suggest that providing a visual that represents a word, and a sentence that explains how it represents the word, may be crucial for helping ELLs understand the word's meaning.

4 Implications and recommendations for writing word problems

This set of key insights from research can be used to help guide the design, modification, and support of word problems and help determine which words should go into glossaries. Below we summarize recommendations for addressing three levels in the language of mathematics word problems:

a) Cultural level: When choosing situations for word problems, background and cultural knowledge is crucial for understanding the context for word problems. Short explanations of the setting or context are often necessary. Since it is not possible to predict what settings, context, or background knowledge all students bring or share, it is important to use common settings, such as school, and to provide short explanations of any setting described in an item.
b) Syntactic level: When considering syntax, as a rule, it is best to keep sentence structure clear and straightforward. However, connective words in longer sentences that help students understand relationships should not be eliminated if they facilitate meaning. Long sentences should be broken into shorter, less complex sentences only if the shorter sentences make the meaning clearer. Avoiding complex noun and verb phrases, using an active voice with an agent, reducing nested constructions and subordinate clauses, using pronouns with clear references, and using sentences with a clear subject would also help students process the problems more easily.
c) Lexical level: When choosing words for word problems it makes sense to avoid the type of words that may cause comprehension problems for ELLs. To facilitate word recognition and understanding, first acknowledge that word meanings are flexible and variable, and that word problems often include polysemous words. If multiple meanings might confuse students, it might be worthwhile to use another word. When there is a choice between words, it makes sense to choose imageable words and cognates. If that isn't possible, a glossary and the context should provide sufficient information to make the word meaning clear.

Infrequent or unusual words, words with complex spellings, vague words, abstract words, idioms, and colloquial or slang phrases should all either be replaced or glossed. Using words in a students' oral language, cognates, concrete words, and easily decodable words will help focus the assessment on the mathematics instead of English proficiency. Uncommon verbs, particularly those that are central to comprehending a word problem, should be glossed for most problems.

Glossaries can be a useful resource, if they are well written, provide visuals, and explain the context of the word. Glossaries are particularly important for polysemous words, multiword phrases (e.g., *standard deviation*), satellite or phrasal verbs known to be problematic for ELLs (e.g. *look up, clean up, turn off, bring up*, etc.); regionalisms (e.g., *bag/sack, soda/pop*), and mathematical terms that differ across countries (e.g., *trapezoid*).

4.1 Designing a glossary

The definitions provided in a glossary should try to explain words in complete, easy-to-read sentences. There is no need to use the terse, convoluted language forms found in typical dictionaries. Example sentences should be used as needed. A picture is often the quickest way to convey the meaning, particularly for concrete nouns. It is much easier to show a picture of a kangaroo than to try to explain what a kangaroo is in words. Since imageability of words can aid recognition and understanding, it makes sense to make those images available to students.

The following are a few guidelines for identifying terms for an English-language glossary:

a) Background knowledge: Provide short explanations or descriptions of the setting or context that may not be familiar to all students; use visuals and pictures to illustrate setting and context.
b) Syntactic level: Write in a clear, straightforward, and cohesive manner; make relationships within the word problem transparent.
c) Lexical level: Use visuals and pictures to illustrate words, particularly concrete nouns; provide glossaries for unfamiliar words, unfamiliar phrases, and unfamiliar connotations of words with multiple meanings (polysemous words); replace words, or provide glossaries for abstract words and words with unusual spelling patterns; provide glossaries for verbs, particularly those that are central to comprehending and completing the word problem.

5 Examples

Several of the word problems we use below as examples illustrate overlapping issues. For instance, infrequent vocabulary is often correlated with lack of background knowledge. However, we have created these revisions from real items, and grouped them as examples of areas of concern.

5.1 Cultural (background knowledge) examples

Example 1 – Original
A company purchases $24,500 of new computer equipment. For tax purposes, the company estimates that the equipment decreases in value by the same amount each year. After 3 years, the estimated value is $9,800.

Which of the following is an explicit function that gives the estimated value of the computer equipment n years after purchase?

It is unlikely that many high school students know about paying taxes and depreciation.

Example 1 – Revised
A company buys $24,500 of new computer equipment. The company estimates that the equipment decreases in value by the same amount each year. After 3 years, the estimated value is $9,800.

Which of the following is an explicit function that gives the estimated value of the computer equipment *n* years after purchase?

Since paying taxes and depreciation are complex ideas to explain, it might be easiest to eliminate the phrase "For tax purposes." This also eliminates the need to explain this complex idea and doesn't affect the word problem. Replace *purchases* with *buys*, a more familiar and frequently used word.

Glossary: Estimate—to guess an answer close to the correct value.

Example 2 – Original
Ms. Olsen is having a new house built on Prospect Road.
She is designing a sidewalk from Prospect Road to her front door.
Ms. Olsen wants the sidewalk to have an end in the shape of an isosceles trapezoid, as shown.

The contractor charges a fee of $200 plus $12 per square foot of sidewalk. Based on the diagram what will the contractor charge Ms. Olsen for her sidewalk?

The idea of having a contractor and designing a sidewalk may be unfamiliar to many students. *Charges* is a polysemous word.

Example 2 – Revised
Ms. Olsen wants her new house to have a new sidewalk that goes from the road to her front door. But, she doesn't want it to have straight lines. Instead, she wants the sidewalk to end in the shape of an isosceles trapezoid, as shown.

The builder charges a fee of $200 plus $12 per square foot of sidewalk. Based on the diagram what will the builder charge Ms. Olsen for her sidewalk?

This revised version provides more context more clearly, without the passive voice or the use of unfamiliar terms and puts *charges a fee* in a glossary.

Glossary: Charges a fee—when someone charges a fee, they are asking the other person to pay that much money.

5.2 Syntactic examples

Example 3 – Original

Jason uses a balance scale to measure the weight of objects. Each of the objects is the same size but has a different weight. Four of the objects have their weights labeled and one does not. Jason is trying to find the weight of the object that is not labeled. He performs the two experiments shown below.

Which could be the weight of the object without the label?

The phrase, "Which could be" uses the conditional tense serves no real purpose and could confuse students. An additional concern is that the word *object* is generic, polysemous, and abstract, and *perform* is also polysemous.

Example 3 – Revised

Jason uses a balance scale to measure the weight of bricks. Each of brick is the same size but has a different weight. Four of the bricks have their weight labeled and one does not. Jason is trying to find the weight of the brick that is not labeled. He does the two experiments shown below.

What is the weight of the brick without the label?

The word *object* could be easily replaced with a word that is more concrete and imageable, like *brick*. A word like *brick*, with a visual image in the glossary, would be more familiar and accessible. The question without the conditional tense is more straightforward and easier to comprehend.

Fig. 1: Visual image for the word *brick* in Example 3. (Image courtesy of iStockphoto | vladakela)

5.3 Lexical examples

Example 4 – Original

A giant gumball machine has the following qualities: It can hold no more than 1,000 gumballs.
- Exactly 7/40 of the gumballs are blue.
- Exactly 4/21 of the gumballs are red.
- There are more white gumballs than blue gumballs.
- There are fewer white gumballs than red gumballs. Which is a possible number of white gumballs in the giant gumball machine?

Infrequent words and multiple meanings

While the word *gumball* is a concrete noun, it is also a low-frequency word in English. The word *qualitie*s and its singular form *quality* have multiple meanings. Students may confuse this sense of the word with the sense that relates to a grade of excellence (e.g., *high quality*).

Example 4 – Revised

A giant gumball machine can hold no more than 1,000 gumballs. Out of those 1,000 gumballs . . .
- Exactly 7/40 of the gumballs are blue.
- Exactly 4/21 of the gumballs are red.
- There are more white gumballs than blue gumballs.
- There are fewer white gumballs than red gumballs. Which is a possible number of white gumballs in the giant gumball machine?

Rewording the sentence to eliminate *qualities* reduces confusion. Adding a link to a picture of a gumball machine in a glossary, or to the item itself, and defining it in a glossary help students access the meaning.

Glossary: A giant gumball machine is a tall machine that gives out brightly colored balls of sugar-coated chewing gums in exchange for coins.

Fig. 2: Visual image for the phrase *gumball machine* in Example 4.

Example 5 – Original
Cara makes scarves in different sizes.
The first scarf has 12 stripes and 5 tassels.
The second scarf has 5 stripes and 8 tassels.

The spelling pattern of *scarves* and *scarf* is irregular and the word *tassels* is infrequent. It is also unlikely that *tassels* is a word in students' oral vocabulary.

Example 5 – Revised
Cara makes pillows in different sizes.
The first pillow has 12 stripes and 5 buttons
The second pillow has 5 stripes and 8 buttons.

A better alternative would be to change *tassels* to more familiar, imageable words like *bedspread*, *pillows*, or *bookmarks* that could have *buttons*, *beads*, *or stickers*. Although these are common words, a picture can facilitate comprehension.

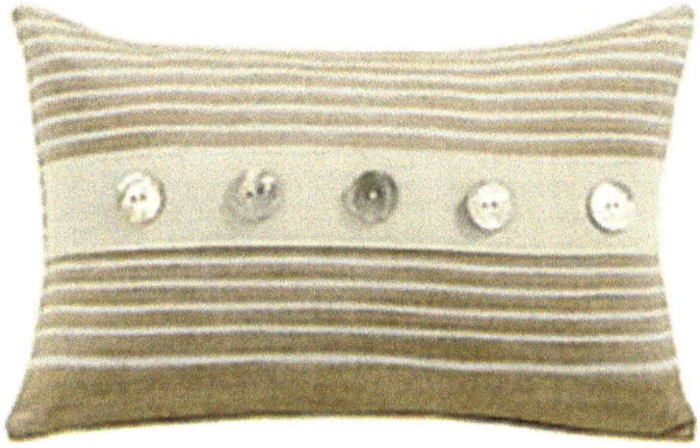

Fig. 3: Visual image for the word *pillow* in Example 5.

Example 6 – Original

The noise level at a music concert must be no more than 80 decibels (dB) at the edge of the property on which the concert is held.
Melissa uses a decibel meter to test whether the noise level at the edge of the property is no more than 80 dB.

- Melissa is standing 10 feet away from the speakers and the noise level is 100 dB.
- The edge of the property is 70 feet away from the speakers.
- Every time the distance between the speakers and Melissa doubles, the noise level decreases by about 6 dB.

Rafael claims that the noise level at the edge of the property is no more than 80 dB since the edge of the property is over 4 times the distance from where Melissa is standing. Explain whether Rafael is or is not correct.

The concept of a *decibel* is highly specific and the word is a low-frequency English word. Understanding the meaning of this word is central to understanding the word problem. In addition, *speakers* and *speaker* have multiple meanings, and could be interpreted as human speakers if the reader doesn't know this context.

Example 6 – Revised

Decibels (dB) are used to measure noise levels. The law says that the noise level at a music concert must be no more than 80 decibels at the edge of the property where the concert is held. Melissa uses a decibel meter to measure the noise level at the edge of the property.

- When Melissa is standing 10 feet away from the sound speakers, the noise level is 100 dB.
- The edge of the property is 70 feet away from the speakers.
- Every time the distance between the speakers and Melissa doubles, the noise level decreases by about 6 dB.

Rafael claims that the noise level at the edge of the property is no more than 80 dB since the edge of the property is over four times the distance from where Melissa is standing. Explain whether Rafael is or is not correct.

This revision adds context to the word problem and defines what a decibel and a decibel meter is within the item. It also disambiguates the word *speaker*. The addition of a picture of *speakers* in a glossary would also help.

Fig. 4: Visual image for the word *speakers* in Example 6.

References

Abedi, Jamal (2002): Standardized achievement tests and English language learners: Psychometrics issues. *Educational Assessment* 8 (3), 231–257.

Abedi, Jamal (2009): Computer testing as a form of accommodation for English language learners. *Educational Assessment* 14 (3/4), 195–211.

Abedi, Jamal, Hofstetter, Carolyn Huie & Lord, Carol (2004): Assessment accommodations for English language learners: Implications for policy-based empirical research. *Review of Educational Research* 74 (1), 1–28.

Abedi, Jamal & Lord, Carol (2001): The language factor in mathematics tests. *Applied Measurement in Education* 14 (3), 219–234.

Adams, Mark James (1990): *Beginning to Read: Thinking about Print*. Cambridge: MIT Press.

Anderson, Richard & Nagy, William (1991): Word meanings. In Barr, Rebecca, Kamil, Michael, Mosenthal, Peter, David Pearson, P. (Eds.): *The Handbook of Reading Research*, Volume II. Mahwah, NJ: Lawrence Erlbaum Associates, 690–724.

Anderson, Richard, Pichert, James, Goetz, Ernest, Diane. L., Schallert, Stevens, Kathleen & Trollip, Stanley (1976): Instantiation of general terms. *Journal of Verbal Learning and Verbal Behavior* 15 (6), 667–679.

August, Diane & Shanahan, Timothy (Eds.) (2006): *Developing Literacy in Second-language Learners: Report of the National Literacy Panel on Language Minority Children and Youth*. Mahwah, NJ: Lawrence Erlbaum Associates.

Beck, Isabel & McKeown, Margaret (1991): Conditions of vocabulary acquisition. In Barr, R., Kamil, M., Mosenthal, P., Pearson, P. (Eds.): *Handbook of Reading Research Volume II*. Mahwah, NJ: Lawrence Erlbaum Associates, 789–814.

Blau, Eileen (1982): The effect of syntax on readability for ESL students in Puerto Rico. *TESOL Quarterly* 16, 517–528.

Cook, Gary & MacDonald, Rita. 2013. Tool to Evaluate Language Complexity of Test Items (WCER Working Paper No. 2013-5). Retrieved from University of Wisconsin– Madison, Wisconsin Center for Education Research website: http://www.wcer.wisc.edu/publications/workingPapers/papers.php

Crowhurst, Marion (1994): *Language and Learning across the Curriculum*. Scarborough, Ontario: Allyn and Bacon.

Davison, Alice & Kantor, Robert (1982): On the failure of readability formulas to define readable texts: A case study from adaptations. *Reading Research Quarterly* 17 (2), 187–209.

Negueruela, Eduardo, Lantolf, James, Jordan, Stefanie & Gelabert, Jaime (2004): The "private function" of gesture in second language speaking activity: A study of motion verbs and gesturing in English and Spanish. *International Journal of Applied Linguistics* 14, 113–147.

Gentner, Deirdre (1982): Why nouns are learned before verbs: Linguistic relativity versus natural partitioning. In Kuczaj, S. (Ed.): *Language Development: Language, Cognition and Culture*. Hillsdale, N.J.: Lawrence Erlbaum Associates Inc, 301–334.

Gentner, Deirdre (2006): Why verbs are hard to learn. In Gentner, D. (Ed.): *Action Meets Word: How Children Learn Verbs*. New York: Oxford University Press, 544–564.

Graves, Michael, August, Diane & Mancella-Martinez, Jeanette (2012): *Teaching Vocabulary to English Langue Learners*. New York: Teachers College Press.

Halliday, Michael (1975). Some aspects of sociolinguistics, in E. Jacobsen (Ed.), *Interactions Between Linguistics and Mathematics Education: Final report of the symposium sponsored by UNESCO, CEDO, and ICMI; Nairobi, Kenya, September 11–14, 1974*. UNESCO Report No. ED-74/CONF.808). Paris, UNESCO, 64–73.

Hiebert, E. H., Scott, J. A., Castaneda, R. & Spichtig, A. (2019): An analysis of the features of words that influence vocabulary difficulty. *Education Sciences* 9 (1), 8.

Hirigoyen, Héctor (1997): Dialectical variations in the language of mathematics: A source of multicultural experiences. In Trentacosta, J., Kenney, M. J. (eds): *Multicultural and Gender*

Equity in the Mathematics Classroom: The Gift of Diversity-1997 Yearbook. Reston, VA: NCTM, 164–168.

Landau, Sidney (1984): *Dictionaries: The Art and Craft of Lexicography*. New York: Charles Scribner's Sons.

Manyak, Patrick (2012): Powerful vocabulary instruction for English learners. In Kame'enui, E., Baumann, J. (Eds.): *Vocabulary Instruction: From Research to Practice, 2nd Ed*. New York: Guilford Press, 280–302.

Martiniello, Maria (2008): Language and the performance of English language learners in math word problems. *Harvard Educational Review* 78 (2), 333–368.

Martiniello, Maria (2009): Linguistic complexity, schematic representations, and differential item functioning for English language learners in math tests. *Educational Assessment* 14 (3), 160–179.

Martiniello, Maria & Wolf, Mikyung Kim. (2012): Exploring ELLs' understanding of word problems in mathematics assessments: The role of text complexity and student background knowledge. In Celedón-Pattichis, S., Ramirez, N. (Eds.): *Beyond Good Teaching: Strategies that are Imperative for English Language Learners in the Mathematics Classroom*. Reston, VA: NCTM, 151–162.

McKeown, Margaret (1993): Creating definitions for young word learners. *Reading Research Quarterly* 28, 16–33.

McNamara, Timothy, Miller, Diana & Bransford, John (1991): Mental models and reading comprehension. In Barr, R., Kamil, M., Mosenthal, P., Pearson, P. D. (Eds.): *The Handbook of Reading Research*, Mahwah, NJ: Lawrence Erlbaum Associates, 490–511.

Mesmer, Heidi, Cunningham, James & Hiebert, Elfrieda (2012): Toward a theoretical model of text complexity for the early grades: Learning from the past, anticipating the future. *Reading Research Quarterly* 47, 235–258.

Miller, George & Gildea, Patricia (1987): How children learn words. *Scientific American* 2573, 94–99.

Moschkovich, Judit (2007): Examining mathematical discourse practices. *For the Learning of Mathematics* 27 (1), 24–30.

Nagy, William (2009): Understanding words and word learning: Putting research on vocabulary into classroom practice. In Rosenfield, S., Berninger, V. (Eds.): *Implementing Evidence-based Academic Interventions in School Settings*. New York: Oxford University Press, 479–500.

Nagy, William, García, G. E., Durgunoğlu, A. Y. & Hancin-Bhatt, B. (1993): Spanish-English bilingual students' use of cognates in English reading. *Journal of Literacy Research* 25 (3), 241–259.

Nagy, William & Hiebert, Elfrieda (2011): Toward a theory of word selection. In Michael Kamil, P., Pearson, David, Moje, Elizabeth, Afflerbach, Peter (Eds.): *Handbook of Reading Research, Volume IV*. New York: Routledge, 388–404.

Nagy, William & Scott, Judith (2000): Vocabulary processes. In Kamil, Michael, Peter Mosenthal, P., Pearson, David, Barr, Rebecca (Eds.): *Handbook of Reading Research: Volume III*. Mahwah, NJ: Erlbaum, 269–284.

O'Halloran, Kay L. (2005): *Mathematical Discourse: Language, Symbolism and Visual Images*. London, England: Continuum.

Peregoy, Suzanne & Boyle, Owen. F. (2000): English learners reading English: What we know, what we need to know. *Theory into Practice* 39 (4), 237–247.

Pimm, David (1987): *Speaking Mathematically: Communication in Mathematics Classrooms*. London: Routledge & Kegan Paul.
Rand Reading Study Group (2002): *Reading for understanding: Toward a research and development program in reading comprehension*. Prepared for the Office of Educational Research and Improvement. Washington, DC: US Department of Education.
Sato, Edynn, Rabinowitz, Stanley, Carole, Gallagher & Huang, Chun-Wei. 2010. *Accommodations for English language learner students: The effect of linguistic modification of math test item sets*. (Final Report. NCEE 2009-4079) National Center for Education Evaluation and Regional Assistance National Center for Education Evaluation and Regional Assistance.
Scott, Judith & Nagy, William (1997): Understanding the definitions of unfamiliar verbs. *Reading Research Quarterly* 32 (2), 184–200.
Shaftel, Julia, Belton-Kocher, Evelyn, Glasnapp, Douglas & Poggio, John (2006): The impact of language characteristics in mathematics test items on the performance of English language learners and students with disabilities. *Educational Assessment* 11 (2), 105–126.
Walkerdine, Valerie (1988): *The Mastery of Reason: Cognitive Development and the Production of Rationality*. London: Routledge.
Zeno, Susan, Ivens, Stephan, Millard, Robert & Duvvuri, Raj. (1995): *The Educator's Word Frequency Guide*. New York: Touchstone Applied Science Associates.

Moritz Herzog, Erkan Gürsoy, Caroline Long, and Annemarie Fritz

Fifth-grade students' production of mathematical word problems

1 Introduction

Mathematical word problems challenge students significantly, as empirical studies have shown (e.g., Bush & Karp, 2013; Lewis & Mayer, 1987). Difficulties mostly arise from two aspects, mathematical characteristics, and linguistic structure. Mathematical characteristics of the word problem, such as number size, number and complexity of required operations, and applicable strategies, increase problem difficulties. While on the linguistic side, semantic as well as syntactical characteristics of word problems add to the difficulty (for an overview, see Daroczy et al., 2015). Besides these factors, it is building a mathematical model based on a situation described in a text that is a main difficulty to identify in empirical research (Jupri & Drijvers, 2016; Leiss et al., 2010; Maaß, 2010).

We use the term "situation" to refer to a context, which serves the purpose of exemplifying a concept or set of related concepts. As a situation is related to a specific mathematical conceptual field, it formulates a mathematical problem that requires a predictive response. Thus, situations go beyond stimuli, which cause a specific behavior, but are rather typical settings in which mathematical concepts become visible. Situations can be given by illustrations and also by contextual descriptions with mathematics concepts embedded. While research on word problems has focused on contextual descriptions of situations, this chapter aims at investigating how children produce word problems from engaging with illustrated situations.

Children encounter word problems that contextualize a more, or less, complex mathematical task in a real-world situation in different ways (Verschaffel et al., 2000). A typical, simple word problem is: "Alex has 3 packages of chocolate. In every package there are 5 pieces. How many pieces of chocolate does Alex have in total?" In this example, the encoded arithmetic task ($3*5=?$) is rather transparent in the word problem, as all numbers are given and the multiplicative structure is highlighted by cue words or phrases (here: "in every") (LeBlanc & Weber-Russell, 1996). Jupri and Drijvers (2016) report that finding all these cue words and phrases is a main obstacle for students while mathematizing a situation. In such tasks, the real-world context often appears to be designed for the task, thereby casting the word problem's authenticity into doubt

https://doi.org/10.1515/9783110661941-018

(Palm, 2009). The main purpose of these so-called "dressed-up" problems is practicing basic operations in real-world contexts (Leiss et al., 2019; Verschaffel et al., 2000). Actual mathematical problem solving requires a significantly more complex mathematizing process (Leiss et al., 2010). Mathematical problem solving is thus a more authentic application of mathematics to real-world problems (Maaß, 2010). Therefore, mathematical problem solving is a necessary part of national curricula (Bush & Karp, 2013; Jupri & Drijvers, 2016; KMK, 2005).

Mathematical word problems are commonly investigated in a manner we will call *receptive mode*: children are given mathematical word problems which they then have to solve (Thevenot & Barrouillet, 2015). In contrast to the receptive mode, Frank and Gürsoy (2014) investigated how grade 5 students create word problem texts in response to a given illustration, which we will call word problem *production*. As both aspects – context and task-relevant structure – are important to design such a text, the visualizations used as stimuli had pictorial properties. As their study had a linguistic research focus on children's awareness of language and multilingualism in mathematics classrooms, the extent to which such tasks can provide information about children's word problem-solving processes is rather limited (Frank & Gürsoy, 2014). This chapter reports on subsequent research which investigated how fifth graders write mathematical word problems in response to given illustrations, pictorial representations of situations, and how that relates to both individual characteristics (mathematical skills and reading comprehension proficiency) and to the features of the illustrated mathematical situations (obviousness of appropriate mathematical model for the illustration).

2 Mathematical modeling

In general, word problems require a complex process of modeling that translates the real-world context into a solvable mathematical task (Borromeo Ferri, 2006). Blum and Leiss (2007) propose a cyclic model of mathematical modeling, comprising cognitive, linguistic, and math-specific processes (Leiss et al., 2019). Fig. 1 shows the phases of mathematical modeling following Blum and Leiss (2007) that are involved in word problem solving that differ in extent and focus, depending on the task (Leiss et al., 2010, 2019).

The problem-solving process of a word problem can generally be divided into two phases. On the modeling side, children need to construct a suitable representation of the situation through identifying the embedded mathematical problem. On the computational side, the mathematical problem has to be solved by applying appropriate strategies and operations (Mayer, 1999). Modeling processes bridge the

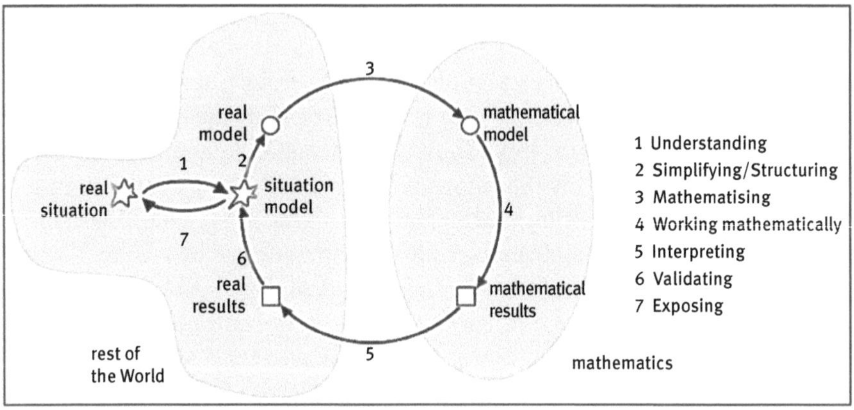

Fig. 1: Modeling cycle (Blum & Leiss, 2007).

gap between these phases and make mathematics applicable for real-world contexts (Leiss et al., 2019).

In the first step (1), children understand the real situation and translate it into a *situation model* of the context. This involves the student reading the text and sometimes providing an illustration containing the necessary information required to solve the task (Leiss et al., 2010). Thus, linguistic aspects are of great importance in this phase (Leiss et al., 2019). Empirical studies have shown that errors in this initial phase often lead to subsequent errors, which highlights the significance of understanding processes in word problems (Clements, 1980; Leiss et al., 2010). Recent research provided insights on linguistic difficulties in this phase (Daroczy et al., 2015; Leiss et al., 2019; Prediger et al., 2013). The information obtained from the word problem is organized and structured in the second step (2), leading to a *real model* (i.e., a model of the real-world problem). In this model, the context is reduced to (mathematically) relevant information and represented more precisely. Particularly in realistic word problems, this step might involve making hypotheses about information not given in the text or not obviously applicable (Leiss et al., 2010). For example, a problem such as "Linda bought 5 liters of juice for her café. How many glasses of 0.4 liters can she sell?" requires a child to keep in mind that you might not have partially filled glasses in some contexts (e.g., when selling them in a café).

At this point, the modeling process switches from real-world representations and considerations to the mathematical aspect of word problems (Mayer, 1999). While mathematizing (3), children transfer the real model into a *mathematical model* that can be solved by mathematical means (Leiss et al., 2010). As children need to know which mathematical model is appropriate, this step requires profound conceptual knowledge of the involved operations (Rittle-Johnson & Schneider, 2015; Daroczy et al., 2015; Leiss et al., 2010). In contrast, the purely mathematical solution process (4) that generates a *mathematical result* is more process-based and draws on strategy choice (e.g., counting, fact retrieval) as well as arithmetic proficiency (Daroczy et al., 2015).

In the next step (5), children interpret the mathematical result by relating the outcome to the real-world contexts. This process entails using conceptual knowledge and bridging the gap between mathematics and the real world again (Leiss et al., 2010; Mayer, 1999). This step leads to a *real result* that solves the mathematical part of the word problem under the given assumptions and with the numbers presented. However, this result has to be validated (6) within the context of the situation model (Leiss et al., 2010). Questions need to be asked: is the result realistic, are the assumptions adequate, and what other parameters besides the mathematical result could be taken into account? Based on these considerations in the situation model, children can finally present and explain (7) their solution to the word problem in terms of the real situation.

Obviously, the construction of a situation model is an integral part of the modeling process (Blum & Leiss, 2007), which can be underpinned by empirical research (Leiss et al., 2019, 2010). While building an adequate situation model of the real situation, children derive the necessary information such as numbers, operations, and their respective relation from the context. A mere combination and manipulation of numbers mentioned in the word problem is not sufficient to find a correct solution (Thevenot & Barrouillet, 2015; Verschaffel et al., 2000). Thus, the situation model allows transferring the not-yet-solved real-world problem into a solvable mathematical problem, if necessary under the constraint of simplification or additional reasonable assumptions; these restrictions are undone or at least discussed during the interpretation phase (Leiss et al., 2019, 2010).

3 Visualization and rewording in word problem solving

While constructing a situation model by understanding and organizing the real-world situation, children often make use of organizing illustrations. Hegarty and Kozhevnikov (1999) differentiate between pictorial illustrations that represent the context of a word problem and visual-schematic representations that organize the given information. For example, to a combinatorial word problem such as "Peter has two trousers and three shirts. How many different outfits can he wear?" a pictorial illustration might show Peter in front of his wardrobe, which does not give any hint of the problem structure or of an appropriate solving strategy. A visual-schematic illustration could show the trousers and shirts, enabling the imagining of combinations more easily. Empirical research revealed that the quality of the illustrations and in particular the use of visual-schematic illustrations is of significant relevance for solving word problems (Hegarty & Kozhevnikov, 1999; Vicente et al., 2008). More recently, Boonen et al. (2013) investigated the relation of the use of visual-schematic representations and reading comprehension, as both skills are discussed as crucial for solving word problems. Boonen et al. (2013) report significant effects of producing visual-schematic representations and reading comprehension. In contrast to visual-schematic visualizations, the use of pictorial illustrations seems to have negative effects on word problem performance (Hegarty & Kozhevnikov, 1999; van Garderen & Montague, 2003). Against this background, some researchers suggest that pictorial representations rely on isolated information derived from single phrases or words, while schematic representations are more likely to integrate the information given in the text (Boonen et al., 2013; van Garderen & Montague, 2003).

As a consequence, a situation cannot be represented by only pictorial illustrations, but requires at least to some extent schematic information that indicates which mathematical concepts are involved. This applies in particular to dressed-up problems, in which the underlying relations are made as obvious as possible, because the intention of such problems is not teaching the modeling process, but practicing arithmetic within real-world contexts (Palm, 2009). Dressed-up problems do not contain an actual problem that has to be solved (e.g., "Linda bought 5 liters of juice for her café. How many glasses of 0.4 liters can she sell?"), but just contextualize a mathematical problem (e.g., "Linda has 2 glasses and gets 3 more. How many does she have now?"). Thus, a dressed-up problem refers to a specific and pre-defined solving strategy and a corresponding operation. In contrast, more complex problems allow for various solution strategies: In case of the given example of Linda's café above, several strategies are equally appropriate (divide 5 by 0.4, add up 0.4 until reaching 5, etc.). Illustrations of dressed-up and more complex problems differ in their obviousness. As dressed-up problems are meant to elicit the pre-defined operation, corresponding illustrations have to be as obvious as possible in order to guide children to that intended operation. A more complex problem, however, can be illustrated less obviously, as there is not one specific intended solution strategy. Naturally, less obvious illustrations of word problems are supposed to elicit more diverse strategies and operations than more obvious illustrations.

De Corte and Verschaffel (1987) highlight the relevance and benefits of rewording word problems as an instructional method in mathematics classrooms. Rewording means that children rephrase the situation of a given word problem in different words. Because children have to understand the structure and situation of the word problem to reword it correctly, it might contribute more to the "concept acquisition function" (De Corte & Verschaffel, 1987: 379) of arithmetic instruction. Therefore, rewording as instructional method might have positive effects on word problem performance (Thevenot & Barrouillet, 2015). As for visualizations, children also benefit from rewordings of the word problem structure. In contrast, situational rewordings that do not refer to the structure of a word problem but just slightly alter the context have no effect on word problem solving. These findings underpin the importance of children's understanding of the situation given in a word problem.

4 Conceptual foundations of multiplication and division

Multiplication and division mark a change in conceptual thinking from addition and subtraction, in that the concepts are two-dimensional rather than one-dimensional. Whereas addition and subtraction can be conceptualized along a number line, multiplication and division require more complex metaphors. Early conceptions of multiplication or grouping can be understood as repeated addition; however, more complex conceptions of multiplication require two-dimensional models such as area models.

In addition and subtraction, quantities of the same type are added or subtracted, for example, 3 candies and 4 candies are added together; however, in multiplication the quantities involved are of different types, for example, 3 jars each with 4 candies per jar, means that there are 12 candies altogether. The division operation, explained as 12 candies divided equally between three jars, will result in four candies per jar (partitive division), or by placing four candies in each jar how many jars are needed will result in three jars (quotative division). From this distinction we see that in addition and subtraction the total number of candies is preserved, while in multiplication and division, the total number is transformed, the 4 candies are taken three times, which makes 12 candies (adapted from Schwartz, 1988: 41).

The developmental path from counting to additive reasoning to multiplicative reasoning is mapped below. We acknowledge that the multiplicative conceptual field incorporates many concepts and that proficiency may be incrementally gained through different pathways; nevertheless, there is a logical mathematical route for the teaching of these concepts that can be determined theoretically, and corroborated or challenged empirically.

5 From counting to additive reasoning to multiplicative reasoning

In the early years of schooling the students move from counting, to addition and subtraction, and to multiplication and division. The reasoning develops from additive reasoning to multiplicative reasoning and finally to this most important construct proportional reasoning, understood to be "the *capstone* of children's elementary school arithmetic and the *cornerstone* of all that is to follow" (Lesh et al., 1988: 93–94).

Counting is the basic concept related to the creation of number concept (Desoete, 2015; Fritz et al., 2018). Counting meaningfully implies that students understand one-to-one correspondence, that is "the situation where there is one item to a set" (Bakker, van den Heuvel-Panhuizen & Robitzsch, 2014: 70). According to Bakker et al. (2014) one-to-many correspondence provides the student with "the awareness that a set has more than one item and the student can count the groups according to number in a set" (p. 70). This concept then leads to the understanding of multiplication and division, which are the base concepts for more complex mathematical concepts in the multiplicative conceptual field such as ratio, fractions, and linear functions (Vergnaud, 1983). Multiplicative reasoning relates to proportional reasoning (Lesh et al., 1988).

Vergnaud's (1983) rationale for clustering the concepts involving multiplication and division as the multiplicative conceptual field is that both operations are related in many everyday situations. As multiplication and division are inverse operations, a problem containing a many-to-many relationship can often be solved by multiplication and by division. In the example task of Linda's café above, multiplication (multiplying 0.4 with increasing numbers until reaching 5) or division (divide 5 by 0.4) is a suitable operation. This conceptualization of the many-to-many relationship from multiplicative concepts to situations supports the rationale for this study.

6 The current study

The aim of this study was to identify how children produce complete word problems for illustrations of given situations. Based on pictures of multiplication or division situations, we investigated children's word problem production, a process which can be interpreted as an inversion within the modeling cycle (Leiss & Blum, 2007): Children start with a situation model (2). To write a mathematical word problem, they create an appropriate real situation (1), a possible real model

(3), and a mathematical model (4). The relationship between writing a full mathematical word problem, numeracy, and literacy, which might be new to many students, can provide insights into fifth-graders' production of word problems.

Characteristics of the given situation are likely to affect the mathematical word problems that children produce, in particular the choice of operations. For example, when a task contains numbers in a specific and salient relation (e.g., numbers that can be divided without remainder such as 12 and 4), it can be translated directly to a suitable mathematical problem. However, if there is no directly translatable operation, with no common factors, (e.g., 13 and 4), the given situation has to be transformed into a possible task by manipulating the given information, and an appropriate word problem might be harder to find.

Mathematical word problems form part of a specific text genre (Frank & Gürsoy, 2014; Hyland, 2007). This context implies that mathematical word problems have particular aspects such as the necessary numerical information, a clearly formulated relation between them, and a mathematical problem (Frank & Gürsoy, 2014). In addition, mathematical word problems can have typical structures, such as "dressed-up" problems. As children are used to such structures from a teaching environment, they might rely on patterns they encounter often in mathematics classrooms. The use of typical word problem structures while writing mathematical word problems might mirror students' approaches to word problem solving.

The scope of the study has been operationalized by the following research questions:

(RQ1) To what extent are fifth graders able to produce word problems from situations given in illustrations with varying level of obviousness regarding an intended strategy?

While reading and solving word problems has been addressed by several studies in the past, children's production of word problems is less well researched (Frank & Gürsoy, 2014). We expect that children have more difficulties with writing word problems that are less obvious in terms of less clearly suggesting a certain multiplicative or divisional relation between the numbers.

(RQ2) How can multiplication and division performance as well as reading fluency predict the successful production of word problems?

Arithmetic performance and reading fluency both were predictors for solving word problems in recent studies (e.g., Leiss et al., 2019; Stephany, this volume). This raises the question, how arithmetic performance and reading fluency can predict children's writing of word problems?

(RQ3) To which operations do students refer in their word problems and to what extent does their choice relate to the illustration?
This research question focusses how illustration type and intended operation in the word problems written by the children relate. Among other representations, the multiplicative conceptual field can be illustrated as distribution contexts, which refer more to division tasks (e.g., Tasks 1 and 2 in Fig. 2), or repeated addition contexts, which refer more to multiplication tasks (e.g., Tasks 3 and 4 in Fig. 2). We expect that more obvious illustrations (Tasks 2 and 3 in Fig. 2) mostly lead to the operations intended by the illustrations.

(RQ4) How do fifth graders verbalize the parts of a complete mathematical word problem (background story, mathematical problem, and mathematical task)?
Word problems consist of different parts, such as a background story contextualizing the problem, a mathematical problem that has to be solved, and a mathematical task that specifies which question has to be solved in the word problem (Frank & Gürsoy, 2014). In addition, these parts have to match (e.g., the mathematical task has to refer to the mathematical problem). This research question aims at investigating how children address the different parts of a word problem and to what extent they can match them.

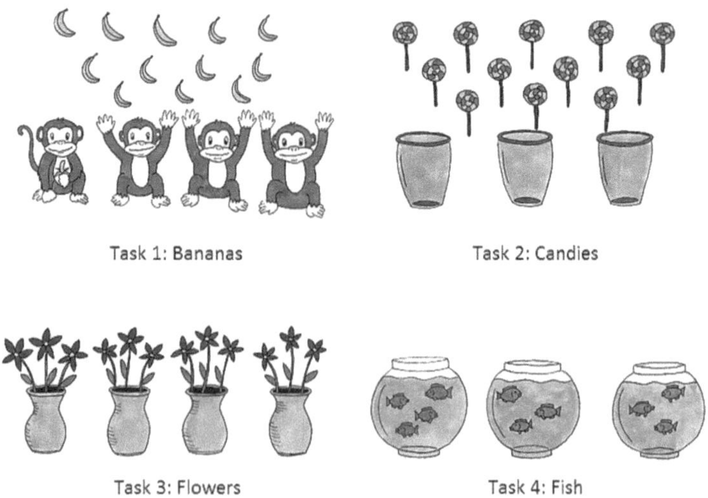

Fig. 2: Illustrations used in the word problem writing tasks (Frank & Gürsoy, 2014).

7 Method

7.1 Sample

Fifth graders ($n = 368$; 226 girls, 61.4%; 142 boys, 38,6%; $m_{age} = 136.0$ months, $SD_{age} = 5.7$ months) from Western Germany participated in the study. Students from 12 different schools were tested at the end of grade 5. In this part of Germany, students are separated based on their academic performance in primary school into three different school levels after grade 4. In this study, 234 (63.6%) students attended the highest school level ("Gymnasium"; preparing for university), 62 (16.8%) the medium school level ("Realschule"), and 72 (19.6%) the lowest school level ("Gesamtschule" and "Sekundarschule").

7.2 Instruments

Writing word problems: In line with Frank and Gürsoy (2014), children were asked to write a mathematical word problem for four given illustrations (Fig. 2). The illustrations included schematic as well as pictorial properties. The objects depicted set up a general context (e.g., monkeys eating bananas). Their arrangement in general was supposed to elicit a multiplication or division task by activating the respective operational understanding (vom Hofe, 1998): While two pictures (monkeys and candies) suggested a distributive context, the other two pictures (flowers and fish) were expected to lead to a compositional understanding.

Two pictures – one multiplication (flowers) and one division task (candies) – contained numbers from the multiplication table (12 and 3) that would elicit a whole number answer. The other two pictures were designed such that the answers would include fractions. As well, one of these illustrations was an intended multiplication task (fish; $3 + 3 + 4$) and one was an intended division (bananas; 13 and 4). In case of number pairs with common factors (flowers and candies), the illustrated situations supported a dressed-up problem clearly. Thus, we will refer to these situations as relatively obvious compared to the other situations, in which number pairs did not support a direct dressed-up problem. We expected that the number pairs that were designed to elicit whole number answers would lead to more multiplication and division tasks than the less obvious number pairs suggesting rational number answers.

Multiplication and division problems: Children were given 27 multiplication and division problems in the number range up to 100 (Crombach's $\alpha = .72$). The problems contained dressed-up word problems as well as pure arithmetic problems. This subtest was not timed, so children solved these problems without any time pressure. In all division problems number pairs could be divided without remainder. In all problems products and dividends were two-digit numbers, while factors, divisors, and quotients were single-digit numbers.

Reading skills: A standardized speed test for reading skills was administered (ELFE 1–6, Lenhard & Schneider, 2006). The test consisted of two subscales: sentence comprehension and text comprehension. In the sentence comprehension subtest, children were asked to complete a given sentence to form a meaningful and grammatically correct sentence. Text comprehension required reading a short text and answering one question on information given in the text.

7.3 Results

Word problem production (RQ1): We evaluated whether the produced word problems were correct as intended in the task ("Write a mathematical word problem connected to the picture"). We considered those word problems as correct that described a situation connected to the picture and contained a problem that could be solved mathematically.

To fulfil the criterion "connected to the situation," it was sufficient to adopt the given context (i.e., bananas and monkeys, candies and jars, flowers and vases, fish and fishbowls) in the word problem. Partial adoption (e.g., only one part: "There are 10 fish to sell at the pet shop. How many are there, after 4 are sold?") as well as additional, not given, information (e.g., "If one fish needs 12 g fish food per week, how much fish food do 10 fish need per week?") was also accepted. The same principle applied to children's counterintuitive interpretations of the pictures (e.g., "Here you see 3 duplication machines [=jars]. Every hour, 12 candies are produced by each machine. How many candies do you have in 1 hour, 13 days, and 4 years?").

The criterion "mathematical word problem" was handled rather strictly. Word problems were considered correct if they demanded a mathematical problem-solving process. That could be implemented by formulating a particular question ("There are 4 vases containing 3 flowers each. How many flowers are there?") or a direct operational instruction ("4 monkeys have 12 bananas. Share them equally").

Evaluated as not correct were word problems that just described the situation without any problem or question ("The jars are empty, but there are 12 candies in the air") and those that just described how to perform a (maybe suitable) operation ("Calculate 4 monkey multiplied by 12") or the solution

("4 candies go in each jar"). Moreover, the word problems did not necessarily have to report the amounts given in the illustration, because children might have anticipated that the illustrations would be part of the word problem (e.g., "How many fish are there?"). As the level of linguistic complexity of a word problem does not matter regarding its solvability, more complex word problems (e.g., "Here you see 3 duplication machines. Every hour, 12 candies are produced by each machine. How many candies do you have in 1 hour, 13 days, and 4 years?") were not distinguished from more simple word problems (e.g., "How many monkeys are there?"). Word problems did not need to include arithmetic operations to be considered correct (e.g., "Which mathematical geometrical shape do the jars have?").

In general, most students were able to correctly write a mathematical word problem suited to the picture. Task 1 (Bananas) was solved by 86.4%, task 2 (Candies) by 88.6%, task 3 (Flowers) by 87.2%, and task 4 (Fish) by 84.8% of the students. There were no significant differences regarding the correct solution rates between all tasks ($\chi^2(3) = 3.257$, $p = .354$). Correct solution rates did not differ between pictures showing a distributive (Tasks 1 and 2) or a multiplicative situation (Tasks 3 and 4) ($\chi^2(1) = .425$, $p = .514$). Contrary to our expectations, pictures directly eliciting multiplication or division tasks did not have higher correct solution rates than less obvious pictures showing situations that cannot be translated directly into either a multiplication or a division task ($\chi^2(1) = 2.757$, $p = .097$).

Reading and mathematical performance (RQ2): Correlation analyses between raw sums of correctly written word problems and number of correctly solved multiplication and division (mathematical performance), as well as sentence and text comprehension tasks, were run. While sentence comprehension correlations were low ($r = .194$, $p < .001$), text comprehension ($r = .322$, $p < .001$) and mathematical performance ($r = .292$, $p < .001$) showed low to medium correlations. This result is mirrored in a regression analysis explaining word problem production ($F(3,363) = 19.914$, $p < .001$, adj. $R^2 = .134$). Only mathematical performance ($\beta = .201$, $p < .001$) and text comprehension ($\beta = .239$, $p < .001$) were significant predictors, while sentence comprehension ($\beta = .027$, $p = .623$) had almost no predictive power. Tab. 1 provides an overview of these results.

Operations in the word problems (RQ3): Written word problems can refer to operations that lead to the solution more or less directly. Very clearly encoded operations are typical of dressed-up problems in which cue words or typical structures indicate the targeted operation (Frank & Gürsoy, 2014; Jitendra et al., 2007). Based on the produced word problems, the encoded operations can be compared. In most word problems (86.3%), a specific encoded operation could be assigned. These were the basic operations and counting. The operations

Tab. 1: Relations between predictor variables and word problem production.

Predictor	Correlation (*r*)	Regression (β)
Sentence comprehension	.194**	.027
Text comprehension	.322**	.239**
Mathematical performance	.292**	.201**

Note. ** = $p < .001$.

might be explicitly mentioned ("There are 4 vases and 3 flowers in each vase. Calculate how many flowers there are. Add them all up") or implicitly suggested by cues ("There are 12 candies and they are supposed to be shared equally among the jars"). In some produced word problems, no mathematical operation was encoded (no operation). In these cases, mostly no mathematical problem was formulated (e.g., "Those are vases," "There are 4 monkeys and one of them eats a banana"). In other word problems a mathematical problem was formulated, the specific operation, however, was not explicated (unclear, e.g., "Calculate how many flowers are in one vase"). Fig. 3 shows the shares of these categories for the four tasks.

In general, word problems encoded operations as intended when designing the illustrations. Tasks 1 (Bananas) and 2 (Candy) mostly lead to division encoding, while in task 3 (Flowers) most word problems encoded a multiplication task. However, the picture is rather fuzzy for task 4 (Fish): About one third of the word problems encoded a multiplication or division task and another third was classified as unclear. In most of these word problems, children described the situation and simply asked for the total sum of fish but indicated no particular operation. As depicted in Fig. 3, there was a greater variety in the encoded operations in the less obvious tasks 1 and 4.

Verbalizations of word problems (RQ 4): A mathematical word problem can be decomposed into three main parts (Frank & Gürsoy, 2014): A background story that describes the situational setting of the word problem (e.g., "Lukas, Emma, and Lilli go to the kiosk to buy some candies"), a mathematical problem ("They bought 12 candies. They wanted to share them fairly"), and a mathematical task ("How many candies does everyone get?"). While the mathematical problem refers to the operational relation of the numbers, the mathematical task refers to the formulation of a prompt to solve the mathematical problem. Obviously, the background story is not necessary to solve the mathematical problem. However, the situational background is relevant when the mathematical result has to be evaluated and explained regarding the real-world context (steps 6 and 7) in the modeling cycle (see Fig. 1, Blum & Leiss, 2007). Note that

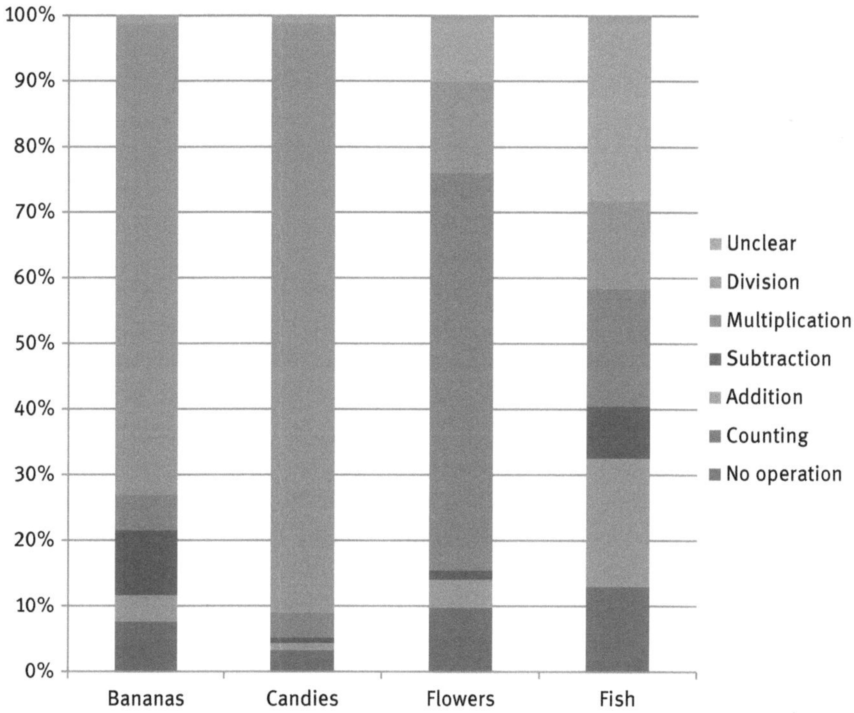

Fig. 3: Operations encoded in the word problems.

the background story has to provide more situational information than describing the items relevant for the mathematical task (e.g., "There are 4 monkeys and 13 bananas"). We claim that a lack of situational contexts is one of the characteristics for "dressed-up" problems, since such word problems do not refer to a realistic context (Verschaffel et al., 2000). Word problems can be solved only if the (numerical) relation of the elements (e.g., candies and children) is specified. This criterion constitutes a mathematical problem. Finally, a word problem is supposed to contain a mathematical task (e.g., "How many candies and jars are there in total?"). As mathematical word problems might have more than one possible task, this part is highly relevant. Obviously, the mathematical problem and the task have to be related in order to be solved. In other words, the mathematical task has to extend the mathematical problem.

In this study, we accepted such background stories that introduced names and specific characters (e.g., parents or zookeepers) or any situation going beyond the mere mention of the depicted items. Any mathematical relation between elements that can be solved by mathematical means was coded as a valid mathematical problem. Finally, mathematical tasks were accepted if they formulated a question that can be solved mathematically in general. Note that mathematical problems and mathematical tasks were evaluated separately and thus did not necessarily have to be related. If the mathematical problem and the mathematical tasks were related, this was evaluated as an independent characteristic of the produced mathematical word problem. In the last step, we checked if the mathematical word problem contained all the information necessary to solve the respective word problem.

Table 2 provides an overview of the percentages of correctly verbalized characteristics as found in the word problems the students produced. In general, there are little differences between the tasks. This indicates that the way children produced word problems was independent of the given picture and refers to children's ideas and skills in general. Children's word problems rarely contained background stories. Chi-square statistics reveal no significant differences in correct response rates between the tasks regarding writing a background story ($\chi^2(3) = 1.894$, $p = .595$), verbalizing a mathematical problem ($\chi^2(3) = 1.232$, $p = .745$), and formulating a mathematical task ($\chi^2(3) = 2.885$, $p = .410$). Only success in matching mathematical problem and mathematical task differed between the tasks ($\chi^2(3) = 27.783$, $p < .001$), obviously due to lower matching rates in the less obvious tasks (bananas and fish).

Tab. 2: Correctly verbalized characteristics of the produced word problems in percent.

	Bananas	Candies	Flowers	Fish
Background story	12,5	14,9	16,0	16,3
Mathematical problem	64,9	63,3	66,6	67,7
Mathematical task	88,6	91,3	92,4	89,4
Match of problem and task	58,4	76,4	71,5	67,7

As they were not explicitly asked to do so and background stories are not obligatory, it cannot be guaranteed that children would produce more background stories if these were made more salient in advance (e.g., by providing an example). Children did not provide mathematical problems in about one third of the word problems. One reason might be that not all students understood the necessity for this integral part of a mathematical word problem.

About one third of the produced word problems did not contain a proper mathematical problem. This result indicates that verbalizing the mathematical problem, which includes describing the relevant context and mentioning the necessary information, was a main obstacle for students in this study.

In contrast, children had fewer difficulties in formulating a mathematical task in their word problems, as the vast majority of produced word problems contained a clearly and explicitly formulated task (e.g., "How many flowers go in each vase?"). One might thus argue that children are rather familiar with the aspect of mathematical task in word problems.

While formulating tasks did not challenge students considerably, matching the task to the mathematical problem was not easy for all students. Notably regarding the less obvious tasks that contained numbers that had no common factors (Bananas and Fish), children more often failed to match the problem and the task. In general, this appears as a second main obstacle to students in writing a mathematical word problem.

8 Discussion

To the best of our knowledge, the vast majority of empirical studies on modeling focus on children's construction of a situation model based on a given text. This study aimed at investigating to what extent students are able to write a word problem to a given situation model. In the context of the cyclic model by Blum and Leiss (2007), this approach means reversing the process of understanding between the real situation and the situation model, which generally implies bidirectional processes (Borromeo Ferri, 2006; Leiss et al., 2010). Although this is not part of the national curriculum standards, children did not show particular difficulties with writing mathematical word problems to given situations, as mirrored throughout in the high correct solution rates of about 85%. The produced word problems mostly were "dressed-up" problems suited to the simple situations, which indicate that students generally know how to encode a word problem directly. Characteristics of the given situations, such as being more multiplicative or more distributive, did not affect correct solution rates significantly. In addition, it had no effect on solution rates if numbers were selected from the multiplication table or not.

In line with previous findings, writing word problems is associated with reading proficiency and mathematical performance (Daroczy et al., 2015; Leiss et al., 2019). Thus, these individual factors affect understanding in both directions. In particular text comprehension has to be highlighted, while sentence comprehension correlated only minimally with writing proficiency. This result strengthens the position that mathematical modeling relies on a comprehensive understanding of the situation and goes beyond cue words or phrases (Thevenot & Barrouillet, 2015). However, correlations and the predictive power of the regression analysis in this study are rather low, which corresponds to the literature on this topic (e.g., Leiss et al., 2019).

Children mostly used the intended operations multiplication and division. Varying the obviousness of the given situation led to more variety in the operations used. This outcome might indicate that fifth graders anticipate a kind of standardized dressed-up problem based on typical situations. This might result from schooling that focusses on this kind of word problems (Verschaffel et al., 2000). How deep children's understanding of the situation in such cases is is doubted (Maaß, 2010). It is rather likely that children produce (and read) such word problems by applying a certain, well-trained verbal frame (Borromeo Ferri, 2006).

Obviously, situations indicating a distribution elicit division more specifically than groupings do for multiplication. A closer look at the visualizations used in classrooms might provide a better understanding of the processes underlying children's use of operations.

Children's focusing on standardized dressed-up problems might also be observed in the few differences between the less obvious (bananas and fish) and more obvious (flowers and candy) situations in the study. First, it has to be noticed that obviousness did not significantly affect performance at all. However, the operations referred to in the word problems were more diverse in the less obvious situations, which indicates that the children's search for a dressed-up problem was hampered by the less obviousness of relatively prime numbers. The less obviousness of tasks 1 and 4 might have resulted in less-well-trained problem-solving processes, in which children used a broader variety of operations. Consequently, not all of these operations could be conceptually supported by the illustrations to the same extent. In this problem-solving process, children did not struggle more with formulating a mathematical problem or a mathematical task than in the more obvious tasks. However, children showed more difficulties in relating the mathematical problem and the task, which could be consequence of their adapted problem-solving process that digressed from the well-trained search for a dressed-up problem.

Children's production of word problems might reflect what they focus on when they encounter word problems in mathematics classrooms. The results show that the mathematical task was particularly salient to the children in this study. The mathematical task is usually the actual question that is asked in word problems. Children obviously are used to focus on that question as they have to answer it and are evaluated based on their answer. However, the mathematical problem, which is often considered as a main aspect of a word problem, was less often given attention. In addition, the children in this study showed considerable difficulties in matching the mathematical problem and task (see Tab. 2). These results suggest that fifth graders pay much attention on the question asked at the cost of understanding the mathematical problem. It would be of great interest to investigate to what extent children are aware of the different parts of a word problem and how they address them.

Such knowledge of children's meta-knowledge of word problems could be useful for instruction. Based on the framework of so-called *genre pedagogy*, producing word problems could help students to understand word problems (Hyland, 2007). The central idea of genre pedagogy is that every discipline has its own typical text types (genres) that have specific properties and follow particular rules. This coherence allows genre-specific teacher scaffolding. A typical "teaching-learning cycle" (Hyland, 2007: 159) covers teaching the purpose and typical use of the genre; identification of key characteristics of the genre as well as possible variations; a joint construction of a typical text type such as a word problem, in which the teacher supports the children by providing appropriate exercises; children's independent construction of such a text; and interrelating the text type to other types (e.g., theorems).

With the teacher scaffolding and co-constructing word problems, children can understand which parts (e.g., given information or targeted result) a word problem has and how these parts can be identified. This instructional method might be a successful addition to existing approaches such as schema-based instruction to promote students' ability to establish a suitable situation model (Frank & Gürsoy, 2014; Hyland, 2007; Jitendra, 2019). Developing such an intervention definitively asks for more detailed and more qualitative research on children's production of word problems in response to given situations.

References

Bakker, Marjoke, van den Heuvel-panhuizen, Marja & Alexander, Robitzsch (2014): First-graders' knowledge of multiplicative reasoning before formal instruction in this domain. *Contemporary Educational Psychology* 39, 59–73.

Blum, Werner & Leiss, Dominik (2007). How do students and teachers deal with modelling problems? In Haines, Christopher (ed.), *Mathematical modelling (ICTMA 12). Education, engineering and economics: Proceedings from the twelfth international conference on the teaching of mathematical modelling and applications* 222–231. Chichester: Horwood.

Boonen, Anton J. H., van der Schoot, Menno, van Wesel, Floryt, de Vries, Meinou H. & Jolles, Jelle (2013): What underlies successful word problem solving? A path analysis in sixth grade students. *Contemporary Educational Psychology* 38 (2013), 271–279.

Borromeo Ferri, Rita (2006): Theoretical and empirical differentiations of phases in the modelling process. *ZDM Mathematics Education* 38 (2), 86–95.

Bush, Sarah B. & Karp, Karen S. (2013): Prerequisite algebra skills and associated misconceptions of middle grade students: A review. *Journal of Mathematical Behavior* 32 (3), 613–632.

Clements, M. A. Ken (1980): Analyzing children's errors on written mathematical tasks. *Educational Studies in Mathematics* 11 (1), 1–21.

Daroczy, Gabriella, Wolska, Magdalena, Meurers, Walt D. & Nuerk, Hans-Christoph (2015): Word problems: A review of linguistic and numerical factors contributing to their difficulty. *Frontiers in Psychology* 6, 1–13.

De Corte, Erik & Verschaffel, Lieven (1987): The effect of semantic structure on first-graders' strategies for solving addition and subtraction word problems. *Journal for Research in Mathematics Education* 18, 363–381.

Desoete, Annemie (2015): Cognitive predictors of mathematical abilities and disabilities. In Kadosh, Roi C., Dowker, Ann (Hrsg.): *The Oxford Handbook of Mathematical Cognition*. Oxford: 2 Medicine UK, 915–932.

Frank, Magnus & Gürsoy, Erkan (2014): Sprachbewusstheit im Mathematikunterricht in der Mehrsprachigkeit – Zur Rekonstruktion von Schülerstrategien im Umgang mit sprachlichen Anforderungen von Textaufgaben. In Ferraresi, Gisella, Liebner, Sarah (Hrsg.): *SprachBrückenBauen*. Göttingen: Universitätsverlag, 29–46.

Fritz, Annemarie, Ehlert, Antje & Leutner, Detlev (2018): Arithmetische Konzepte aus kognitiv-psychologischer Sicht. *Journal für Mathematik-Didaktik* 39, 7–41.

Hegarty, Mary & Kozhevnikov, Maria (1999): Types of visual-spatial representations and mathematical problem solving. *Journal of Educational Psychology* 91, 684–689.

Hyland, Ken (2007): Genre pedagogy: Language, literacy and L2 writing instruction. *Journal of Second Language Writing* 16, 148–164.

Jitendra, Asha K. (2019): Using schema-based instruction to improve students' mathematical word problem solving performance. In Annemarie Fritz, Vitor G. Haase, Räsänen, Pekka (eds.): *The International Handbook of Mathematical Learning Difficulties*. New York: Springer, 595–609.

Jitendra, Asha K., Griffin, Cynthia, Haria, Priti, Leh, Jayne, Adams, Aimee & Kaduvetoor, Anju (2007): A comparison of single and multiple strategy instruction on third grade students' mathematical problem solving. *Journal of Educational Psychology* 99, 115–127.

Jupri, Al & Drijvers, Paul (2016): Student diffculties in mathematizing word problems in algebra. *EURASIA Journal of Mathematics, Science, & Technology Education* 12 (9), 2481–2502.

KMK (Kultusministerkonferenz) (2005): *Bildungsstandards im Fach Mathematik für den Primarbereich. Beschluss vom 15. 10.2004*. München: Luchterhand Verlag.

LeBlanc, Mark & Weber-Russell, Sylvia (1996): Text integration and mathematical connections: A computer model of arithmetic word problem solving. *Cognitive Science* 20, 357–407.

Leiss, Dominik, Plath, Jennifer & Schippert, Knut (2019): Language and mathematics – Key factors influencing the comprehension process in reality-based tasks. *Mathematical Thinking and Learning* 21 (2), 131–153.

Leiss, Dominik, Schukajlow, Stanislaw, Blum, Werner, Messner, Rudolf & Pekrun, Reinhard (2010): The role of the situation model in mathematical modelling – task analyses, student competencies, and teacher interventions. *Journal Für Mathematik-Didaktik* 31, 119–141.

Lenhard, Wolfgang & Schneider, Wolfgang (2006): Ein Leseverständnistest für Erst- bis Sechstklässler. Göttingen: Hogrefe.
Lesh, Richard, Post, Thomas & Behr, Merlyn (1988): Proportional reasoning. In Hiebert, James, Behr, Merlyn (eds.): *Number Concepts And Operations in the Middle Grades*. Reston, Virginia: Lawrence Erlbaum, 93–118.
Lewis, Anne B. & Mayer, Richard E. (1987): Student's miscomprehension of relational statements in arithmetic word problems. *Journal of Educational Psychology* 79, 363–371.
Maaß, Katja (2010): Classification scheme for modelling tasks. *Journal für Mathematik-Didaktik* 31 (2), 285–311.
Mayer, Richard E. (1999): *The Promise of Educational Psychology Vol. I: Learning in the Content Areas*. Upper Saddle River, NJ: Merrill Prentice Hall.
Palm, Torulf (2009): Theory of authentic task situations. In Verschaffel, Lieven, Greer, Brian, van Dooren, Wim, Mukhopadhyay, Swapna (eds.): *Words and worlds. Modelling verbal descriptions of situations*. Rotterdam: Sense, 3–20.
Prediger, Susanne, Renk, Nadine, Büchter, Andreas, Gürsoy, Erkan & Benholz, Claudia (2013): Family background or language disadvantages? Factors for underachievement in high stakes tests. In Lindmeier, Anke, Heinze, Aiso (Hrsg.), *Proceedings of the 37th Conference of the International Group for the Psychology of Mathematics Education* 4,49–56. Kiel: PME.
Rittle-Johnson, Bethany & Schneider, Martin (2015): Developing conceptual and procedural knowledge of mathematics. In Kadosh, Roi C., Dowker, Ann (eds.): *The Oxford Handbook of Numerical Cognition*. New York: Oxford University Press, 1118–1134.
Schwartz, Judah (1988): Referent preserving and referent transforming operations on qualities. In Hiebert, James, Behr, Merlyn (eds.): *Number Concepts and Operations in the Middle Grades*. Vol. 2, Reston, Virginia: Erlbaum/N.C.T.M, 41–52.
Thevenot, Catherine & Barrouillet, Pierre (2015): Arithmetic word problem solving and mental representations. In Kadosh, Roi Cohen, Dowker, Ann (eds.): *The Oxford Handbook of Numerical Cognition (pp. 158–179)*. Oxford, UK: Oxford University Press.
van Garderen, Delinda & Montague, Marjorie (2003): Visual–spatial representation, mathematical problem solving, and students of varying abilities. *Learning Disabilities Research & Practice* 18, 246–254.
Vergnaud, Gerard (1983): Multiplicative structures. In Lesh, Richard, Landau, Marsha (eds.): *Acquisition Of Mathematics Concepts and Processes*. New York: Academic Press, 127–174.
Verschaffel, Lieven, Greer, Brian & De Corte, Erik (2000): *Making Sense of Word Problems*. Lisse: Swets & Zeitlinger.
Vicente, Santiago, Orrantia, Josetxu & Verschaffel, Lieven (2008): Influence of situational and mathematical informationon situationally difficult word problems. *Studia Psychologica* 50 (4), 337–356.
Vom Hofe, Rudolf (1998): Generation of basic ideas and individual images. In Sierpinska, Anna, Kilpatrick, Jeremy (eds.): *Mathematics Education as a Research Domain*. Dordrecht: Kluwer Academic, 316–332.

Sabine Stephany

The influence of reading comprehension on solving mathematical word problems: A situation model approach

1 Introduction

Tasks presenting mathematical information as text, known as word problems, are one of the key components in the teaching of mathematics in primary school. Still, they are also one of the most difficult ones. Studies have revealed that word problems are solved up to 30% less successfully than tasks in numerical notation (Duarte et al., 2011). This discrepancy indicates that besides mathematical competence aspects of language play a significant role in the processing of word problems as well (Duarte et al., 2011; Gürsoy, 2016; Heinze et al., 2011; Prediger et al., 2015; Verschaffel et al., 2000). In order to solve a word problem, students not only need to perform the necessary mathematical operations, they also need to understand the text of the task.

There are two intertwining factors central to understanding the text of a word problem: the task text itself and the problem solver's individual reading competence. Both can cause difficulties. In recent years, increasing attention has been paid to the task text and its linguistic characteristics that are considered to challenge the processing of the task, such as academic language (Abedi & Lord, 2001; Gürsoy, 2016; Haag et al., 2015; Martinello, 2008; Prediger et al., 2015). However, there have only been a few studies that focus on reading skills as a factor in explaining students' difficulties in dealing with word problems. Since reading and understanding the text of a task are fundamental for solving word problems, it can be assumed that, in addition to the linguistic characteristics of the task text itself, reading competence also has an important influence on the solution process.

Hence, the present study examines the impact of reader characteristics on the handling of word problems in primary school and explores aspects of reading competence that may be relevant for their solution. Putting mathematical content and processes aside, word problems are primarily texts that have to be read and understood. In addition to the mathematical perspective on solving word problems (Section 2), the reading process is therefore examined from a cognitive-psychological perspective to transfer relevant conclusions to mathematics (Section 3). This may lead to important conclusions on ways of promoting

https://doi.org/10.1515/9783110661941-019

mathematical reading skills. The empirical part (Sections 4 and 5) describes the study carried out on the relationship between various aspects of reading competence, mathematical competence, and the probability of solving word problems. Finally, the results regarding their impact on teaching mathematical word problems are discussed (Section 6).

2 Solving mathematical word problems

Mathematical word problems are mathematical exercises whose main characteristic is the embedding of mathematical relations in a text rather than in a mathematical notation (Duarte et al., 2011), for example, "Peter has 3 marbles and Ann has 5 marbles. How many marbles do they have altogether?" The context of the tasks varies and ranges from excerpts from students' everyday lives to facts and figures and fantasy worlds. Word problems pursue three goals: the application of mathematics, the development and expansion of problem-solving skills, and the exploration of the environment by means of mathematics (Franke & Ruwisch, 2010).

Solving word problems is complex and includes several stages (Blum & Leiss, 2007; Verschaffel et al., 2000). The modeling cycle of Blum and Leiss (2007) shown in Fig. 1 illustrates the cognitive operations during the solution process.

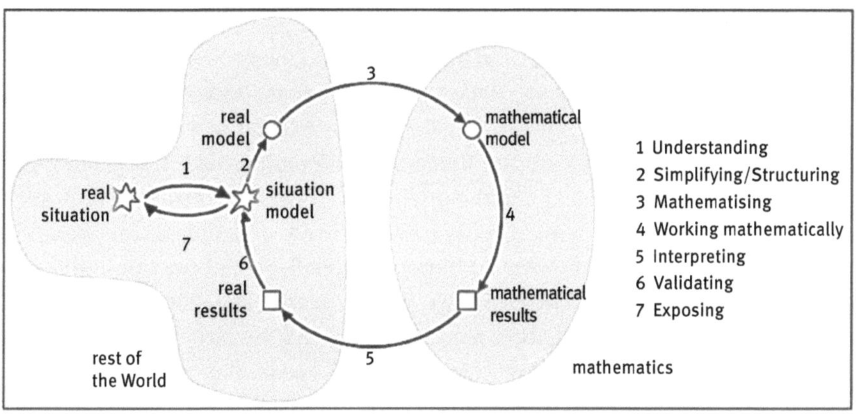

Fig. 1: Modeling cycle (Blum & Leiss, 2007).

At the beginning of the process, the word problem has to be read and (1) the problem situation has to be understood. During the reading process, a situation model, that is, a mental representation of the initial situation described in the text, is generated. In a next step, a (2) real model has to be formed through processes of simplification and structuring. The real model contains only certain features essential for processing the task, meaning that every information is excluded from the situation model that is not necessary for computation. (3) Mathematizing transforms the real model into a mathematical model, which contains relations between relevant elements, numbers, variables, and so forth and is created by incorporating mathematical concepts. This process is not a one-to-one translation between the real model and the mathematical model, but rather a constructive act that depends, among other things, on the objectives and mathematical knowledge of the problem solver (Franke & Ruwisch, 2010). Finally, by (4) working mathematically, that is, performing mathematical operations, the result is calculated, and interpreted with respect to real-life situation (5) and validated with respect to the situation model previously constructed (6). The modeling cycle ends with the (7) exposition and explanation of the solution. Metacognitive strategies "check" the processes for plausibility throughout the entire modeling process, and lead, if necessary, to a restart of the cycle or individual sub-processes.

Empirical evidence suggests that the construction of a situation model is crucial for processing word problems successfully (Hegarty et al., 1995; Kintsch, 1998; Reusser, 1989; Thevenot, 2010).

Depending on the text and the problem solver's individual characteristics such as goals, mathematical knowledge, language and reading skills, and metacognitive abilities, the processing of tasks can be less linear than described above. Difficulties can occur in all sub-processes. In particular, the construction of a situation model and a real model can be a major obstacle for learners (Greefrath et al., 2013; Hegarty et al., 1995; Verschaffel et al., 2000). Therefore, often inadequate strategies are used that usually do not result in a correct solution. Frequently used strategies are the immediate calculation with the given numbers without reading the text and the orientation toward alleged keywords such as "more" or "less," which are directly translated into a mathematical operation that seemingly fits the keyword (addition, respectively subtraction), simultaneously ignoring the particular context. Consequently, key sub-processes such as the construction of a situation model are skipped (Hegarty et al., 1995; Verschaffel et al., 2000). Thus, Hegarty et al. (1995) speak of a "direct translation strategy" (p. 18) in contrast to the "problem model strategy" (p. 18).

However, the processes involved in building a situation model are not the primary concern of mathematical research, which is why conclusions about the situation model concerning word problems remain somewhat vague.

Building a situation model is a decisive process not only for solving word problems but also for reading comprehension in general. As word problems are texts that have to be read and understood, it is crucial to give an in-depth examination of the construction of the situation model from a text comprehension perspective to gain insights into possible reasons for difficulties during problem solving.

There are only a few studies that deal with the relationship between reading comprehension and word problems and even less that focus on the construction of a situation model. Nevertheless, these studies provide evidence that reading competence plays an important part in solving word problems (Boonen et al., 2016; Capraro et al., 2012; Fuchs et al., 2015; Jordan et al., 2003; Leiss et al., 2019; van der Schoot et al., 2009). So far, however, little attention has been paid to the sub-processes involved in the construction of the situation model while reading word problems. Thus, the aim of the present study is to shed light on these sub-processes and their relevance for the solution of word problems.

Hence, in the next section, the processes involved in building a situation model will be closely examined from the psychology of reading point of view, in order to derive indications for factors affecting reading and understanding word problems for the empirical study.

3 Text comprehension

The complex process of reading is composed of several sub-processes. At word level, these include the basic processes of letter and word recognition as well as the acquisition of word meaning. At sentence and text level, syntactic and semantic relationships between words and sentences have to be established and transformed into a coherent meaning of the text by integrating previous knowledge (Christmann & Groeben, 1999; Richter & Christmann, 2009).

Recognizing the meaning of words is undeniably crucial for reading comprehension. However, understanding texts is more than just decoding single words. Current theoretical approaches assume that comprehension requires the construction of multiple mental text representations (Graesser et al., 1994; Kintsch, 1998; Schnotz, 2006). Three main levels of text processing are distinguished:[1] the surface level of the text, the text base, and the level of the situation model (Kintsch, 1998; Schnotz, 2006). As the construction of a situation model is the core process of text comprehension, the following section focusses on these processes without considering basic reading processes.

1 In current research, additionally a communication and a genre level are assumed. Schnotz and Dutke (2004) refer to the latter as meta-levels since they do not represent facts described in the text, but rather characteristics of the situation in which a text is received.

3.1 Construction of a situation model

The *surface level* represents the entire linguistic information of a text, for example, the literal wording and syntactic constructions (Graesser et al., 1997). A mental representation of the text surface arises from processes of recoding and parsing. Semantic processing does not yet take place on this level. The following example of a pseudo-sentence with fake words illustrates these processes: "The ploor proy yegged." When processing this pseudo-sentence, the grapheme-phoneme correspondence is established, and syntactical parsing is carried out. Thus, for instance, even without understanding the meaning, one can easily identify the noun in the sentence ("proy"). On this level of word processing, it is therefore possible to reproduce sentences literally without understanding their meaning. It is assumed that the mental representations of the text surface form the structural basis for higher semantic representations (Schnotz, 2006).

Cognitive theories agree that the semantic information of the text is transferred into a mental representation, the so-called *text base*. The text base represents the semantic content of a text (Kintsch, 1998; van Dijk & Kintsch, 1983). As texts consist of single pieces of semantic information, semantic relations between these pieces must be established in the process of text comprehension, so that a mental network is created. At the level of the text base these connections are established by means of inference that are text-based and do not require any extra-textual knowledge, such as creating co-reference with pronouns (Graesser et al., 1997). The emerging mental network is at best locally coherent because there is no top-down processing integrating the reader's background knowledge on this level.

The use of prior knowledge is essential for global text comprehension since texts do not explicitly represent all the information necessary for a complete reconstruction of the meaning (Schwarz, 2001). The missing references have to be established by the recipient taking into account his or her background knowledge. Hence, the construction of a text base allows only for shallow comprehension without establishing global connections and a deeper meaning and is therefore insufficient for the understanding of texts (Schnotz, 2006). The integration of textual information with the reader's knowledge into a coherent mental representation of the text, a so-called situation model is crucial for text comprehension.

A *situation model* is "a cognitive representation of the events, actions, persons, and in general the situation that a text is about" (van Dijk & Kintsch, 1983: 11). The situation model built on the sentences "Peter has three marbles and Ann has five marbles. How many marbles do they have altogether?" could be the imagination of two kids sitting on the pavement combining their marbles. The construction of a situation model is an integrative and step-by-step process: On the one hand, relations between sentences are established during reading, and on the other hand, the situation model constructed so far forms the context for the interpretation of the next sentence. Thus, an extended new model is created, which in turn provides the context for the interpretation of the next passage of text. During this process, explicit information from the text is integrated with the reader's prior knowledge from long-term memory (Garnham & Oakhill, 1996; Kintsch, 1998). This includes linguistic knowledge as well as knowledge of the world or expert knowledge (Nussbaumer, 1991).

The successful drawing of *inferences* is crucial for the construction of a situation model. Two types of inferences are particularly relevant here: "Local Cohesion Inferences" and "Global Coherence Inferences" (Graesser et al., 2007; Oakhill et al., 2015), the former often referred to as "bridging inferences" (Graesser et al., 2007; Kintsch, 1998). "Local Cohesion Inferences" are used to create references at a local level within and between sentences. Anaphora like pronouns and connectives are important linguistic means for establishing local connections. "Global Coherence Inferences" establish global coherence by connecting larger parts of the text in the situation model. They are more important for text comprehension than "Local Cohesion Inferences." Global inferences are knowledge-based; that is, they cannot be drawn without extra-linguistic prior knowledge. Since inference making depends on a number of different impact factors, different readers might create situations models of the same text that might vary widely in character and complexity (Kintsch, 1998).

3.2 Impacts on the construction of a situation model

Text comprehension is a complex process involving the interaction of different mental sub-processes. Individual differences in mastering these processes lead to difficulties in constructing mental representations. Parameters affecting the construction of a situation model are mainly inference skills and metacognitive skills such as the ability to monitor one's own understanding (Oakhill & Garnham, 1988). These aspects will be examined in more detail below.

3.2.1 Inference skills

Oakhill and Garnham (1988) assume that individual differences in inference making are responsible for differences in the construction of situation models and thus for text comprehension. In a longitudinal study, Oakhill and Cain (2012) investigated factors influencing text comprehension in eleven-year-old children.

They were able to show that the ability to build inferences during reading of seven- and nine-year-olds is the critical predictor of later text comprehension. Decoding ability and implicit syntactic knowledge, however, had no predictive power on text comprehension in eleven-year-olds. The ability to draw inferences during the reading process develops over the course of schooling, independently of the learners' prior knowledge (Barnes et al., 1996; Oakhill & Garnham, 1988). Klicpera and Gasteiger-Klicpera (1993) demonstrated that third graders had difficulties in drawing inferences despite a high level of prior knowledge. The ability to draw necessary inferences during the reading process not only distinguishes younger from older children, but also differentiates decisively between competent and less competent readers. Studies reveal that children with lower text comprehension skills draw fewer inferences than good readers, despite having the same prior knowledge. Difficulties in drawing inferences and thus in constructing a coherent situation model can be the result of a lack of prior knowledge or insufficient retrieval of knowledge, poor processing of anaphors and connectors, limited working memory capacity, and limited vocabulary knowledge.

Vocabulary knowledge is essential for text comprehension and also for drawing inferences. A broad mental lexicon supports the construction of a situation model (Oakhill et al., 2015; Oakhill & Garnham, 1988; Schnotz & Dutke, 2004). However, vocabulary is not the only factor in determining text comprehension.

A large number of studies have found that a possible explanation of individual differences in drawing inferences is rooted in the extent of *prior knowledge* (McNamara et al., 2011). Regardless of their reading skills, children who had the most background knowledge about the topic of the text they had to read scored best in these studies. Children with low-level reading skills but a high level of prior knowledge outperformed children with high-level reading skills but low-level prior knowledge in the number of correct inferences (Adams et al., 1995; Recht & Leslie, 1988; Schneider & Körkel, 1989). Further studies revealed that prior knowledge influences mainly the situation model level and hardly the text base. Readers with little background knowledge do not construct a global situation model during reading, but rather build up a text base (Dutke, 1993; Fincher-Kiefer et al., 1988). This is because many pieces of information remain unconnected in long-term memory. Kintsch (1998) refers to "many different unconnected islands" (p. 232), because the necessary knowledge to establish connections is missing. However, a lack of prior knowledge does not completely explain the problems that occur during inference making. Even if poor readers have background knowledge, they often seem to be unable to retrieve this knowledge sufficiently from long-term memory to draw inferences or fail to integrate prior knowledge into information drawn from the text. Thus, poor readers may not know when to draw inferences (Cain & Oakhill, 1999; Cain et al., 2001).

Anaphora resolution and understanding *connectives* are crucial for drawing local inferences. The ability to resolve pronouns while reading develops during primary school is critical. For example, children up to the age of 10 still find it difficult to use the context when resolving syntactically ambiguous pronouns (Oakhill et al., 2015). In particular, children with lower-level reading skills often have difficulties in resolving pronouns. They frequently tend to interpret the noun closest to the pronoun as an antecedent (Megherbi & Ehrlich, 2005). This may be due to lower working memory capacity. The distance between pronouns and antecedent thus plays a role in pronominal resolution and in the process of comprehension (Daneman & Carpenter, 1980; Oakhill et al., 2015). Since in German not only semantic but also syntactic clues are relevant for the resolution of pronouns, such as the consistency of the grammatical gender, this can also result in difficulties, especially for second-language learners.

Connectives are gradually acquired while children are still of primary school age. Regarding the English language, acquisition patterns reveal that additive connectives are acquired first, followed by temporal, causal, and finally adversative connectives (Bloom et al., 1980). As regards the German language, Dragon et al. (2015) demonstrated throughout different studies that second and third graders, whose first or second language was German, processed temporal and causal connectives with high frequency significantly better than concessive connectives, regardless of the language spoken at home. Differences between children only speaking German at home and those who spoke another language with family members were, if at all, only marginally present and therefore of no practical relevance.

Working memory is relevant for both global and local comprehension. Information that is spread across the entire text must be maintained in memory and integrated into the situation model created so far. At local level, for example, the referential words for pronouns have to be maintained. A lower working memory capacity leads to a reduced ability to link information from the text to prior knowledge which hinders the creation of a coherent global situation model. The capacity and processing efficiency of the working memory develop throughout childhood. In children with below-average text comprehension skills, it has been found that working memory performance is a distinguishing factor as compared to good readers, especially when the information which has to be integrated does not occur in adjacent sentences, but is located further apart (Oakhill et al., 2005).

3.2.2 Metacognition

Children with good and lower-level text comprehension skills differ in the strategies they consider appropriate for reaching specific reading goals. Similarly, below-average readers often have different individual theories about reading compared to good readers (Cain, 1999). "They tend to view reading as a word decoding activity rather than one of meaning-making" (Oakhill & Cain, 2007: 67). Even those having good decoding ability tend to focus on word reading. Thus, their reading aim is more directed toward understanding individual words. The establishment of references and the understanding of the text as a whole are not focused (Garner, 1981; Oakhill & Cain, 2007; Oakhill et al., 2015; Oakhill & Garnham, 1988). Oakhill et al. (2015) accordingly state, "If reading is all about 'getting the words right' then a high standard for comprehension will not be set" (p. 105). Contrary to good readers, children who read a text word by word do not expect to build up a coherent situation model of the text when reading it. Oakhill and Cain (2007) refer to this as a lower "standard for coherence – caring that a text makes sense" (p. 67). A low "standard for coherence" leads to less or no inferring and, thus, no global situation model is established. According to Schnotz (1994), in this case, one can speak of an "illusion of understanding" (p. 208). Therefore, less efficient readers are not aware of their comprehension problems at all. Thus, they regard further activities aimed at optimizing comprehension as pointless, although a coherent situation model has not yet been established. Consequently, there is hardly any monitoring of the reading process; otherwise, problems of understanding would be noticed (Cain, 1999; Schnotz, 1994). However, constant monitoring of one's own reading process is critical to ensure text comprehension. New information has to be compared with the previously formed situation model and checked for inconsistencies and plausibility.

4 The current study

Empirical studies demonstrate that the construction of a situation model is crucial for solving word problems. However, the situation model is not the focus of current research when it comes to students' difficulties in solving word problems. As word problems are texts that have to be read and understood, in line with the theoretical background presented, it is to be assumed that the aspects relevant for the comprehension of non-mathematical texts also have an effect on the construction of a situation model when reading mathematical texts. It is important to examine more closely, which reading skills contributing to the construction of a situation model also contribute to solving mathematical word problems and how these skills are related. This knowledge is vital to shed light on non-mathematical aspects that make it challenging to solve word problems and to purposefully promote "mathematical" reading comprehension. In order to do so, the study at hand includes measures of different skills of text comprehension relevant to the construction of a situation model, rather than a global assessment of reading competence.

The study addresses the following questions: Does the ability to construct a situation model influence the solution of word problems? Do inference skills make a contribution to the solution process of word problems? Are comprehension monitoring and standard for coherence also important parameters for dealing with word problems? How do mathematical processes and reading processes interact when solving word problems?

The relationships examined here should apply at least to children without special educational needs, regardless of their language or reading skills since the cognitive processes behind reading competence are not fundamentally different in first- and second-language learners. Therefore, there are no differences expected to be observed between students with German as first and German as second language. That is not to assume that there are no differences in the level of reading competence and the solving of word problems between those students, but these are not further examined in the present study, as this study focuses on relationships between reading subskills and solving word problems.

4.1 Method

The design of the study is correlational. Nevertheless, some hypotheses about the direction of relations were formulated and being tested by means of a structural equation model. Structural equation modeling is a statistical procedure that incorporates the relationship of latent variables and confirms hypotheses about the validity of measurements. This particular model is based on the theoretical findings and assumptions described in the overview above. The goodness of fit of the parameters was decisive for the evaluation of the model. A total of three variants of the model were calculated, one overall model with all participants and one model each for L1 and L2 learners in order to compare these subgroups.

The reliability of the different measures was assessed by calculating Cronbach's alpha, whenever this measure was applicable. The present study is part of a broader project on the relationship between language and mathematical word problems (Stephany, 2018).

4.2 Participants

A total of 381 fourth graders from seven schools participated in the study. Children with special educational needs, diagnosed dyslexia or dyscalculia as well as children who had only been learning German for a few weeks were excluded from the analysis. The resulting sample comprised 352 students aged between eight and eleven years ($M = 9.05$; $SD = 0.47$), 47.9% were girls, 39.3% of the children spoke German as their second language. The schools were located in a range of lower- to middle-class urban areas.

4.3 Assessments and material

4.3.1 Reading skills

Some tasks were used from standardized tests and others were developed to measure specific components of comprehension skills notably necessary for the construction of a situation model in the three following areas: inference skills, comprehension monitoring, and the standard for coherence. Additionally, word-reading ability was measured.

4.3.1.1 Inference skills

The ability to establish local and global references was measured with the ELFE 1–6 subtest "text comprehension" (Lenhard & Schneider, 2006). Children read short stories, and for each one they had to select one out of four statements that fits the text the best. It was evaluated how well children established anaphoric references or build inferences. The reliability of the subtest "text comprehension" was $\alpha = .76$. A factor value "inference skill" was calculated with both variables.

4.3.1.2 Comprehension monitoring

An inconsistency detection task was used to measure comprehension monitoring. Children read two short stories containing internal inconsistencies. One line at the beginning and at the end of each of the stories contained contradictory information (Oakhill et al., 2005). Children were asked to underline any parts in the stories that did not make sense. Inconsistencies between parts of texts can be detected only by a continuous comparison with the situation model constructed so far. Correlation between the inconsistency items in both stories was $r = .30$ ($p < .001$).

4.3.1.3 Standard for coherence

Based on the considerations of Garner (1981), Oakhill and Cain (2007), and Oakhill and Garnham (1988), a questionnaire was constructed in order to find out if children tend to be "word readers" with a low "standard for coherence." The term "word readers" is used for children who consider reading to be exclusively a decoding activity, that is, who read a text word by word. "[They] manage written language as bits and pieces, not as textual wholes" (Garner, 1981: 161). A questionnaire with seven items was developed. Each item consisted of two conflicting statements: one reflected the strategy of word by word reading, and one statement focused on text comprehension. Some of these statements were used several times in different combinations. The polarity of the statements was randomized. The children were asked to choose one of the statements. All items were assigned to three superordinate questions: "What makes a good reader?" "What makes a text difficult to read?" and "When are you satisfied with yourself when reading?" Fig. 2 shows an excerpt from the questionnaire. Cronbach's alpha for all seven items was $\alpha = .53$. This is only moderate but sufficient for scientific purposes.

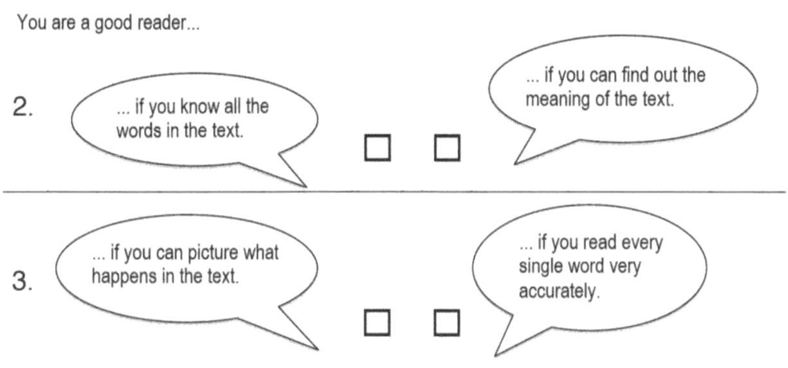

Fig. 2: Excerpt from the reading questionnaire.

4.3.1.4 Word-reading ability

Word-reading ability was controlled in the study. All children completed the subtest "word comprehension" of the German reading comprehension test ELFE 1–6 (Lenhard & Schneider, 2006). Children had to underline the word that matched a picture. The reliability of the test was high ($\alpha = .94$).

4.3.2 Mathematical skills

4.3.2.1 Mathematical ability

The subtest "Arithmetic" of the German Mathematics Test DEMAT 3+ (Roick et al., 2004) was used to assess arithmetic competence. The reliability of the subtest "arithmetic" was $\alpha = .77$. In addition, the grades of the participating students in mathematics were elicited.

4.3.2.2 Mathematical reading competence

To assess if children tend to use a direct-translation strategy in contrast to a problem-model strategy a test consisting of three word problems was developed, which contained additional numerical data that was irrelevant for answering the questions. The children first had to circle the numbers they needed for the calculations and then they had to solve the tasks. Cronbach's alpha was $\alpha = .89$. To calculate a value "mathematical reading competence" the circled numbers and the numbers actually used for calculating were combined with factor analysis.

4.3.2.3 Word problems

To assess the ability to solve word problems, four word problems with two different grades of text coherence were developed ("koala," "tortoise," "swallow," "ant eater"). This approach controls for the influence of text coherence on the development of a situation model. All tasks used the topic "records in the animal world." Fig. 3 shows one version of the word problem "tortoise."

Riesenschildkröten sind die ältesten Tiere der Welt. Sie werden häufig über 200 Jahre alt. Die älteste bekannte Riesenschildkröte Adwaita lebte 140 Jahre in einem indischen Zoo. Sie wurde aber erst im Alter von 116 Jahren gefangen. Wie alt wurde sie?

Giant tortoises are the oldest animals in the world. They often grow over 200 years old. The oldest known giant tortoise, Adwaita had lived for 140 years in an Indian zoo. However, it was captured only at the age of 116 years. How old did it get?

Fig. 3: Word problem "tortoise" in German and translated into English.

With regard to their mathematical content, all word problems corresponded to the curricular requirements in the field of elementary arithmetic for the fourth grade. When creating the tasks, it was ensured that in no case a "direct translation strategy," that is, an exclusive orientation toward numbers and supposed key words, could lead to the correct solution.

4.3.3 Situation model

A central aspect of the present study was the assessment of the situation model. The few studies available in mathematics assess the construction of a situation model indirectly either by rating the tasks (Leiss et al., 2010) or by using the correct solution as an indicator. In contrast to them, in this study the situation model was measured via images and statements matching the word problem. Thus, the situation model was measured independently of mathematical processes.

In order to obtain a value for the construction of a situation model, three images and four statements for each word problem were developed. One image represented the global topic of the word problem (attractor); two images served as distractors. Fig. 4 shows the images of the word problem "tortoise." Children were asked to select the pictures depicting the word problem. In order to select the corresponding image, the construction of a situation model of the task text is required, which needs to be validated against the content of the images (Schnotz & Dutke, 2004).

Furthermore, children were asked to decide whether statements about the word problem were true or false. For instance, the correct answer required the inference of local ("The giant tortoise Adwaita was the oldest animal in the world.") and global ("The time in freedom and the time in the zoo altogether add up to the age of the tortoise Adwaita.") connections in the text of the task. The correct evaluation of the global statement required the construction of a coherent situation model of the word problem. To calculate a value "situation model" for each word problem students' answers on images and statements were summarized with the help of a factor analysis.

4.4 Procedure

Data collection took place in seven schools after the summer break in a classroom assessment. The tests were conducted by the author on different days. The total assessment time per student was 75 min. Each student completed all four word problems, two high and two low coherent versions. Images and statements had to be processed after solving each word problem without viewing the

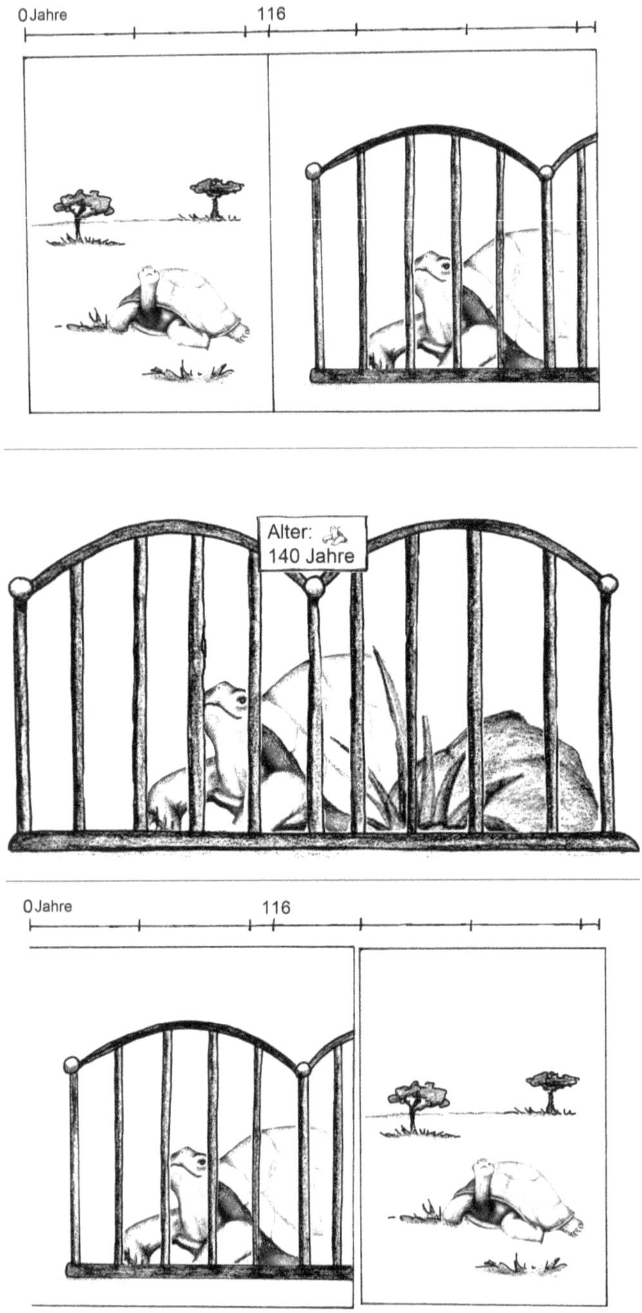

Fig. 4: Image to measure the situation model "tortoise.".

text. Thus, the situation model was supposed to be captured at the time of task processing without being distorted by re-reading the text. In order to rule out the possibility that word problems are not solved solely due to a lack of prior knowledge on the topic of the tasks, this likelihood was controlled by activating and building vocabulary and knowledge on the topic.

5 Results

The hypothesized relationships among variables were evaluated by means of structural equation modeling. Since not all variables could be transformed into a latent construct, manifest variables are also represented in the model. Fig. 5 depicts the model for all participants. The model's paths and path directions were derived from a variety of studies and theoretical expectations. To evaluate the fit of the model the root-mean-square error of approximation (RMSEA) was used. The results indicate that the fit of the proposed model was good RMSEA = .068; Chi-square was 146.15 (df = 56), $p < .001$. These results suggest acceptance of the proposed model as the most parsimonious. The standardized and unstandardized regression weights and the significance levels of these variables are depicted in Tab. 1.

5.1 Confirmatory part of the model

Before evaluating the structure model, it is investigated whether the manifest variables make an actual contribution to the respective latent construct. The "situation model" was assessed through the pictures and statements concerning the four word problems. These variables (pictures and statements) were combined into a single factor entering the structural equation model. The confirmatory part of the structural equation model reveals a moderate fit of these four variables to the latent construct "situation model." The loads of the single variables ranged between .24 and .74. The variables measuring "mathematical performance," arithmetical skills and math grade, loaded between .73 and .86 and thus showed a good fit. Although the factor was constructed by mathematical skills, it is named "performance" since in the modeling cycle, problem solvers have to perform math based on their skills. The latent factor "solution" measured by the variables correct solution path and result (.97 each) also demonstrates a good model fit.

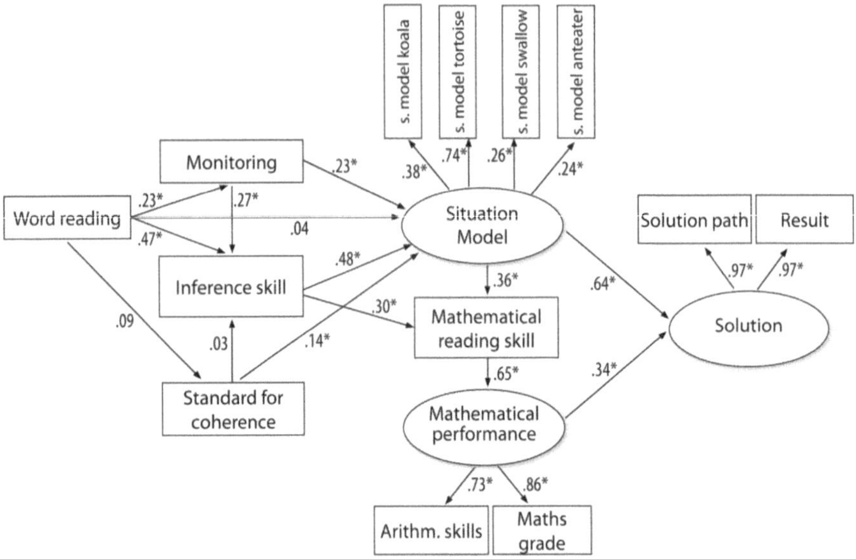

Fig. 5: *Structural equation model for the processing of word problems – all students.* Rectangles represent measured variables; circles represent latent factors. Measurement errors and structural errors were removed from the figure. Significant and non-significant paths are displayed; * = path coefficient $p < .05$; s. model = situation model.

Tab. 1: Standardized and unstandardized regression weights and their levels of significance.

			Unstandardized regression weight	Standardized regression weight	S.E.	C.R.	p
Word reading	→	Monitoring	0.19	.23	0.042	4.428	<.001
Word reading	→	Standard for coherence	0.10	.09	0.057	1.755	.079
Word reading	→	Inference skill	0.50	.47	0.048	10.235	<.001
Monitoring	→	Inference skill	0.36	.27	0.061	5.943	<.001
Standard for coherence	→	Inference skill	0.03	.03	0.044	0.656	.512
Standard for coherence	→	Situation model	0.10	.14	0.040	2.574	.010
Monitoring	→	Situation model	0.23	.23	0.059	3.821	<.001
Word reading	→	Situation model	0.03	.04	0.050	0.643	.520

Tab. 1 (continued)

			Unstandardized regression weight	Standardized regression weight	S.E.	C.R.	p
Inference skill	→	Situation model	0.35	.48	0.053	6.670	<.001
Inference skill	→	Mathematical reading	0.28	.30	0.071	3.926	<.001
Situation model	→	Mathematical reading	0.46	.36	0.117	3.916	<.001
Mathematical reading	→	Mathematical performance	0.54	.65	0.048	11.286	<.001
Mathematical performance	→	Solution	0.56	.34	0.095	5.913	<.001
Situation model	→	Solution	1.12	.64	0.128	8.727	<.001

Notes. S.E. = Standard error, C.R. = critical ratio.

5.2 Structural part of the model

In consideration of the structural part of the model, there was a high connection between the construction of a situation model and the solution ($\beta = .64$). The relationship between mathematical performance and the solution was considerably smaller ($\beta = .34$). In this model, the ability to solve word problems was composed of both the skill to construct a situation model and mathematical performance. However, the construction of the situation model was a much better predictor (and prerequisite) for correct solutions. Although the variable solution model explained around 41% of the total variance in solving word problems, mathematical performance explained only 11.6% of it.

The differences in the ability to build a situation model when solving word problems were explained by 42% of the variance in inference making, monitoring the reading process and the standard for coherence. Furthermore, monitoring the reading process had a moderate, yet significant, impact on inference making ($\beta = .27$) and a direct path to the situation model ($\beta = .23$). Higher-level reading processes are, therefore, responsible for building an adequate situation model. Word reading had an impact on monitoring ($\beta = .23$) and inference making ($\beta = .47$). The direct path from word reading to the latent factor situation model was somewhat weaker ($\beta = .04$, n.s.) than the path from inference making – situation model ($\beta = .48$), meaning higher-order reading skills have a higher impact on the construction of a situation model than basic reading skills, at least when reading word problems.

According to the model, 35% of the variance in mathematical reading could be explained by inference skills ($\beta = .30$) and the situation model ($\beta = .36$). Mathematical reading, on the other hand, correlated strongly with mathematical performance ($\beta = .65$), and thus explained 42% of the variance in mathematical performance.

5.3 Differential aspects of learning German as first or as second language

To analyze the differential aspects between L1 and L2 learners, two different models were calculated. It was assumed that the interconnections in the L2 model are not substantially different from those in the L1 model, because the cognitive processes underlying reading competence are not fundamentally different in first- and second-language learners. A closer look at both models (Fig. 6) shows that this assumption can largely be supported since only marginal differences are revealed.

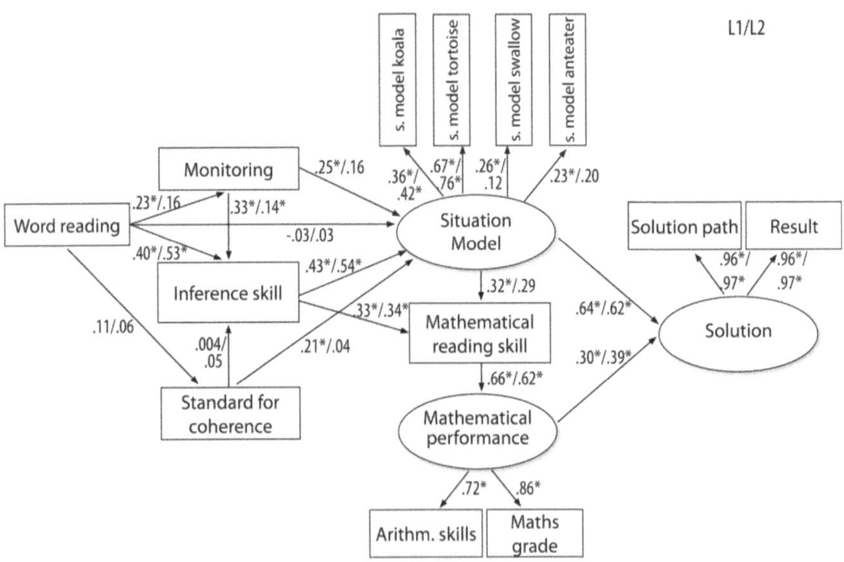

Fig. 6: *Structural equation model for the processing of word problems – German as a first and as a second language.* Rectangles represent measured variables; circles represent latent factors; measurement errors and structural errors were removed from the figure. Significant and non-significant paths are displayed; * = path coefficient $p < .05$; s. model = situation model; values show L1/L2.

The core idea of this paper referred to the connection between the construction of a situation model and the solution of word problems. The part of the model that refers to this relationship shows no substantial differences between L1 and L2 learners. However, minor variations in the overall model were found concerning the variables involved in the construction of a situation model. It was found that the ability to decode words had the same minor effect on the construction of a situation model in L1 and L2 learners. The influence of inference making and monitoring the construction of a situation model was slightly different in both groups. However, these differences are marginal and are not of any practical relevance. The only substantial difference was found in the connections between the "standard for coherence" and the "situation model" among L1 and L2 learners (higher correlation for L2 than for L1). This could be due to only a moderate reliability of the questionnaire or a poor understanding of the items.

It was revealed that the effect of the variables "monitoring," "inference making," and "standard for coherence" on the construction of an situation model are composed slightly differently for L1 and L2 learners. However, this is hardly of any practical relevance. The variance of the situation model explained by the three variables monitoring, inference making, and standard for coherence was about 39% both for L1 and L2 learners. Hence, the joint influence of these variables was the same for both groups. In conclusion, the relevance of an adequate situation model for solving word problems is equally high for both groups.

Both SEM showed a good fit of the RMSEA ($RMSEA_{L1} = .077$, $n = 214$; $RMSEA_{L2} = .045$, $n = 138$). Chi-square was significant for L1 but not for L2 ($Chi\text{-}square_{L1} = 122.39$ ($df = 56$), $p < .001$; $Chi\text{-}square_{L2} = 1{,}024$ ($df = 56$), $p = .085$). Due to the smaller number of L2 participants in contrast to the L1 model, more paths do not reach significance.

6 Discussion

In this contribution, word problems were presented as texts that have to be read. To solve a word problem, students not only need to perform the necessary mathematical operations, they also need to read and understand the text of the task. Therefore, it was assumed that one reason for students' difficulties with word problems relates to reading processes, in particular to processes involved in the construction of a situation model. In recent years, an increasing number of studies focused on language-related reasons for complications with word problems. In doing so the main focus was on the task text itself and its linguistic features. Only a few studies examined students' reading competences and even fewer focused on the problem solvers' ability to construct a situation model, in order to explain difficulties and give guidance to foster the solving of word problems in the classroom. The present study sheds light on the sub-processes necessary for building a situation model known from psychological reading research.

With the help of various standardized as well as specifically developed measurement tools, a structural equation model was created to provide answers to four questions concerning the role of the situation model and the factors involved in its construction in the context of mathematical word problems.

It has been revealed that the situation model has a strong direct effect on the solution of word problems. This effect is even much stronger than the influence of mathematical performance on the correct solution. This indicates that not only mathematical competence but also the construction of a situation model is crucial for the solution of mathematical word problems. If no situation model is constructed, as in the "direct translation strategy," it is usually not possible to solve the task. Since the present model confirms the relevance of the situation model also for mathematical word problems, it is even more important to closely examine the factors influencing the construction of a situation model. The ability to draw inferences has the strongest influence; that is, children who draw few inferences also construct a situation model less well, and the frequency of solutions is correspondingly lower.

Furthermore, reading-related metacognitive strategies were considered. The model shows a significant influence of comprehension monitoring and the "standard for coherence" on the construction of a situation model. The latter turns out to be lower than assumed. However, this may also be related to the merely average reliability of the measurement tool developed for the study and would have to be re-examined in a further investigation. Still, there is an influence of both variables. For instance, children who monitor their reading process less and tend to read texts word by word rather than looking at the text as a whole are less successful in building up a situation model even regarding mathematical word problems. A low standard for building a mental coherence structure, therefore, also plays a role in solving word problems.

The results of the study demonstrate that especially higher-order processes of inference making and metacognition have an effect on the construction of a situation model and thus on the solution process. Competence in word reading, on the other hand, only indirectly affects the situation model via the ability to draw inferences. In the model, the situation model also affects mathematical reading competence: Only when a situation model has been built are adequate solution strategies applied instead of relying on substitute strategies, such as focusing on numbers without considering the context.

Overall, it can be stated that the processes of understanding non-mathematical texts and mathematical word problems are comparable. Influencing factors that play a role in reading texts and have been examined in detail in this study are also relevant for dealing with word problems. Inadequate strategies, in which the situation model is omitted, such as in the case of the direct translation strategy, might be a student's attempt to handle the lack of understanding of a task text. This indicates that reading promotion must also play a part in mathematics lessons. Language classes can do this only to a limited extent since word problems are a subject-specific genre with its specific characteristics. Accordingly, Leiss et al. (2010) refer to mathematical reading competence and explicitly demand its promotion in mathematics lessons. The results of the present study can provide starting points for such support, suggesting that the promotion of reading in mathematics lessons should start with processes of inference making. Particularly with mathematical word problems, the establishment of references on a local and global level is essential. For example, underlining keywords is not helpful if references cannot be established at all. Therefore, support should include, for instance, exercises for making references in mathematical texts. In this study prior knowledge was controlled for, so no conclusion can be made about its influence. Nevertheless, it seems to be reasonable to put word problems in mathematics lessons in a common thematic context and to build up the necessary prior knowledge and vocabulary before working on the actual task. A stronger focus on comprehension monitoring should also be part of the teaching of mathematics, for example, by detecting inconsistencies in task texts. Further studies must examine to what extent such methods are effective and whether they particularly support students with reading difficulties in solving word problems.

As expected, the observed relationships do not differ substantially between L1 and L2 learners. The ability to draw inferences and the underlying processes, as well as standard for coherence or comprehension monitoring, are language independent. Nonetheless, individual difficulties of some children in this particular area may still be due to low basic reading skills, limited vocabulary, or unfamiliarity with certain connectives. However, this can affect L1 learners as well as L2 learners. Support targeting inference skills should therefore be effective for all students regardless of their first language.

References

Abedi, Jamal & Lord, Carol (2001): The language factor in mathematics tests. *Applied Measurement in Education* 14 (3), 219–234.

Adams, Beverly C., Bell, Laura C. & Perfetti, Charles A. (1995): A trading relationship between reading skill and domain knowledge in children's text comprehension. *Discourse Processes* 20 (3), 307–323.

Barnes, Marcia A., Dennis, Maureen & Haefele-Kalvaitis, Jennifer (1996): The effects of knowledge availability and knowledge accessibility on coherence and elaborative inferencing in children from six to fifteen years of age. *Journal of Experimental Child Psychology* 61, 216–241.

Bloom, Lois, Lahey, Margaret, Hood, Lois, Lifter, Karin & Fiess, Kathleen (1980): Complex sentences: Acquisition of syntactic connectives and the semantic relations they encode. *Journal of Child Language* 7, 235–261.

Blum, Werner & Leiss, Dominik (2007): How do students and teachers deal with mathematical modelling problems? The example sugarloaf. In Haines, Christopher, Galbraith, Peter, Blum, Werner, Khan, Sanowar (eds.): *Mathematical Modelling: Education, Engineering, and Economics*. Chichester: Horwood, 222–231.

Boonen, Anton J. H., de Koning, Björn B., Jolles, Jelle & Van der Schoot, Menno (2016): Word problem solving in contemporary math education: A plea for reading comprehension skills training. *Frontiers in Psychology* 7, 191.

Cain, Kate (1999): Ways of reading: How knowledge and use of strategies are related to reading comprehension. *British Journal of Developmental Psychology* 17, 295–312.

Cain, Kate & Oakhill, Jane (1999): Inference making ability and its relation to comprehension failure in young children. *Reading and Writing: An Interdiciplinary Journal* 11, 489–503.

Cain, Kate, Oakhill, Jane, Barnes, Marcia A. & Bryant, Peter E (2001): Comprehension skill, inference-making ability, and their relation to knowledge. *Memory & Cognition* 29 (6), 850–859.

Capraro, Robert M., Capraro, Mary M. & Rupley, William H (2012): Reading-enhanced word problem solving: a theoretical model. *European Journal of Psychology of Education* 27, 91–114.

Christmann, U. & Groeben, N. (1999): Psychologie des lesens. *Handbuch lesen* 2, 145–223.

Daneman, Meredith & Carpenter, Patricia A. (1980): Individual differences in working memory and reading. *Journal of Verbal Learning and Verbal Behaviour* 19, 450–466.

Dijk, Teun A. van & Kintsch, Walter (1983): *Strategies of Discourse Comprehension*. New York: Academic Press.

Dragon, Nina, Berendes, Karin, Weinert, Sabine, Heppt, Birgit & Stanat, Petra (2015): Ignorieren Grundschulkinder Konnektoren? – Untersuchung einer bildungssprachlichen Komponente. *Zeitschrift für Erziehungswissenschaft* 18, 803–825.

Duarte, Joana, Gogolin, Ingrid & Kaiser, Gabriele (2011): Sprachlich bedingte Schwierigkeiten von mehrsprachigen Schülerinnen und Schülern bei Textaufgaben. In Prediger, Susanne, Özdil, Erkan (eds.): *Mathematiklernen unter Bedingungen der Mehrsprachigkeit*. Münster: Waxmann, 35–53.

Dutke, Stephan (1993): Mentale Modelle beim Erinnern sprachlich beschriebener räumlicher Anordnungen. Zur Interaktion von Gedächtnisschemata und Textrepräsentation. *Zeitschrift für experimentelle und angewandte Psychologie* 40, 44–71.

Fincher-Kiefer, Rebecca, Post, Timothy A., Greene, Terry R. & Voss, James F (1988): On the role of prior knowledge and task demands in the processing of text. *Journal of Memory and Language* 27, 416–428.
Franke, Marianne & Ruwisch, Silke (2010): *Didaktik des Sachrechnens in der Grundschule*. Heidelberg: Spektrum Akademischer Verlag.
Fuchs, Lynn S., Fuchs, Douglas, Compton, Donald L., Hamlett, Carol L. & Wang, Amber Y (2015): Is word-problem solving a form of text comprehension? *Scientific Study of Reading* 19 (3), 204–223.
Garner, Ruth (1981): Monitoring of passage inconsistency among poor comprehenders: A preliminary test of the "piecemeal processing" explanation. *The Journal of Educational Research* 74, 159–162.
Garnham, Alan & Oakhill, Jane (1996): The mental models theory of language comprehension. In Britton, Bruce K., Graesser, Arthur C. (eds.): *Models of Understanding Text*. Mahwah, N. J.: Erlbaum, 313–339.
Graesser, Arthur C., Louwerse, Max M., McNamara, Danielle S., Olney, Andrew, Cai, Zhiqiang & Mitchell, Heather H (2007): Inference generation and cohesion in the construction of situation models: Some connections with computational lingusitics. In Schmalhofer, Franz, Perfetti, Charles (eds.): *Higher Level Processes in the Brain: Inference and Comprehension Processes*. Mahwah, NJ: Erlbaum, 289–310.
Graesser, Arthur C., Millis, Keith K. & Zwaan, Rolf A (1997): Discourse comprehension. *Annual Review of Psychology* 48, 163–189.
Graesser, Arthur C., Singer, Murray & Trabasso, Tom (1994): Constructing inferences during narrative text comprehension. *Psychological Review* 101 (3), 371–395.
Greefrath, Gilbert, Kaiser, Gabriele, Blum, Werner & Borromeo Ferri, Rita (2013): Mathematisches Modellieren – Eine Einführung in theoretische und didaktische Hintergründe. In Ferri, Rita Borromeo, Greefrath, Gilbert, Kaiser, Gabriele (eds.): *Mathematisches Modellieren für Schule und Hochschule. Theoretische und didaktische Hintergründe*. Wiesbaden: Springer Spektrum, 11–37.
Gürsoy, Erkan (2016): *Kohäsion und Kohärenz in mathematischen Prüfungstexten türkisch-deutschsprachiger Schülerinnen und Schüler. Eine multiperspektivische Untersuchung*. Münster: Waxmann.
Haag, Nicole, Heppt, Birgit, Roppelt, Alexander & Stanat, Petra (2015): Linguistic simplification of mathematics items: Effects für language minority students in Germany. *European Journal of Psychology of Education* 30 (2), 145–167.
Hegarty, Mary, Mayer, Richard E. & Monk, Christopher A (1995): Comprehension of arithmetic word problems: A comparison of successful and unsuccessful problem solvers. *Journal of Educational Psychology* 87 (1), 18–32.
Heinze, Aiso, Herwartz-Emden, Leonie, Braun, Cornelia & Reiss, Kristina (2011): Die Rolle von Kenntnissen der Unterrichtssprache beim Mathematiklernen. In Prediger, Susanne, Özdil, Erkan (eds.): *Mathematiklernen unter Bedingungen der Mehrsprachigkeit*. Münster: Waxmann, 11–33.
Jordan, Nancy C., Hanich, Laurie B. & Kaplan, David (2003): A longitudinal study of mathematical competencies in children with specific mathematics difficulties versus children with comorbid mathematics and reading difficulties. *Child Development* 74, 834–850.
Kintsch, Walter (1998): *Comprehension. A Paradigm for Cognition*. Cambridge: Cambridge University Press.

Klicpera, Christian & Gasteiger-Klicpera, Barbara (1993): *Lesen und Schreiben – Entwicklung und Schwierigkeiten. Die Wiener Längsschnittuntersuchungen über die Entwicklung, den Verlauf und die Ursachen von Lese- und Schreibschwierigkeiten in der Pflichtschulzeit.* Bern: Huber.

Leiss, Dominik, Plath, Jennifer & Schwippert, Knut (2019): Language and mathematics – key factors influencing the comprehension process in relity-based tasks. *Mathematical Thinking and Learning* 21 (2), 131–153.

Leiss, Dominik, Schukajlow, Stanislaw, Blum, Werner, Messner, Rudolf & Pekrun, Reinhard (2010): The role of the situation model in mathematical modelling: Task analyses, student competencies, and teacher interventions. *Journal für Mathematik-Didaktik* 31 (1), 119–141.

Lenhard, Wolfgang & Schneider, Wolfgang (2006): *ELFE 1–6. Ein Leseverständnistest für Erst- bis Sechstklässler. Manual.* Göttingen: Hogrefe.

Martiniello, Maria (2008): Language and the performance of English-language learners in math word problems. *Harvard Educational Review* 78 (2), 333–368.

McNamara, Danielle, Ozuru, Yasuhiro & Floyd, Randy G. (2011): Comprehension challenges in the fourth grade: The roles of text cohesion, text genre, and readers' prior knowledge. *International Electronic Journal of Elementary Education* 4 (1), 229–257.

Megherbi, Hakima & Ehrlich, Marie-France (2005): Language impairment in less skilled comprehenders: The online processing of anaphoric pronouns in a listening situation. *Reading and Writing* 18, 715–753.

Nussbaumer, Markus (1991): *Was Texte sind und wie sie sein sollen: Ansätze zu einer sprachwissenschaftlichen Begründung eines Kriterienrasters zur Beurteilung von schriftlichen Schülertexten.* Tübingen: Max Niemeyer.

Oakhill, Jane & Cain, Kate (2007): Issues of causality in children's reading comprehension. In McNamara, Danielle S. (ed.): *Reading Comprehension Strategies: Theories, Interventions, and Technologies.* Mahwah, NJ: Erlbaum, 47–72.

Oakhill, Jane & Cain, Kate (2012): The precursors of reading ability in young readers: Evidence from a four-year longitudinal study. *Scientific Studies of Reading* 16 (2), 91–121.

Oakhill, Jane, Cain, Kate & Elbro, Carsten (2015): *Understanding and Teaching Reading Comprehension. A Handbook.* London: Routledge.

Oakhill, Jane & Garnham, Alan (1988): *Becoming a Skilled Reader.* Oxford: Blackwell.

Oakhill, Jane, Hartt, Joanne & Samols, Deborah (2005): Levels of comprehension monitoring and working memory in good and poor comprehenders. *Reading and Writing* 18, 657–686.

Prediger, Susanne, Wilhelm, Nadine, Büchter, Andreas, Gürsoy, Erkan & Benholz, Claudia (2015): Sprachkompetenz und Mathematikleistung – Empirische Untersuchung sprachlich bedingter Hürden in den Zentralen Prüfungen 10. *Journal für Mathematik-Didaktik* 36 (1), 77–104.

Recht, Donna R. & Leslie, Lauren (1988): Effect of prior knowledge on good and poor readers. *Journal of Educational Psychology* 80, 16–20.

Reusser, Kurt (1989): *Vom Text zur Situation zur Gleichung. Kognitive Simulation von Sprachverständnis und Mathematisierung beim Lösen von Textaufgaben.* Zürich: Universität Zürich.

Richter, Tobias & Christmann, Ursula (2009): Lesekompetenz: Prozessebenen und interindividuelle Unterschiede. In Groeben, Norbert, Hurrelmann, Bettina (eds.): *Lesekompetenz. Bedingungen, Dimensionen, Funktionen.* Aufl. (Vol. 3). Weinheim: Juventa, 25–58.

Roick, Thorsten, Dietmar, Gölitz. & Hasselhorn, Marcus (2004): *DEMAT 3+. Deutscher Mathematiktest für dritte Klassen*. Göttingen: Beltz.

Schneider, Wolfgang & Körkel, Joachim (1989): The knowledge base and text recall: Evidence from a short-term longitudinal study. *Contemporary Educational Psychology* 14, 382–393.

Schnotz, Wolfgang (1994): *Aufbau von Wissensstrukturen: Untersuchungen zur Kohärenzbildung beim Wissenserwerb mit Texten*. Weinheim: Beltz.

Schnotz, Wolfgang (2006): Was geschieht im Kopf des Lesers? Mentale Konstruktionsprozesse beim Textverstehen aus der Sicht der Psychologie und der kognitiven Linguistik. In Blühdorn, Hardarik, Breindl, Eva, Waßner, Ulrich H. (Hrsg.): *Text – Verstehen. Grammatik und darüber hinaus*. Berlin: de Gruyter, 222–238.

Schnotz, Wolfgang & Dutke, Stephan (2004): Kognitionspsychologische Grundlagen der Lesekompetenz. Mehrebenenverarbeitung anhand multipler Informationsquellen. In Schiefele, Ulrich, Artelt, Cordula, Schneider, Wolfgang, Stanat, Petra (Hrsg.): *Struktur, Entwicklung und Förderung von Lesekompetenz. Vertiefende Analysen im Rahmen von PISA 2000*. Wiesbaden: VS Verlag für Sozialwissenschaften, 61–100.

Schwarz, Monika (2001): Kohärenz: Materielle Spuren eines mentalen Phänomens. In Bräunlich, Margret, Neuber, Baldur, Rues, Beate (Hrsg.): *Gesprochene Sprache – transdisziplinär. Festschrift zum 65. Geburtstag von Gottfried Meinhold*. Frankfurt: Peter Lang, 151–159.

Stephany, Sabine (2018): *Sprache und Sprache und mathematische Textaufgaben. Eine empirische Untersuchung zu leser- und textseitigen sprachlichen Einflussfaktoren auf den Lösungsprozess*. Münster: Waxmann.

Thevenot, Catherine (2010): Arithmetic word problem solving: Evidence for the construction of a mental model. *Acta Psychologica* 133 (1), 90–95.

Van der Schoot, Menno, Annemieke H., Bakker Arkema, Horsley, Tako M., Lieshout, Van & Ernest, C. D. M (2009): The concistency effect depends on markedness in less successful but not successful problem solvers: An eye movement study in primary school children. *Contemporary Educational Psychology* 34 (1), 58–66.

Verschaffel, Lieven, Greer, Brian & de Corte, Eric (2000): *Making Sense of Word Problems*. Lisse: Swets & Zeitlinger.

Farran Mackay, Jantien Smit, Arthur Bakker, and Ronald Keijzer

Supporting teachers to scaffold students' language for mathematical learning

1 Introduction

Teachers often are unaware of language issues and avoid linguistic challenges in their classrooms to focus on mathematics (e.g., Van Eerde et al., 2008). Specifically, teachers typically do not attend to the language students need for mathematical learning, and rarely know how to support the development of subject-specific language required for mathematical learning (e.g., Hajer & Norén, 2017). Yet students, especially those with low language proficiency, require support from teachers within this subject because shortcomings in subject-specific language can impede their development of mathematical understanding (Moschkovich, 2010). Despite the importance of improving language-responsive teaching, there is a profound lack of opportunity for teachers to develop the required teaching practices, especially in mathematics (Essien et al., 2016). The required teaching practices integrate language learning and mathematics in a domain-specific way (Van Eerde & Hajer, 2009). Although there are some insights into the professional development of secondary school teachers (e.g., Prediger, 2019), relatively little is known about how to support primary school teachers in realizing language-responsive teaching. This chapter provides insights into how teachers can be supported within a professional development program (PDP), focusing on genre awareness and scaffolding students' language for mathematical learning.

2 Theoretical background

To specify our approach to supporting teachers in a PDP, we first address what the literature considers essential learning goals for teachers (2.1) before we characterize our approach to designing and evaluating our PDP (2.2).

https://doi.org/10.1515/9783110661941-020

2.1 Learning goals for teachers: Genre awareness and scaffolding language

Key to participating in mathematical discourse is access to the subject-specific language required for learning mathematics (e.g., Prediger, 2019). Each domain has not only its own vocabulary but also phrases which learners need to recognize and adequately use to participate successfully (Moschkovich, 2010; Schleppegrell, 2007). Consequently, a PDP should help teachers pay attention to the required language and establish an environment that allows learners to interact and communicate at a mathematical level (e.g., Lampert & Cobb, 2003). Genre pedagogy is a promising approach to explicitly address the language required for learning in that it provides learners with metalinguistic knowledge about how (both spoken and written) language is structured and used to achieve particular goals (e.g., describing or persuading) (Martin, 2009). The notion of genre is typically associated with certain literary forms, such as poem or novel. In genre pedagogy, the notion of genre is mainly used for academic text types used throughout the curriculum. Six key genres have been distinguished (e.g., narratives, reports). However, for mathematics education, a more domain-specific investigation and identification of genres for mathematical learning are needed to centralize linguistic competency explicitly (Moschkovich, 2010).

Using genre pedagogy in mathematics education, Smit et al. (2016) formulated linguistic and structure features needed to identify the mathematical language to describe and interpret line graphs (see Fig. 1). The linguistic features included, for example, subject-specific vocabulary and phrases, and also the use of an expression of gradation steepness (as in "the graph descends *gradually*"), as well as general academic language to be employed when interpreting the graph (e.g., *the number* of people *increases*). The structure features comprised the stages of students' reasoning about graphs. For example, students are expected to interpret each part of the graph (e.g., "it was less busy") with a description related to the course of the graph ("you can tell as the graph shows a steep fall").

Linguistic and structure features of genres can provide teachers with a lens through which they can identify the language for mathematical learning (Smit et al., 2016). Through this identification, teachers can become more aware of the language required for mathematical discourse, and they can then better support their students express their thoughts in a mathematically accepted way (e.g. Schleppegrell, 2007).

Next to awareness of genres, it is considered useful for teachers to learn how to scaffold their students' language (Gibbons, 2002). Scaffolding, in short, is temporary, adaptive support, fostering students' independence regarding a

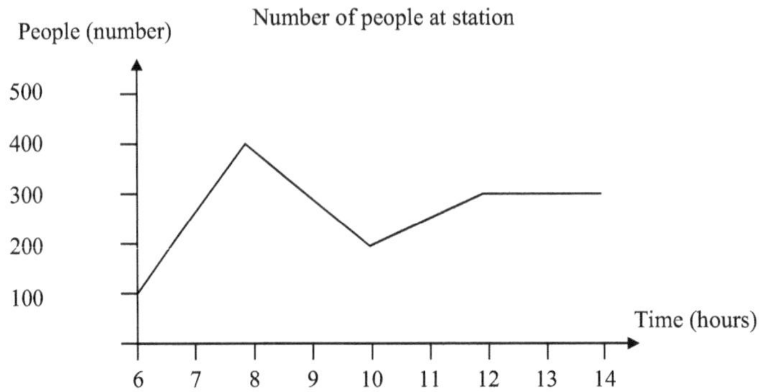

At 06:00 there are 100 people at the station. Between 06:00 and 08:00 the number of people increases, as the graph ascends. Between 08:00 and 10:00 the number of people decreases; the graph descends. Between 10:00 and 12:00 the number of people slowly increases as the graph gradually ascends again. From 12:00 the number of people remains the same. The graph is constant.

Fig. 1: Line graph and exemplary text from a domain-specific genre of interpreting line graphs.

particular topic (Maybin et al., 1992). For whole-class scaffolding, Smit et al. (2013) formulated three key characteristics. The first, *diagnosis*, is the assessment of the learners' level and needs. Second, *responsiveness* is the adaptation of support to learner's needs based on a diagnosis. The third characteristic is *handover to independence* which is the fading of support as the learner's independence increases.

The literature suggests that teachers can enact several scaffolding strategies to offer adaptive support (e.g., Gibbons, 2002). Our PDP's goal was that teachers would learn to use the scaffolding strategies in Tab. 1 (Smit et al., 2016) in response to the diagnoses of students' language proficiency as part of a language-responsive approach informed by genre pedagogy.

2.2 Teacher learning within adaptive professional development

For conceptualizing and analyzing teacher learning in our study, we used a framework developed by Bakkenes et al. (2010). It distinguishes four main categories of teacher learning: changes in knowledge and beliefs, intentions for practice, changes in practice, and emotions.

Tab. 1: Strategies for scaffolding language for mathematical learning.

Strategies	**Example**
Reformulating students' utterances into more academic wording	[In response to *the graph goes higher*] *Indeed, the graph rises steeply*
Ask students to be more precise or improve their spoken language	*What do you mean by "it"?*
Repeat correct student utterances	*Indeed, the graph descends slowly*
Refer to features of the text type	*Into how many segments can we split the graph?*
Use gestures or drawings to support verbal reasoning	Gesturing a horizontal axis when discussing a topic
Remind students (by gesturing or verbally) to use a designed scaffold (i.e., word list or writing plan) as a supporting material	*The word you are looking for is written down for you here*
Ask students how written text can be produced or improved	*How can we rewrite this using mathematical wording?*

To design a PDP based on previous research on genre pedagogy and scaffolding language, we capitalized on several sources. First, we drew on four learning activities identified by Bakkenes et al. (2010) as key elements for developing teacher expertise in the context of educational renewal: (1) learning by experimenting; (2) learning in interaction with others; (3) using external sources (e.g., publications); and (4) consciously reflecting on one's teaching practices. Second, we drew on insights concerning the promotion of language-responsive teaching (e.g., Hajer & Norén, 2017). Prediger (2019) gained insights into the learning pathways and obstacles of secondary school teachers in their development of language-responsive teaching practice concerning the identification of language for mathematical language. Despite these insights, there are still ongoing challenges in guiding teachers in the identification and therefore support of language for mathematical learning. One such challenge, as noted by Lyon (2013), is that teachers need to view language as inherent to one's classroom culture and not merely a technical issue to control. As such, the guidance of teachers in the identification of language for mathematical learning is required to be adaptive to teachers' contexts and needs.

A third source used for developing our PDP formed insights on adaptivity – generally assumed to be an essential characteristic of PDP as teacher learning depends on numerous factors (Putnam & Borko, 2000). Individual teacher learning may be different, and variations across settings need to be acknowledged (e.g., Goldsmith et al., 2014). If a PDP is adaptive, it provides the opportunity for participant teachers to take ownership of the content (e.g., Davis, 2002). To date, little is known on how adaptivity to teachers' individual needs can be achieved, which specific steps must be made in the design of the PDP to allow for this, and what teacher learning takes place in the context of these efforts.

The aim of the study reported here was to gain insights into how primary school teachers in an adaptive professional development program (2.2) can be supported in developing genre awareness and the scaffolding of students' mathematical language (2.1). We ask two questions:

1. What adaptations to the program were necessary to support teacher learning?
2. What did a case-study teacher learn from participating in the program?

3 Methods

We used design research to design and evaluate the PDP. Design research can be characterized as an interventionist approach to which prediction of and reflection on learning processes are central, and where the design of an intervention and the actual research are intertwined (McKenney & Reeves, 2018; Prediger et al., 2015). The design and redesign of sessions of the PDP were carried out by the second author. She acted as both researcher and teacher educator within the PDP, with a larger team acting as a soundboard. The PDP comprised seven monthly group sessions of 2.5 h each.

To achieve adaptivity to the teachers' needs, several steps were taken. Before the PDP, the participants completed questionnaires, and a mathematics lesson was observed to determine the starting point concerning language-responsive teaching. Throughout the PDP participants completed electronic logs, which the researcher-educator could access. The role of the participant logs was twofold. First, the participant logs consisted of open questions designed to promote teachers' reflection on their learning and classroom practice (Bakkenes et al., 2010) in the context of the PDP. Second, the participant logs gave insights into teachers' self-reported learning. Inspired by hypothetical learning trajectories (Simon, 1995), a researcher log was written before each session with the intentions and expectations for the session. After each session, a reflection document was created where the researcher log was compared with what actually happened in the session. The reflection document plus the most recent participant logs formed the basis for the researcher log of the next session. As such, the goals of the PDP were addressed while being adaptive to participants' needs and levels of understanding.

3.1 The five participants

One participating teacher (female, 25 years of experience), who taught grade 3 (students aged 8–9), agreed to be a case study to allow for more in-depth insight into the learning processes and development of the participants occurring within the PDP. The second participant (female, 13 years of experience) taught grade 4. A third participant (female, 27 years of experience) taught mathematics in a one-on-one setting supporting special education in a mainstream primary school. The final two participants came from the same school where one (female, 29 years of experience) and the other (male, 10 years of experience) taught grade 6.

3.2 Instruments and data collection

Apart from the aforementioned participant and researcher logs, and reflection documents for each session, the data collection consisted of completed exercises by the participants and verbatim transcription of the interaction between the researcher-educator and participants from video recordings of each group session.

To gain more in-depth insight into the responses in the participants' logs and PD sessions, two semi-structured interviews of the case-study teacher were conducted: one between the fourth and fifth group session and the second after the final session. The interview timing was chosen to allow for the characterization of the teacher's learning throughout the whole PDP. In the interviews, we asked the teacher to elaborate on points of interest mentioned in her logs. Audio recordings of both semi-structured interviews were transcribed verbatim.

3.3 Data analysis

3.3.1 Enactment and adaptation of the PDP

Participants' progress in terms of the learning goals mentioned in Section 2.1 was analyzed to evaluate the need for adaptation of the PDP. To detect these progressions within the PDP, the researcher logs and reflection documents were analyzed in the following way. We compared our expectations with what happened and traced how decisions were made based on new insights. For constructing the narrative presented in 4.1, we focused on the most apparent discrepancies between expectations and actual teacher learning. For example, we used Smit et al. (2016) genre of interpreting line graphs to illustrate what a domain-specific genre with its linguistic and structure features could look like. However, we quickly noted that teachers had little affinity with that domain. Hence we needed to adapt our initial plans and find another example that was closer to teachers' own practices. For more details, see Bakker et al. (2019). The first author created the theme-oriented narratives, after which the researcher-educator reviewed them; minor updates for clarification were made.

3.3.2 Teacher's perceived learning within the PDP

To gain insight into teachers' learning during the adaptive PDP, the semi-structured interviews with Mary (pseudonym) were analyzed based on the aforementioned main categories of Bakkenes et al. (2010) framework for teachers' self-reported learning.

Tab. 2: Coding scheme for reported learning outcomes.

Code	Global description	Example
CKB	Change in knowledge/beliefs: The teacher reports on growing awareness acquired knowledge, or the teacher reports on the confirmation of already existing beliefs	I am more aware of the role language plays in the mathematics classroom.
CP	Change in practice: The teacher reports that things have changed in his/her way of teaching or students' participation in the mathematics lessons.	I now prepare my lessons with a language goal in mind.
IP	Intention for practice: The teacher reports that he/she wants to teach differently in the future or reports that he/she wants to hold on to certain teaching practices	I want to spend more attention to reasoning steps in the future.
E	Emotion: Teacher reports on emotions related to using the knowledge from the PDP in the classroom, or reports on being surprised	I was surprised to see the level of interaction increased in the class.

All utterances in the transcripts of the interviews, in which Mary explicitly reports on a learning outcome, were identified by the observing researcher. Two independent raters then coded the 126 utterances to measure the inter-rater reliability of the coding process. The coding by the first and second coders resulted in an agreement of 95.2% and a Cohen's kappa of .91, meaning the four categories could be reliably distinguished.

The utterances were placed in chronological order by category to gain insight into the nature of the reported learning outcome. The utterances were analyzed, and the observing researcher generated a summary of changes in the four categories. The researcher-educator read the ordered data to validate the conclusions drawn from the analysis; minor updates for clarification were made.

4 Results

4.1 Enactment and adaptation of the PD

Analysis of the enactment of the PDP through the researcher and participant logs, and reflection documents, yielded two separate chronological narratives focused on the two main themes of the program: genre awareness and scaffolding language. For the latter theme, relatively minor adaptations were made to the PDP. Therefore we focus in this chapter on the adaptations made for the theme of genre awareness sharing the chronological narrative.

4.1.1 Narrative of genre awareness

In the first session, the participants were introduced to the identification of language required for mathematical learning through the concept of genre. The researcher-educator presented the concept of genre by using an example from the domain of line graphs (Smit et al., 2016). However, based on the reactions of the participants, the researcher-educator diagnosed that since line graphs are not regularly taught in primary school, this example was not close enough to the teachers' own practices. As such, the line graph example did not provide sufficient support for the teachers to understand the idea of genres in the domain of mathematics.

In the second session, the researcher-educator responded by characterizing two other genres for domains that were closer to the participants' teaching: estimation and expanded column method for subtraction. During the analysis of these genres, she drew the attention of the participants not only to the linguistic features (i.e., general academic language and subject-specific language) but also to the structure features (i.e., the required ordering of the steps by students to give mathematical meaning). While reviewing the in-session completed tasks, the researcher-educator diagnosed that the participants were still struggling with the concept of genre. This diagnosis was corroborated when two participants contacted the researcher-educator to report that they could not grasp how to complete the homework assignment related to the estimation genre. The researcher-educator concluded that the notions of linguistic and structure features of genres were not as fully understood as the researcher-educator had anticipated they would be at this stage in the PDP and that the term "genre" was a stumbling block for the participants. During this conversation, the researcher-educator explained the concept of genre in the context of language for mathematical learning as the text that includes the specific language and reasoning that is particular to that domain. It was in this conversation that the term "domain text" was first coined as a concretization to the more technical concept of a genre, where domain text was considered to be a prototypical text for a particular domain.

In the third session, the researcher-educator made three fundamental adaptations to the PDP. The first was to replace genre by *domain text*. Rather than using the more abstract and unfamiliar notion of genre, the researcher-educator thus responsively reframed the language for mathematical learning as *domain texts* that represent typical language usage in the different mathematical domains (deploying particular words and phrases). The second was to shift focus from identifying the structure features of the spoken or written mathematical text to identify the *reasoning steps* needed by students to solve mathematical problems. This new focus was better aligned with participating teachers' existing views on how students can solve mathematical problems. The third adaptation was to shift identifying the language for mathematical learning for a particular mathematical *domain*, to that for solving a particular mathematical *problem* within a domain. This adaptation was regarded as crucial by the participants and researcher-educator, as each mathematical problem, even within the same domain, requires its language to be identified in association with the particular reasoning steps.

In the fourth session, the participants used a domain text preparation template that was developed as a response to the diagnosis that the participants needed a scaffolding device for the identification of domain texts based on the completed homework assignments of the participants. The preparation template included the identification of reasoning steps and the required language components for solving the mathematical problem. The participants first had to identify the reasoning steps needed to solve the problem and then identify the language components of the domain text. Finally, the participants had to write how a student should articulate the solution. While the participants were analyzing videos of language-focused mathematics classes and the accompanying domain text, the participants reported that the dual analysis of the video and the domain text helped with their understanding of the identification of language for mathematical learning. In response, the researcher-educator and participants agreed that in the following sessions, every participant should present for peer feedback at least one video of their language-focused mathematics classes and the associated domain text.

In the fifth session, the teacher-educator led a discussion with the participants on the differences between a teacher's instruction (procedural steps) and students' thinking (reasoning steps) required to solve a mathematical problem. The addition of the discussion was based on the analysis of the participants' homework assignment to develop a domain text. The researcher-educator diagnosed an increase in independence as the participants were beginning to identify words and formulations for their chosen domain. However, as part of their development of domain texts, three of the five participants were still unable to centralize student thinking in their identification of reasoning steps and instead focused on procedural steps. To initiate the discussion and emphasize the importance of centralizing students' thinking in a lesson, the researcher-educator asked one participant who had already perceived this difference between procedural steps and reasoning steps, to explain the difference to the other participants. The discussion clarified the concept of reasoning steps for the other participants. The participants made comments such as "reasoning steps stimulate thinking," "maybe we give too little attention to reasoning steps," and "normally language in the mathematics lesson is focused on the mathematical procedures, not the reasoning steps of a student." By the end of the sixth session, most of the participants showed some form of independence concerning reasoning steps during the session: "You get closer to the thinking of the children" and, concerning language and reasoning steps, "[language and reasoning steps] support each other. You can see the thought process in the children."

In sum, responsive adaptations to genre pedagogy could be characterized as from a more global to a more local orientation on the one hand, and from more abstract (i.e., genre with language and structure features for a domain) to more concrete notions (i.e., language and reasoning steps for a problem) on the other hand.

4.2 Teacher's perceived learning within the PDP

The distribution of utterances among the categories of reported learning is shown in Tab. 3. With 125 utterances in total, the majority of the reported learning outcomes fell into the category changes in practice.

Tab. 3: Distribution of reported learning outcomes among three categories.

	Mid-interview (Mid)	Post-interview (Post)
Changes in knowledge and beliefs	15	16
Changes in practice	30	31
Intentions for practice	13	20

In the following subsections, quotes from the interviews (from which interview is cited in parentheses) and observations are presented to illustrate Mary's development from the mid- to post-interview for each category of reported learning.

4.2.1 Changes in knowledge and beliefs

Based on the analysis of the semi-structured interviews, the changes in knowledge and beliefs primarily concerned the overarching intention of the PDP for scaffolding language for students' mathematical learning. The most noticeable change in Mary's beliefs was the importance of language in the mathematics classroom and that "language in a mathematics lesson is not separate, mathematics and language belong together" (Mid). For Mary, the concept of the scaffolding language seemed easy to understand as early in the PDP she had formulated for herself that "they are strategies I can use to help the students use more precise language" (Mid).

However, with the theme of identification of genres or language required for mathematical learning, Mary had some conceptual issues. Mary struggled in shifting her view from considering that the language required for mathematical learning consists only of vocabulary to viewing it as a means to allow a student to articulate their reasoning steps. Mary suggested that her difficulties in comprehending the concept of reasoning steps may have been due to her choice of domain. The domain of measuring – at the age of Mary's class – does not involve a large amount of reasoning knowledge but does involve substantial procedural knowledge of how to measure. For Mary, this was highlighted when she reflected on another participant's domain text where she could immediately identify the reasoning steps.

Mary reported that the learning structure within the PDP where participants spent time analyzing their assignments and video recordings of their classroom interaction as a group was for her valuable and crucial to her learning as "You cannot do that alone. And also not by reading a book" (Post).

4.2.2 Changes in practice

Mary reported several changes in practice as a result of the PDP. She changed the preparation of her mathematics lessons to include a language goal. Specifically, she now considers "what language is important to focus on in the lesson and how they (learners) can use it" (Post).

By the end of the PDP, Mary reported how several scaffolding language strategies had become embedded in her class: "how can we say that using mathematical language? That is a standard sentence I use often" (Mid) and "discussing with each other how to word it more precisely" (Post). She also recognized the benefits of not only confirming that her student had used the correct articulation but asking another student to also try as "it may help the student who had not yet understood it" (Post).

The focus in Mary's mathematics class shifted from getting the answer to allowing the students to articulate their reasoning steps:

> The students articulate their reasoning steps. You then know: the answer is not correct but in how you tried to get there one small step is missing. If you can explain something, then you understand it. They previously had not been able to articulate that. (Mid)

Mary also observed changes in the form of interaction she had with the students: "I noticed I have a lot more dialogue with the students" (Mid); "they use the required terms and often say 'oh we need to say that using mathematical language' " (Post).

4.2.3 Intentions for practice

In the mid-interview, Mary's intentions for practice were focused on scaffolding students' language required for mathematical learning. In the post-interview, Mary reported the intention of continuing to work on components of the domain text, the identification of the articulation of the reasoning steps. Mary also intended to use the knowledge she gained in the PDP, not only in the mathematics class but to extend scaffolding of language to other subjects: "I think that I will try it in other subjects" (Post).

In the post-interview, Mary's reported intentions were not limited to her learning and practice but also included the dissemination of knowledge to her colleagues. Mary also noted that the dissemination of the knowledge related to the scaffolding of language required for mathematical learning is not something that can be achieved in the short term "but slowly as this is something that must be absorbed. This cannot be done in a couple of months. This is something where you need a year to become more confident about it" (Post).

5 Discussion

The presented study aimed to gain insights into how a professional development program (PDP) can adaptively support teachers to gain awareness of genres relevant to learning mathematics and scaffold language required for mathematical learning within their lessons. To fulfil this aim, two aspects of the PDP have been analyzed. The first was the adaptations made to the PDP and the observations that triggered these adaptations. The second aspect was the learning reported by a participating teacher.

The researcher-educator had the opportunity to adapt the contents of the sessions to the needs of the group and individual participants by analyzing the development of the teachers in between PDP sessions through reflection documents and participant logs. Similar to findings in other studies (e.g. Prediger, 2019), the identification of language for mathematical language proved to be challenging. For example, the notion of "genre" proved too abstract for the participants to identify the language for mathematical learning. This led us to use the notion of "domain text" instead – a move that can be considered practicalizing principled theoretical knowledge (Bakker et al., 2019; Janssen et al., 2015). The shift from the abstract to concrete was achieved through using a narrower focus of language needed for reasoning about and solving a particular mathematical problem (called "reasoning steps" – "denkstappen" in Dutch). In line with the approach taken with secondary school teachers (Prediger, 2019), focusing on a mathematical problem, as opposed to an entire mathematical domain, was closer to teachers' own practical experiences. This practicalization of principled theoretical knowledge was enacted both in the design of the course activities (e.g., the preparation template) and in interaction with the participants during sessions.

We give this as an example to underpin a broader call to action for researchers and educators. Practicalization requires flexibility on the side of the researchers and educators, a willingness to adapt original plans where the input and knowledge from the participants are employed to judge what works in the context of their teaching practice. By doing so, researchers and educators have the opportunity to concretize the methods and strategies, aligned with the theory in question, that the teachers can successfully employ in their classrooms. However, such adaptations should not lead to "lethal mutations" (Brown & Campione, 1996).

Second, we analyzed what a teacher perceived to have learned from participating in the PDP. The case-study teacher reported an increase in awareness in the connection between mathematics and language. She also reported that the most noticeable impact of the PDP was the change in social norms concerning mathematical language during the interaction that occurred within her class, both between herself and the students and between the students themselves, in line with often reported observations (e.g., Yackel & Cobb, 1996). She reported how a number of the language scaffolding strategies had become embedded in her class with the effect that students became accustomed to using the language for mathematical learning as well as verbalizing their reasoning steps.

A limitation of our study is that we only analyzed the participants within the PDP itself and relied on self-report for their teaching in the classroom. The impact of the PDP on the participating teachers' classroom enactment and students' mathematical performance was not observed or analyzed, and the extent of the handover to independence could thus not be fully monitored.

What we as authors have learned about our national case is that when promoting a particular vision on mathematics education, professional development has to be taken much more seriously. For example, the ideas from Realistic Mathematics Education have been influential both nationally and internationally (Treffers & van den Heuvel-Panhuizen, 2020) but proved hard to realize in practice in the Netherlands. The general ideas and curricular materials were insufficient for implementing Realistic Mathematics Education in education (Gravemeijer et al., 2016), pointing to a lack of investment in professional development. Thus there should be a focus not only on the education characteristic in question (in our case language-responsiveness) but also on continued investment in the professional development that is adaptive to the local context.

References

Bakkenes, Inge, Vermunt, Jan D. & Wubbels, Theo (2010): Teacher learning in the context of educational innovation: Learning activities and learning outcomes of experienced teachers. *Learning and Instruction* 20 (6), 533–548.

Bakker, Arthur, Mackay, Farran, Smit, Jantien & Keijzer, Ronald (2019): Practicalizing principled knowledge with teachers to design language-oriented mathematics lessons: A design study. In Graven, M, Venkat, H., Essien, A, Vale, P. (eds.): *43rd Conference of the International Group for the Psychology of Mathematics Education*. 2, Pretoria, South Africa: PME, 57–64.

Brown, Ann L. & Campione, Joseph C. (1996): Psychological theory and the design of innovative learning environments: On procedures, principles, and systems. In Schauble,

L., Glaser, R. (eds.): *Innovations in learning: New environments for education*. Mahwah, NJ: Lawrence Erlbaum, 289–325.

Davis, Kathleen S. (2002): "Change is hard": What science teachers are telling us about reform and teacher learning of innovative practices. *Science Education* 87, 3–30.

Essien, Anthony A., Chitera, Nancy & Planas, Núria (2016): Language diversity in mathematics teacher education: Challenges across three countries. In Barwell, R., Clarkson, P., Halai, A., Kazima, M., Moschkovich, J., Planas, N., Setati-Phakeng, M., Valero, P., Villavicencio Ubillús, M. (eds.): *Mathematics Education and Language Diversity*. Cham: Springer, 103–119.

Gibbons, Pauline (2002): *Scaffolding Language, Scaffolding Learning: Teaching Second Language Learners in the Mainstream Classroom*. Portsmouth, NH: Heinemann.

Goldsmith, Lynn T., Doerr, Helen M. & Lewis, Catherine C. (2014): Mathematics teachers' learning: A conceptual framework and synthesis of research. *Journal of Mathematics Teacher Education* 17 (1), 5–36.

Gravemeijer, Koeno, Bruin-Muurling, Geeke, Kraemer, Jean-Marie & Van Stiphout, Irene (2016): Shortcomings of mathematics education reform in the Netherlands: A paradigm case? *Mathematical Thinking and Learning* 18 (1), 25–44. doi:10.1080/10986065.2016.1107821.

Hajer, Maaike & Norén, Eva (2017): Teachers' knowledge about language in mathematics professional development courses: From an intended curriculum to a curriculum in action. *EURASIA Journal of Mathematics Science and Technology Education* 7b (13), 4087–4114.

Janssen, Fred, Westbroek, Hanna & Doyle, Walter (2015): Practicality studies: How to move from what works in principle to what works in practice. *Journal of the Learning Sciences* 24 (1), 176–186.

Lampert, Magdalena & Cobb, Paul (2003): Communication and language. In Kilpatrick, J., Martin, W. G., Schifter, D. (eds.): *A Research Companion to Principles and Standards for School Mathematics*. Reston, VA: National Council of Teachers of Mathematics, 237–249.

Lyon, Edward G. (2013): What about language while equitably assessing science?: Case studies of preservice teachers' evolving expertise. *Teaching and Teacher Education* 32, 1–11.

Martin, James R. (2009): Genre and language learning: A social semiotic perspective. *Linguistics and Education* 20 (1), 10–21.

Maybin, Janet, Mercer, Neil & Stierer, Barry (1992): 'Scaffolding' learning in the classroom. In Norman, K. (ed.): *Thinking Voices: The work of the National Oracy Project*. London: Hodder and Stoughton, 185–195.

McKenney, Susan & Reeves, Thomas C. (2018): *Conducting Educational Design Research (2nd Ed.)*. London: Routledge.

Moschkovich, Judit N. (2010): *Language and Mathematics Education: Multiple Perspectives and Directions for Research*. Charlotte, NC: IAP.

Prediger, Susanne (2019): Investigating and promoting teachers' expertise for language-responsive mathematics teaching. *Mathematics Education Research Journal* 31 (4), 367–392.

Prediger, Susanne, Gravemeijer, Koeno & Confrey, Jere (2015): Design research with a focus on learning processes: An overview on achievements and challenges. *ZDM* 47 (6), 877–891.

Putnam, Ralph T. & Borko, Hilda (2000): What do new views of knowledge and thinking have to say about research on teacher learning? Educational Researcher 29 (1), 4–15.

Schleppegrell, Mary J. (2007): The linguistic challenges of mathematics teaching and learning: A research review. *Reading & Writing Quarterly* 23 (2), 139–159.
Simon, Martin A. (1995): Reconstructing mathematics pedagogy from a constructivist perspective. *Journal for Research in Mathematics Education*, 26 (2), 114–145.
Smit, Jantien, Bakker, Arthur, Van Eerde, Dolly & Kuijpers, Maggie (2016): Using genre pedagogy to promote student proficiency in the language required for interpreting line graphs. *Mathematics Education Research Journal* 28 (3), 457–478.
Smit, Jantien, Van Eerde, Henriëtte A. A & Bakker, Arthur (2013): A conceptualization of whole-class scaffolding. *British Educational Research Journal* 39 (5), 817–834.
Treffers, Adri & Van den Heuvel-panhuizen, Marja (2020): Dutch Didactical Approaches in Primary School Mathematics as Reflected in Two Centuries of Textbooks. In van den Heuvel-panhuizen, M. (ed.): *National Reflections on the Netherlands Didactics of Mathematics*. Cham: Springer, 77–103.
Van Eerde, Dolly & Hajer, Maaike (2009): The integration of mathematics and language learning in multiethnic schools. In César, M., Kumpulainen, K. (eds.): *Social Interactions in Multicultural Settings*. Rotterdam/Taipei: Sense, 269–296.
Van Eerde, Dolly, Hajer, Maaike & Prenger, Joanneke (2008): Promoting mathematics and language learning in interaction. In Dean, Jeannie, Haijer, Maaike, Koole, Tom (eds.): *Interaction in Two Multicultural Mathematics Classrooms. Processes of Inclusion and Exclusion*. Amsterdam: Aksant, 31–69.
Yackel, Erna & Cobb, Paul (1996): Sociomathematical norms, argumentation, and autonomy in mathematics. *Journal for Research in Mathematics Education* 27, 458–477.

About the editors

Annemarie Fritz was full professor of psychology at the University of Duisburg-Essen, Germany, since 1998. She graduated in Psychology, habilitated in psychology of special education and rehabilitation. From 2015 to 2020 she was Distinguished Visiting professor at the University of Johannesburg.

In the past 25 years she did research on numerical cognition applied to math-learning difficulties. She is interested in the development of mathematical competencies from kindergarten age to secondary level. The main focus of her scientific work was the empirical validation of a development model of key numerical concepts and arithmetic skills from age 4 to 8. Based on this model some diagnostic assessments (MARKO-Series) and training programs for preschool and elementary school children were developed. Recently, her interest turned to assessments for older children in math and math anxiety.

She is now emerita and heads the "Akademie Wort und Zahl," an international organization for intercultural education.

Erkan Gürsoy is in charge of the project "ProDaZ" at the University of Duisburg-Essen in Germany (funded by the Mercator Foundation from 2010 to 2022), which aims to ensure that teaching and learning contents in formal and nonformal education can be improved through language learning across the subject areas.

He is researcher at the Institute for German as a Second and Foreign Language (University of Duisburg-Essen) and obtained his PhD in Second Language Acquisition from the same university in 2014.

His work and research interests focus on multilingual and subject-based language education, heritage language education, relation of languages and learning mathematics in formal and nonformal education, and teaching newly arrived children/young people and their families. Recently, his interest turned to intersectional perspectives on educational linguistics.

Moritz Herzog works as postdoctoral researcher in special needs education at the University of Wuppertal (Germany). During his studies for teacher education, he worked as a dyscalculia therapist. He worked at the University of Duisburg-Essen from 2015 to 2020. He obtained his PhD in Psychology from the University of Duisburg-Essen in 2020. From 2017 to 2020, he was research associate at the University of Johannesburg.

His research interests are the development of arithmetic concepts, and the diagnosis, prevention, and intervention of mathematical learning difficulties. This covers emotional and affective aspects of mathematical learning, influences of language resources, and the design of effective intervention for children with mathematical learning difficulties.

https://doi.org/10.1515/9783110661941-021

List of authors

Ashkenazi, Sarit The Seymour Fox School of Education, The Hebrew University of Jerusalem, Mount Scopus, Israel

Bahnmueller, Julia Centre for Mathematical Cognition, Loughborough University, Leicesterchire, UK; LEAD Graduate School & Research Network, University of Tuebingen, Germany

Bakker, Arthur ESoE, Eindhoven University of Technology, Utrecht, The Netherlands

Bos, Samantha E. Department of Special Education, The University of Texas at Austin, Texas, USA

Buyle, Margot Institute of Psychology and Institute of Neuroscience, Université Catholique de Louvain, Louvain-la-Neuve, Belgium

Ceulemans, Annelies Department of Experimental Clinical and Health Psychology, University of Ghent, Ghent, Belgium

Crollen, Virginie Institute of Psychology and Institute of Neuroscience, Université Catholique de Louvain, Louvain-la-Neuve, Belgium

Desoete, Annemie Department of Experimental Clinical and Health Psychology, University of Ghent, Ghent, Belgium

Ceulemans, Annelies Artevelde University of Applied Sciences, Ghent, Belgium

Dowker, Ann Department of Experimental Psychology, Oxford University, Oxford, UK

Dresen, Verena Institute of Psychology, UMIT – The Health and Life Science University, Tirol, Austria

Everett, Caleb Anthropology Department, University of Miami, Miami, USA

Fischbach, Antoine Department of Education and Social Work, University of Luxembourg, Luxembourg

Fritz, Annemarie Faculty of Education Sciences, Department of Psychology, University of Duisburg-Essen, Essen, Germany; Center for Educational Practice Research, Faculty of Education, University of Johannesburg, Johannesburg, South Africa

Gürsoy, Erkan Institute of German as a Second/Foreign Language/DaZ/DaF, University of Duisburg-Essen, Essen, Germany

https://doi.org/10.1515/9783110661941-022

Haase, Vitor Geraldi Department of Psychology, Federal University of Minas Gerais, Belo Horizonte, Brazil

Hartmann, Julia Faculty of Education Sciences, Department of Psychology, University of Duisburg-Essen, Essen, Germany

Herzog, Moritz Institute of Educational Research, School of Education, University of Wuppertal, Wuppertal, Germany

Jung, Stefanie Leibniz-Institut fuer Wissensmedien, Tuebingen, Germany

Kaufmann, Liane Department of Psychiatry and Psychotherapy A, General Hospital Hall in Tirol, Tirol, Austria

Keijzer, Ronald Freudenthal Institute, iPabo University of Applied Sciences, Amsterdam, The Netherlands

Klein, Elise Université de Paris, LaPsyDÉ, CNRS, Sorbonne Paris Cité, Paris, France
Leibniz-Institut fuer Wissensmedien, Tuebingen, Germany

Klein, Helga Department of Special Education, Ludwigsburg University of Education, Ludwigsburg, Germany

Lin, Xin Department of Special Educaion, The University of Texas at Austin, Texas, USA

Long, Caroline Faculty of Education Science, University of Johannesburg, Johannesburg, South Africa

Lopes Silva, Júlia Beatriz Department of Psychology, Federal University of Minas Gerais, Belo Horizonte, Brazil

Mackay, Farran ESoE, Eindhoven University of Technology, Utrecht, The Netherlands

Mähler, Claudia Institute of Psychology, University of Hildesheim, Hildesheim, Germany

Mark-Zigdon, Nitza Special Education Department, Levinsky College of Education, Tel Aviv, Israel; The School of Education, Tel Aviv University, Israel

Marlair, Cathy Institute of Psychology and Institute of Neuroscience, Université Catholique de Louvain, Louvain-la-Neuve, Belgium

Martini, Sophie Faculty of Humanities, Education and Social Sciences, University of Luxembourg, Luxembourg

Moeller, Korbinian Centre for Mathematical Cognition, Loughborough University, Leicesterchire, UK; Leibniz-Institut fuer Wissensmedien, Tuebingen, Germany

Moschkovich, Judit Department of Education, University of California, Santa Cruz CA, USA

Moser Opitz, Elisabeth Institute of Education, University of Zurich, Zurich, Switzerland

Moura, Ricardo Department of Basic Psychological Processes, University of Brasília, Brasília, Brazil

Pixner, Silvia Institute of Psychology, UMIT – The Health and Life Science University, Tirol, Austria

Powell, Sarah R. Department of Special Education, The University of Texas at Austin, Texas, USA

Prediger, Susanne Institute for Development and Research in Mathematics Education, TU Dortmund University, Dortmund, Germany

Rocha de Freitas, Fernanda Department of Psychology, Federal University of Minas Gerais, Belo Horizonte, Brazil

Roelants, Mathieu Catholic University of Leuven KUL, Leuven, Belgium

Rousseau, Sofie Ziama Arkin Infancy Institute, Baruch Ivcher School of Psychology, Interdisciplinary Center (IDC), Herzliya, Israel

Schiltz, Christine Department of Behavioural and Cognitive Sciences, University of Luxembourg, Luxembourg

Schindler, Verena University for Teacher Education, Berne, Switzerland

Schuchardt, Kirsten Institute of Psychology, University of Hildesheim, Hildesheim, Germany

Scott, Judith Department of Education, University of California, Santa Cruz CA, USA

Smit, Jantien Research Centre for Education, Hogeschool Utrecht University of Applied Sciences, Utrecht, The Netherlands

Stephany, Sabine Mercator-Institute for Literacy and Language Education, University of Cologne, Cologne, Germany

Teixeira Batista, Luana Institute of Biological Sciences, Federal University of Minas Gerais, Belo Horizonte, Brazil

Ugen, Sonja Faculty of Humanities, Education and Social Sciences, University of Luxembourg, Luxembourg; Observatoire national de la Qualité Scolaire, Luxembourg

Uribe, Ángela Faculty of Mathematics, TU Dortmund University, Dortmund, Germany

Bei Fragen zur Produktsicherheit wenden Sie sich bitte an:
If you have any questions regarding product safety,
please contact:

Walter de Gruyter GmbH
Genthiner Straße 13
10785 Berlin
productsafety@degruyterbrill.com